T3-BNI-879

The Energy Controversy

Friends of the Earth
Principal offices

124 Spear Street
San Francisco, California 94105

620 C Street SE
Washington, DC 20003

72 Jane Street
New York, New York 10014

9 Poland Street
London W1V 3DG, England

THE ENERGY CONTROVERSY

Soft Path Questions & Answers

By Amory Lovins & his critics

Edited by Hugh Nash

Friends of the Earth San Francisco

Design by JoEtta May, San Rafael, and Roosevelt Studios, Berkeley.
Index by Deanne and Tom Holzberlein.
Manufactured in the U.S.A.

Most of the material in this book originally appeared in *Alternative Long-Range Energy Strategies,* Interior Committee Serial No. (94-47)(92-137), Government Printing Office, 1976, 1977.

We are grateful to the following individuals and organizations for permission to quote from their material: Ian A. Forbes; Charles B. Yulish and Charles B. Yulish Associates, Inc.; Hans A. Bethe; National Economic Research Associates, Inc.; Aden and Marjorie Meinel; George Pickering; Arnold E. Safer; Bertram Wolfe; and the Editors of *Foreign Affairs* and *Science.*

Library of Congress Cataloging in Publication Data
Main entry under title:

The Energy controversy.

Includes index.
1. Energy policy. 2. Power resources.
I. Lovins, Amory B., 1947- II. Nash, Hugh, 1923-
HD9502.A2E54357 333.7 78-67907
ISBN 0-913890-21-9
ISBN 0-913890-21-7 pbk.

Distributed in Canada by Hurtig Publishers, Edmonton

CONTENTS

FOREWORD

You don't expect a magazine article to make much of a stir, ordinarily. But occasionally, one does. One that did is Amory B. Lovins's 'Energy Strategy: The Road Not Taken?" published by *Foreign Affairs* in October 1976.

The article was catapulted into prominence by people who loathed it. Favorable critics, of whom there were many, could have made "Energy Strategy" a success of the seven-days' wonder variety. But it took vociferously hostile critics, of whom there were also many, to make the article a storm center around which controversy still rages more than two years later.

Congressional reaction was immediate. In early December, Lovins was testifying about his article and his energy theories before a joint hearing of two U.S. Senate committees. The hearing record contains not only testimony offered in person, but also published and unpublished critiques of the *Foreign Affairs* article, news items, editorials, comments received from interested parties, and the like—plus, when he felt moved or goaded to respond, Lovins's rebuttals. By the time a second volume had followed the first, the hearing record totalled more than two thousand pages.

Publication of such a voluminous record might seem to leave nothing more to be said on the subject. And indeed, professionally or profoundly interested readers may find the complete hearing record indispensable. But for the "general reader," the record can be improved upon. For most of us, it contains too much in too hard-to-read a form. The text was reproduced photographically from original copy that was all too often typed single spaced on a worn ribbon with handwritten interlineations, then reduced for economy to perhaps one-fourth its original size. Much of my editorial work was performed with magnifying glass in hand and imprecations on my lips. This book's goal is to make the official hearing record's very substantial virtues more accessible to a wider audience.

When David Brower asked me to distill a readable book from the record for publication by Friends of the Earth, my first thought was to concentrate on pro-Lovins material. Why provide an audience for anti-Lovins critics who, from the viewpoint of the book's publishers, were plainly wrong? But the deeper I delved into the record, the less useful such an approach seemed. Lovins is his own ablest advocate, and he had stated his case admirably in both the *Foreign Affairs* article and in his testimony; pro-Lovins material by other authors tends not to make his argument any more understandable and plausible than it was already. Gradually it dawned on me: the people who had done the most to bolster Lovins's case were not his intellectual allies but his most implacable critics.

Few of Lovins's detractors lay claim to dispassionate detachment. Some attack his ideas with the wholehearted enthusiasm of people who feel that their own sacred cows have been gored; others, who clearly feel that Lovins's ideas are heretical to the point of blasphemy, sound like modern-day inquisitors who would like to see his notions (if not their author himself) burned at the stake. Lovins's critics are a highly motivated lot, and one must suppose that they demolished his "soft energy" argument as thoroughly as they possibly could. Which makes it all the more striking that they didn't demolish it at all. When a concept emerges stronger than ever after being plunged into the crucible of antagonistic criticism, it must be accepted, provisionally, at least, as a basically sound hypothesis deserving further investigation.

Having decided that the book should consist of adverse criticism coupled with Lovins's reactions to it, I found that assembling the material still posed problems. Both critiques and rebuttals were in monologue form, and one could hardly be expected to keep all of a critic's points in mind while reading Lovins's attempts to refute them. So I created synthetic dialogues, interrupting critics' commentaries in order to place Lovins's rebuttals in close juxtaposition to the points rebutted. That, in essence, is the structure of this book.

(In a few places, at the insistence of an author, we had to abandon the basic pattern. A few people refused to have their prose interrupted, in which case we resorted to a parallel-column format. Several authors refused permission for us to reprint their work, or set conditions that were unacceptable to us. In a couple of instances of this kind, where Lovins's rebuttals were of such interest that we were unwilling to jettison the whole exchange, a critic's comments were paraphrased rather than quoted. As editor, I am solely responsible for the accuracy of paraphrasing, so far as it goes; sometimes I omitted material that was non-controversial.)

So copious is the hearing record that much material suitable for inclusion in this book was necessarily omitted. The final selection was based on my admittedly subjective impression that some critiques were more "representative" than others, or would somehow contribute more than others to the overall quality of the book. Two or three other books of equal size and comparable quality could be assembled from the hearing record, however, and to an extent, my selection of material was undeniably arbitrary. But I did try to give Lovins's bitterest critics another chance to be heard, for it is one of the purposes of this book to demonstrate that *none* of his critics has made any real dent in Lovins's case for a soft energy path.

Critics naturally tend to seize upon the same alleged flaws in Lovins's arguments, so their commentaries sound like variations on a theme. Some very fine music has been built on this plan, and I hope it is an appropriate one for this book. I could have edited out much of the repetition, of course, but not without compromising the integrity of arguments on both sides. Much of the repetition, in any case, may serve to emphasize important points.

In order to meet the hearing record's deadlines, Lovins confronted the nearly impossible task of responding to dozens of critics more or less simultaneously (on top of his normal heavy schedule). All things considered, his replies to critics read surprisingly well as originally written. But there were inevitably things a writer of Lovins's ability would want to smooth out, and I saw no reason to deny him the privilege. It will be apparent to readers, I feel sure, that relatively few and minor editorial improvements could not have tilted the balance unfairly against Lovins's adversaries.

Some critics, it is evident, somehow feel entitled as a matter of right to have the last word. I can't imagine why, unless they consider that Lovins's *Foreign Affairs* article was an attack on them. But the article was scholarly and impersonal; it was the critics who struck the first blow, some of them saying things that could be personally and professionally damaging to Lovins if taken seriously. To give critics the last word in this book would be to give them two whacks at Lovins as opposed to only one counter-whack by him. It would give the "aggressors" a two-to-one advantage over the aggressed-against, and in my view, would be totally unjustifiable.

Lovins wanted to update the material in this book by means of footnote references to recent developments. He had the misfortune, however, to bump into an editor with a strong bias against footnotes. Instead of footnotes, I have put the updatings, surrounded by brackets, in the body of the text. I have indulged my bias also by omitting many footnote references by Lovins and others in the hearing record, wherever the importance of documentation to a few readers seemed to me outweighed by its unimportance to the rest. Lovins suffers more than anyone else from this idiosyncracy of mine, since most other authors used footnote references sparingly if at all.

I'm a Lovins partisan not merely because we both work for Friends of the Earth, but also because his energy strategy seems to me environmentally, economically, socially, culturally, and politically superior to any other in sight. It also has the virtue (which others lack) of permanent sustainability. But partisan as I am, I don't think Lovins covers all bases as well as he's capable of doing. Not quite.

One concept essential to the understanding of Lovins's arguments is perhaps too much taken for granted: the concept of "margin" and "marginal costs." Loosely defined, the margin separates the past and present (which we cannot influence) from the future (whose shape we can determine in large measure). The marginal cost of electricity, for example, is the cost of electricity produced by the next power plant to be planned and built from scratch.

Marginal costs tend to be very much higher than historic or current costs because the cheap and easy things have already been done. One can't put another dam where Hoover Dam already is, and when cheap Persian Gulf oil is exhausted, we can't replace it with more cheap oil from the Persian Gulf.

Confusion is often caused (sometimes deliberately) by comparing the historic cost of one alternative with the marginal cost of another—an example of what Lovins calls asymmetrical analysis. It is misleading, for instance, to compare the cost of solar heating, an emerging technology, with the current cost of price-controlled natural gas. A sound energy strategy must rest upon symmetrical analysis, comparing the marginal costs of all available alternatives with each other.

The margin is where an energy strategist's head must be at. Except as a mental exercise, one can't usefully devise an energy strategy for the 1950s or 1960s. If you want to follow Lovins's thought processes, forget about the past and present except insofar as they inevitably affect the future. Be future-oriented; think in terms of energy options available from now on, and their relative costs (economic, environmental, and social). You'll find, I think, that Lovins consistently thinks in terms of what can be done at the margin, while his critics frequently seem somewhat muddled about past, present, and future energy options.

Lovins has much to say about solar heating, which is all to the good. But it seems to me he talks too little of *passive* solar systems, those in which solar energy is collected, stored, and gradually released by the structure itself. Perhaps he was trapped into this by his critics, whose concept of solar heating tends to be overly gadget-ridden. Active solar systems may be the only way to fit solar heating into many existing structures, and they certainly have much to recommend them. In fact, more than half the 50,000 buildings in the U.S. that have passive solar systems were retrofitted. And at the margin—in new buildings that can be planned from the ground up as solar structures—the passive approach ought usually to be simpler, cheaper, and more elegant than active solar systems, not to mention non-solar alternatives.

One point that Lovins emphasizes again and again, but which can hardly be emphasized enough, is the importance of having tightly-built and well-insulated buildings—however they are heated. A house that lost virtually no heat could be heated (or overheated) by its lights, appliances, and human occupants. Today's conventional house, however, is almost at the opposite extreme: heating it is a step toward heating the great outdoors. One of the infuriatingly common mistakes that many energy analysts make is to poor-mouth solar energy because it cannot adequately heat uninsulated or poorly insulated houses that leak air like sieves leak water. Pouring heat into such a house is like pouring water into a bucket with no bottom. Whatever form of energy you use to heat your home, you'll be wasting most of it unless you've caulked its seams and insulated it well.

Critics deride Lovins for suggesting that phasing out civilian nuclear power would put a decisive end to nuclear proliferation. Of course, Lovins suggests nothing of the sort. He points to the distinction between that which is necessary and that which is sufficient. It is necessary for me to breathe, but I must also eat, drink, find clothing and shelter, perspire, find a purpose in life, and so on. The fact that breathing alone isn't sufficient to maintain life doesn't tempt me to stop breathing. Phasing out civilian nuclear power is not rendered unnecessary by being insufficient, but we must do other things as well to denuclearize the world. Lovins tells us what else he thinks needs to be done. I happen to believe that his own prescription falls into the necessary-but-not-sufficient category; nothing less than a world federation would be *sufficient*, in my opinion. Be that as it may, it's shoddy criticism to ignore most of Lovins's denuclearization recommendations, and then, having ignored them, to accuse him of naivete. And that is precisely what many of his critics have done.

The criticism that Lovins is probably tiredest of is the contention that a nation can simultaneously follow the hard and the soft energy paths. These paths haven't been defined for you yet, so I won't have a great deal to say on this subject. But it's a crucial issue, and Lovins addresses himself to it cogently. Cogently, but perhaps not at great enough length. I'd like to see him write another article, this one explaining in elaborate detail why one can't march north and march south at the same time.

Obviously, say Lovins's critics, a solar house (soft) can be built on the sunny side of a nuclear reactor (hard); so hard and soft technologies can coexist. They're right, obviously. (Whether it's safe for anyone to live in that particular solar house is not so obvious.) But having said this, we haven't said much. Hard and soft technologies can exist in the same world just as an Australian prime minister and an Australian bushman can exist on the same continent. So what?

Technologies do not battle with one another like monsters in a science fiction film. Hard and soft *technologies* are not mutually exclusive in a technical sense. But Lovins argues that hard and soft *paths* are mutually exclusive, and I find this totally convincing. Classical and country music can coexist. But one can't reasonably set out to write music that appeals equally to classicists and to country-music fans. Nor can a society aspire to be both conspicuously consumptive and elegantly frugal. Lovins says that the hard and the soft paths are culturally and institutionally antagonistic, and furthermore, compete for the same limited resources. I believe this. But this seems to be the issue on which critics most unanimously insist that Lovins is wrong. I have to take their unanimity seriously. Not that I think they are right, but it does seem that Lovins (and his adherents) need to work harder to be convincing on this point.

I can imagine Lovins groaning: "What am I supposed to do? Take time out to argue that a society cannot dedicate itself simultaneously to monotheism and polytheism, to vegetarianism and cannibalism? That one cannot mount two horses and ride off in opposite directions? That people cannot ardently consume and equally ardently conserve? . . ."

Well, yes. If people haven't been convinced that following the hard path precludes following the soft one, believers in a soft energy strategy must find ways to persuade them. Patience, Amory; keep at it. You're winning. That's what makes your critics so frantic: you're winning!

Hugh Nash
Napa, California
December, 1978

GLOSSARY

ACES.	Annual cycle energy system.
ad-hocracy.	Improvisational decision-making
AFBC.	Atmospheric fluidized-bed combustion
APAE.	ERDA's office for planning and analysis.
APS.	American Physical Society.
ASHRAE.	American Society of Heating, Refrigeration, and Air Conditioning Engineers.
— bar.	Barometric pressure. 16 bar = 16 times barometric pressure.
bbl.	Barrel (42 gallons or 159.2 liters).
BCURA.	British Coal Utilization Research Association.
BPA.	Bonneville Power Authority.
BT.	Bankers Trust Company.
BTU.	British thermal units.
CANDU.	Canadian fission reactor using unenriched uranium.
CEI.	Council for Energy Independence.
CEQ.	Council on Environmental Quality, Executive Office of the President.
CONAES.	Committee on Nuclear and Alternative Energy Systems (of the National Academy of Sciences/National Research Council).
CONAES-CLOP.	Consumption, Locations and Occupational Panel (also known as The "lifestyles panel," chaired by Laura Nader), of CONAES.
COP.	Coefficient of performance, a measure of heat-pump efficiency.
DAF.	Canadian manufacturer of wind-electric machines, Dominion Aluminum Fabricating, Ltd.
DOE.	(U.S.) Department of Energy.
dq/dt.	Rate of change in energy use, measured in Quads, over time.
(e).	Electric, as in kW(e) = kilowatts (electric).
EEI.	Edison Electric Institute, a trade organization.
EPA.	Environmental Protection Agency.

EPP. Energy Policy Project (of the Ford Foundation)

EPRI. Electric Power Research Institute, utility research center.

ERDA. Energy Research and Development Administration (succeeded by DOE).

ERG. Energy Research Group, private energy consultants.

FA. *Foreign Affairs*, a quarterly; sometimes used here to mean Amory Lovins's article in *FA*, October 1976.

FBC. Fluidized-bed combustion.

FCB. Fives-Cail-Babcock, manufacturers of Ignifluid boilers.

FEA. Federal Energy Agency.

First-Law efficiency. A measure of the fraction of the energy supplied to a device or system that it delivers in its output, based on the First Law of Thermodynamics.

g. Gravity, as in 8g = eight times the acceleration of gravity.

GESMO. General Environmental Impact Statement of Mixed Oxide fuels (i.e., on recycling plutonium).

GJ. Giga-joule = billion joules, a measurement of energy.

GNP. Gross national product.

GW. Giga-watt = billion watts, a measurement of power.

h. Hour.

Heat pump. An electric appliance used to pump heat from one place to another; used to heat houses, or reversed, to air condition them. More efficient than electric resistance heating.

HTGR. High temperature gas reactor, a type of fission reactor.

IEA. Institute for Energy Analysis (at Oak Ridge).

IFIAS. International Federation of Institutes for Advanced Study.

IIASA. International Institute for Applied Systems Analysis.

Insolation. Sunlight falling on a surface (directly, or via cloud scattering, etc.).

J. Joule, a small unit of energy, equal to a watt-second or 1/1055 BTU.

k.	Thousand.
kW.	Kilowatts.
kWp.	Kilowatts of peak power.
LBL.	Lawrence Berkeley Laboratory.
LMFBR.	Liquid metal fast breeder reactor.
LNG.	Liquefied natural gas.
LWR.	Light water reactor, the type of fission reactor most common in the U.S.
m.	Meter, a bit more than one yard.
M.	Million (mega); though some, confusingly, use M for thousand and MM for million.
m^3/m^2.	Cubic meters per square meter, as in cubic meters of heat storage per square meter of collector surface in solar heating systems.
MITRE.	A research corporation that does energy and military studies for the U.S. government.
MJ.	Mega joule = million joules.
MW.	Megawatt = million watts.
NAS/NRC.	National Academy of Sciences/National Research Council.
NASA.	National Aeronautics and Space Administration.
NPDI.	Net Private Domestic Investment.
NRC.	Nuclear Regulatory Commission. Also National Research Council.
NSF.	National Science Foundation.
NSPS	New Source Performance Standards, air pollution control standards administered by EPA.
OCR.	Office of Coal Research (succeeded by DOE).
OECD.	Organization for Economic Cooperation and Development, which includes most of the industrialized nations.
O&M.	Operation and maintenance.
OMB.	Office of Management and Budget.
OPEC.	Organization of Petroleum Exporting Countries.
ORAU.	Oak Ridge Associated Universities.
ORNL.	Oak Ridge National Laboratory.
OTA.	Office of Technology Assessment.
PSC.	Public Service Commission.
PUC.	Public Utilities Commission.

q. Quad = quadrillion Btus, a very large amount of energy.

RD&D. Research, development, and demonstration.

Recuperator. A device that captures and recycles waste heat.

Retrofit. To install new equipment in an existing building, or modify some other existing device.

RFF. Resources for the Future, a private research organization.

RSS. Reactor Safety Study, also known as "The Rasmussen Report."

SEP. *Soft Energy Paths: Toward a Durable Peace,* a book by Amory Lovins, published by FOE/Ballinger, 1977; Penguin, 1977, and Harper & Row (Colophon), 1979. All editions are paginated identically.

SIPI. Scientists in the Public Interest, a private organization.

SIPRI. Stockholm International Peace Research Institute.

SRI. Stanford Research Institute.

Synfuels. Synthetic fuels (e.g., liquid fuels from coal).

(t). Thermal, as in kW(t) = kilowatts (thermal).

T&D. Transmission and distribution (in electric systems).

W. Watt, a unit of power (electric or thermal).

y. Year.

1974 $, 1976 $ Money fluctuates in value over time, owing to inflation, revaluations, etc., so, in order to give true comparisons, costs are expressed in dollars at their value for a particular year. The general 7% annual inflation over the last decade means that a 1974 dollar is worth about $1.145 in 1976 dollars. Conversions between different years' dollars can be made with this general "GNP deflator" or with more specialized indices showing the inflation rates in particular kinds of costs—such as power-plant construction.

Introduction: SENATOR NELSON

A congressional hearing on alternative long-range energy strategies was conducted jointly by two U.S. Senate committees on December 9, 1976, the Select Committee on Small Business and The Committee on Interior and Insular Affairs. Senator Gaylord Nelson, chairman of the former, presided. His opening statement follows.

The Committee on Small Business and the Committee on Interior and Insular Affairs are holding a joint public meeting this morning to hear distinguished experts discuss their views and their differences on energy strategy.

In the field of "energy strategy," we are concerned with such questions as these—among others:

Where are our energy assumptions, policies and programs taking us, in domestic development and international relations?

Might we prefer to go somewhere else?

And, if so, how do we change course?

The way our society answers these questions will affect employment, lifestyles, wealth, equity, war, and peace.

And not just for ourselves. Choices we make in our lifetimes, even within the lifetime of the incoming [Carter] administration, can have repercussions felt by descendants of ours who will be born on dates so far in the future, if our species lasts that long, as to be barely imaginable.

Indeed, some choices we have already made, perhaps irrevocably, will have repercussions over incredible stretches of time.

Let me illustrate. Neanderthal man, the last race of cavemen before Cro-Magnon man, our most immediate ancestor, occupied most of Europe for some 100,000 years, then disappeared rather suddenly toward the end of the Pleistocene epoch, some 25,000 to 30,000 years ago. Cro-Magnon man was himself incredibly ancient, of course—prehistoric, prestone age; but we are talking about his forbears. Neanderthal man's time on earth began roughly 125,000 years ago and ended roughly 25,000 years ago.

The period of time during which deposits of the manmade element plutonium must be separated from the biosphere is 250,000 years—twice as far away in the future as the beginnings of Neanderthal man are distant in the past.

The reason is that this extremely dangerous element, lethal to man and other living things in quantities far too small to see and almost too small to measure, has a radioactive half life for 24,400 years. That means, as a practical matter, that some 250,000 years must pass before it decays into a safe—or at least insignificantly dangerous—state.

One of our energy-strategy options—one that has wide support at this time—involves moving into what is termed the "plutonium economy."

The name stems from the fact that, under that option, plutonium would be one of the large energy sources, and we would be making, transporting, storing and using large quantities of this material. We would also be involved in finding safe ways to store away permanently—meaning 250,000 years—relatively small quantities.

Proponents of the plutonium economy insist that the longevity of the waste disposal problem is exaggerated, because most of the plutonium is consumed in the reactor or recycled for other reactors, while most of the radioactive materials that we are still trying to find ways to store permanently have much shorter half lives.

Dr. Ralph Lapp, for example, has written these reassuring words:

The waste radiation hazard is primarily the thirty-year half-lived strontium-90 and cesium-137 fission products; on a basis of allowing twenty half-lives to reduce the radioactivity to insignificant proportions, this means a containment or isolation period of 600 years—by no means a very great challenge.

[Senator Nelson is non-conservative in saying that plutonium must be separated from the biosphere for 250,000 years, which is about ten half-lives of plutonium. The usual rule of thumb is the one cited by Dr. Lapp: 20 half-lives, equivalent to a decay to slightly less than one-millionth of initial potency. Twenty half-lives of plutonium-239, the main isotope, amount to 488,000 years. Plutonium-239 decays into uranium-235, a slightly radioactive bomb material with a much longer half-life.]

Compared to 250,000 years, of course, 600 years is not very long at all. Going backward 600 years takes us not to pre-Neanderthal man but only to the end of the reign of Richard II in England, only to 116 years before the first voyage of Columbus and 55 years before the burning of Joan of Arc. By any standard that is an impressive period to store and assure the safety of large quantities of dangerous radioactive materials.

However, even the staunchest proponents of the nuclear economy admit that some amounts of plutonium itself will be around, safely or unsafely stored, adequately or inadequately marked in the language of some era or other, for the full 250,000 years it must be isolated.

That is simply unavoidable. There is some already in our temporary storage depositories we are now using while we wait to invent permanent storage or disposal means.

Each nuclear powerplant of 1,000 megawatts electrical generating capacity can be expected to create 220 kilograms of plutonium each year of normal operation.

By 2000, a plutonium power industry would require in the United States alone eight factories where loose plutonium will be fabricated into nuclear fuel. To recover the plutonium from used nuclear fuels will require about five large reprocessing plants.

And all of this plutonium must be transported from place to place. The Nuclear Regulatory Commission—NRC—estimates that if plutonium is used for fuel, or is continued to be used, during the next 25 years, trucks carrying this cargo will travel some 142 million miles and trains will travel some 82 million miles. Each mile presents some risk of escape of plutonium into the biosphere—most temporarily but some permanently—as a result of accidents.

The present generation implications of the plutonium economy may merit some thought. The prestigious Stockholm International Peace Research Institute—SIPRI—created by the Swedish Parliament, has recently published a compendium of papers entitled "Armaments and Disarmament in the Nuclear Age—A Handbook." That publication has resulted in statements in the New York Times and Parade magazine to this effect. [Quoting from Parade:]

"Within nine years, about 35 countries will be able to manufacture atomic weapons and nuclear war will become inevitable. . . . The (SIPRI) publication said 35 nations would achieve nuclear capability by 1985 as a by-product of peaceful nuclear programs."

In America, our own Office of Technology Assessment—OTA—is not so openly pessimistic as some of the SIPRI papers' authors; but it has launched an intensive study of the problems of nuclear proliferation.

In short, the picture that emerges is something like this:

To meet the transitory military competition and civilian electricity needs, or perceived needs, of some three to six generations in our own tiny speck of historic time, those generations are laying a radioactive curse on future generations. In the case of strontium-90, the curse will be deadly for small or large numbers of people in "only" about 150 generations.

In the case of plutonium-239, the number of generations affected, to some degree or other, will be around 10,000.

[Strontium-90 has a half-life of 28 years, and according to the 20-half-lives rule of thumb, requires 560 years to decay to a nominally safe level. This is 19 generations (taking a generation to be 30 years). On the same illustrative basis, plutonium will be unsafe for 16,267 generations.]

One way or another, if we are to go to our graves knowing we have created such military and energy blessings as strontium-90 and plutonium-239, we must, if we are to sleep well, have rationalized our ingenuity very, very well.

We must, at a minimum, be convinced that the social goals we are pursuing with our nuclear weapons and nuclear electricity are not only essential, vital goals, but also that we have absolutely no alternative means of attaining them.

As we all know, many respected scientists and statesmen, and a present majority of the American electorate, have rationalized our military and energy strategies to their own satisfaction; but others have not. Consequently, discussion quite properly continues. It is continuing this morning, at this hearing.

The members of our witness panel today are of uncommon intellect, education, and attainments. They have been trained in high technology. They have thought long and deeply about the questions of energy strategy. And they have come up with significantly different answers. We are indebted to them for their willingness to appear together to discuss those differences.

We have asked them to focus their remarks—to some extent, at least, not exclusively—on the stakes in energy strategy held by three large groups:

- Industry, especially those enterprises that now do and in future might comprise the energy industries;
- Small business, both as users of energy and as present and potential future participants in the energy industries; and
- The public—society at large.

Any such discussion must include the questions with which science and technology are concerned, but it neither begins nor ends with those questions.

Technology and science ask the questions and provide the answers about means. In energy strategy, we are primarily—almost exclusively—concerned about ends.

We tend to lose sight of this fact, but energy itself is not an end but a means.

Therefore, all discussion of how much energy we need and how best we can supply it must be interwoven with a larger question: To what end?

What kind of society are we trying to evolve?

Or are we not trying to evolve at all, but only to preserve the society we have, on a magnified scale?

A new draft report by the Stanford Research Institute—SRI—provides some fresh insights on this subject. It suggests that the predominant attitudes of society may be shifting in ways that might actually force decentralized solar energy technologies to be commercialized much faster.

It also suggests that philosophical and political factors—that is, the attitudes of people about the kind of society they want to live in and the kind of planet they want to leave their children—may be more important ultimately than the scientific, technical, and even economic factors in the shaping of America's choice among various energy options.

1

PREPARED TESTIMONY OF AMORY B. LOVINS

A Target Critics Can't Seem to Get in Their Sights

The first witness to testify before the joint hearing was Amory Lovins, representative of Friends of the Earth in the United Kingdom. His statement, slightly edited, follows.

I am glad to have this opportunity to explore a hopeful view of our energy future on which many energy strategists in many countries have recently begun to converge. To outline this convergence I shall sketch two hypothetical, illustrative energy paths which the United States (or, by analogy, other countries) might follow over the next 50 years. These paths are neither forecasts nor precise recommendations, but rather a qualitative vehicle for ideas of broad international relevance.

This introduction to a complex set of arguments will be brief and selective. A fuller account has been published in the October [1976] issue of *Foreign Affairs* as an annotated paper entitled "Energy Strategy: The Road Not Taken?" Among the many citations supplying technical support for that paper is a companion-piece entitled "Scale, Centralization, and Electrification in Energy Systems," which I prepared for an Oak Ridge symposium in October and which gives the details of calculations and arguments summarized in the *Foreign Affairs* paper. I understand that both papers will appear in the record of this hearing. [Edited versions of both papers, plus much new matter, make up Mr. Lovins's FOE/Ballinger book, *Soft Energy Paths: Toward a Durable Peace.*] For the moment I shall therefore omit most of the technical and numerical details in order to concentrate on fundamental concepts.

Most people and institutions responsible for past U.S. energy policy consider that the future should be the past writ large: that the current energy problem is how to expand supplies (especially domestic supplies) of energy to meet the extrapolated demands of a dynamic economy, treating those demands as homogenous—that is, as aggregated numbers representing total energy in a given year. Solutions to this problem have been proposed by, among others, ERDA, FEA, the Department of the Interior,

Exxon, and the Edison Electric Institute. A composite of the main elements of their proposals might resemble Figure 1. This diagram shows a policy of Strength Through Exhaustion—that is, pushing hard on all fuel resources, whether oil, gas, coal, or uranium. Fluid fuels (oil and gas) are obtained from present fields, the Arctic, offshore, and imports. More fluid fuels are synthesized from coal, whose mining is enormously expanded both for that purpose and to supplement vast amounts of nuclear electricity. In essence, more and more remote and fragile places are to be ransacked, at ever greater risk and cost, for increasingly elusive fuels, which are then to be converted into premium forms—fluids and electricity—in ever more costly, complex, gigantic, and centralized plants. As a result, while population rises by less than a fifth over the next few decades, our use of energy would double and our use of electricity would treble. Not fulfilling such prophecies, it is claimed, would mean massive unemployment, economic depression, and freezing in the dark.

FIGURE 1

AN ILLUSTRATIVE, SCHEMATIC FUTURE FOR U.S. GROSS PRIMARY ENERGY USE

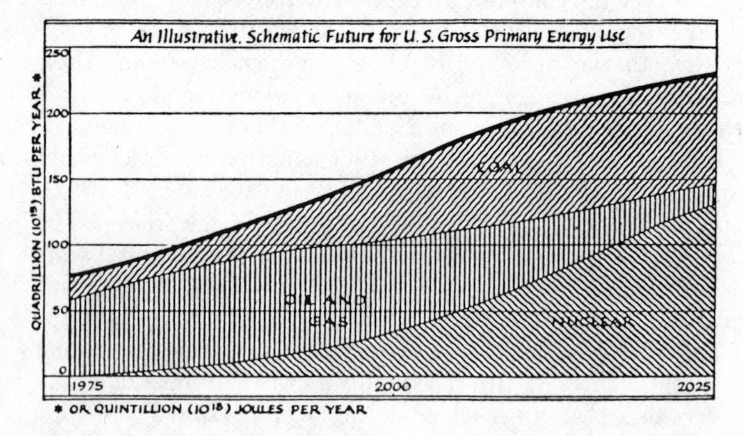

I shall consider in a moment whether such energy growth is necessary. But first let me suggest that it is unwise and unworkable. I do not mean by this that it is *politically* unworkable, though I think that is true; most Americans affected by offshore and Arctic oil operations, coal-stripping, and the plutonium economy have greeted these enterprises with a comprehensive lack of enthusiasm, because they correctly perceive the prohibitive political and environmental costs. Nor do I mean that Figure 1 is *technically* unworkable, though there is mounting evidence that we lack the skills, industrial capacity, and managerial ability to sustain such rapid expansion of untried and unforgiving technologies. I mean rather that Figure 1 is *economically* unworkable.

The first six items in Figure 2 show why. They state the capital investment typically required to build various kinds of complete energy systems

FIGURE 2

Approximate marginal capital investment for complete energy systems delivering 1 bbl/day [~67 kW(t)] enthalpy to US consumers

Energy system	1976 $/ (bbl · day)	Form supplied
Hard technologies:		
Traditional direct fuels, 1950s–1960s or direct coal, 1970s	2–3,000	F
North Sea oil, late 1970s	10,000	F
US frontier oil and gas, 1980s	10–25,000	F
Synthetics from coal or shale, 1980s	20–40,000+	F
Central coal-electric + scrubbers, 1980s	170,000	E
Nuclear-electric (LWR), mid-1980s	235,000+	E
Technical fixes to improve end-use efficiency:		
New commercial buildings	-3,000	HE(?)
Common industrial & architectural leak-plugging; better home appliances	0–5,000	HE
Most heat-recovery systems	5–15,000	H
Bottoming cycles; better motors	20,000	E
Worst-case, very thorough building retrofits	30,000	H
Transitional fossil-fuel technologies:		
Coal-fired fluidized-bed gas turbine with district heating and heat pumps (COP = 2), early 1980s	30,000	H
Most industrial cogeneration, late 1970s	60,000	EH
Soft technologies:		
Passive solar heating (≤ 100%)	< 0–20,000	H
Retrofitted 100% solar space heat, no backup needed, ~ 100-unit neighborhood	~ 20–40,000	H
Same, single house, mid-1980s	~ 50–70,000	H
300°C solar process heat, 1980	120,000	H
Collection and bioconversion of farm and forestry wastes to fuel alcohols, 1980	< 15–25,000+	F
Pyrolysis of municipal wastes, 1980	30,000	F
Microhydroelectric plants	30–140,000	E
Solar pond and Rankine engine	120,000	E
Wind-electric plants	~ 70–185,000	E

to increase delivered energy supplies by the heat equivalent of one barrel of oil per day. As we move from traditional oil and gas and direct-burning coal systems to frontier (Arctic and offshore) oil and gas and to synthetic fuels made from coal, the capital intensity of the system rises by about a factor of ten. As we move from those systems in turn to electrical systems, the capital intensity rises by about a further factor of ten. Such a hundredfold increase in capital intensity has led many analysts (for example, the Shell Group planners in London) to conclude that no major country outside the Persian Gulf can afford to build these big, complex, high-technology systems on a truly large scale—large enough to run a country. The cash flow that central-electric and synthetic-fuel systems generate is so unsustainable (even for a national treasury) that they are starting to look like future technologies whose time has passed.

Because Figure 1 [the "hard energy path"] relies mainly on the most capital-intensive systems, its first ten years—1976 through 1985—typically entail a total capital investment of over one trillion of today's dollars, three-quarters of it for electrification. This sum implies that the energy sector would consume its present quarter of all new private investment in the U.S.—plus about two-thirds of all the rest. We couldn't even afford to build the things that were supposed to *use* all that energy. Later the burden would grow even heavier.

These astronomical investments would give us so many power stations that, as Professors von Hippel and Williams show in the December [1976] *Bulletin of the Atomic Scientists,* we would have to use most of the electricity for very wasteful and economically unjustifiable purposes. Yet because of slow and imperfect substitution of electricity for oil and gas, we would still be seriously short of these indispensable fuels. Even worse, *at least half* of our total energy growth in the next few decades would be *lost* in conversions from one kind of energy to another before it ever got to us. The efficiency of the fuel chain would plummet. Serious shortages of capital, labor, skills, and materials, as these resources were diverted to the energy sector, would exacerbate inflation. And every "quad" (quadrillion BTU) of primary energy fed into new power stations would *lose* some 75,000 net jobs (about 4,000 per 1000-megawatt station) because power stations produce fewer jobs per dollar, directly and indirectly, than virtually any other investment in the whole economy.

The massive diversion of scarce resources into the energy sector would thus worsen, not correct, the economic problems it was intended to prevent. At the same time it would create serious social and political problems. Reallocating scarce resources to priorities that the market is plainly unwilling to support would require a strong central authority with power to override any objections. Political and economic power would concentrate in politically unaccountable oligopolies and bureaucracies. A bureaucratized technical elite, politically remote from energy users, would operate the complex systems and say who could have how much energy at what price. Centralized energy systems would allocate energy and its side-effects to different groups of people at opposite ends of the transmission line or pipeline. As the energy went to Los Angeles or New York and the social costs to Appalachia, Navajo country, Montana, or the Brooks Range, inequity and alienation would fuel new tensions, pitting region against region, even as a wider and more divisive form of centrifugal politics pitted central siting and regulatory authority against local autonomy. [In mid-1978, there were more than 60 such "energy wars" going on in the U.S.]

Energy supplies would depend increasingly on centralized distribution systems vulnerable to disruption by accident or malice. Electrical grids, in particular, distribute a form of energy that cannot readily be stored in bulk, supplied by hundreds of large and precise machines rotating in exact synchrony across a continent, and strung together by a frail network of aerial arteries that can be severed by a rifleman or disconnected by a few strikers. This inherent vulnerability could be reduced only by stringent social controls, similar to those required to protect plutonium from theft, LNG terminals from sabotage, and the whole energy system from dissent. The difficulty, too, of making decisions about compulsory technological hazards that are disputed, unknown, or (often) unknowable would increase the risk that citizens might reject official preferences. Democratic process would thus have to be replaced by elitist technocracy, "we the people" replaced by "we the experts." And over all these structural and political problems would loom the transcendent threat of nuclear violence and coercion in a world dependent on international commerce in atomic bomb materials measured in tens of thousands of bombs' worth per year. The impact of human fallibility and malice on nuclear systems would, I believe, quickly corrode humane values and could destroy humanity itself.

These many side-effects of Figure 1 would be even worse in interacting combination than singly. Continued energy waste implies continued U.S. dependence on imported oil, to the detriment of the Third World, Europe, and Japan. We pay for the oil by running down domestic stocks of commodities, which is inflationary; by exporting weapons, which is inflationary, destabilizing, and immoral; and by exporting wheat and soybeans, which inverts Midwestern real-estate markets, makes us mine groundwater unsustainably in Kansas, and increases our food prices. Exported American wheat diverts Soviet investment from agriculture into military activities, making us increase our own inflationary military budget, which we have to raise anyhow to defend the sea-lanes to bring in the oil and to defend the Israelis from the arms we sold to the Arabs. Pressures increase for energy- and water-intensive agribusiness, creating yet another spiral by impairing free natural life-support systems and so requiring costly, energy-intensive technical fixes (such as desalination and artificial fertilizers) that increase the stress on remaining natural systems while starving social investments. Poverty and inequity, worsened by the excessive substitution of inanimate energy for people, increase alienation and crime. Priorities in crime control and health care are stalled by the heavy capital demands of the energy sector, which itself contributes to the unemployment and illness at which these social investments were aimed. The drift toward a garrison state at home, and failure to address rational development goals abroad, encourages international distrust and domestic dissent, both entailing further suspicion and repression. Energy-related climatic shifts could jeopardize marginal agriculture, even in our own Midwest, endangering an increasingly fragile world peace.

If it were true, as the proponents of Figure 1 insist, that there is no alternative to it, then the prospects for a humane and sustainable energy future would be bleak indeed. But I believe there is another way of looking at the energy problem that can lead us to a very different path: one that is quicker, cheaper, more beneficial to the economy, and politically and environmentally far more attractive. Such an energy path, which I shall call a "soft" path, is illustrated in Figure 3. It has three main technical components, which I shall discuss in turn:

- greatly increased efficiency in using energy:
- rapid deployment of "soft" technologies (which I shall define presently); and
- the transitional use of fossil fuels.

These three components mesh to make a whole far greater than the sum of

FIGURE 3

AN ALTERNATE ILLUSTRATIVE FUTURE FOR U.S. GROSS PRIMARY ENERGY USE

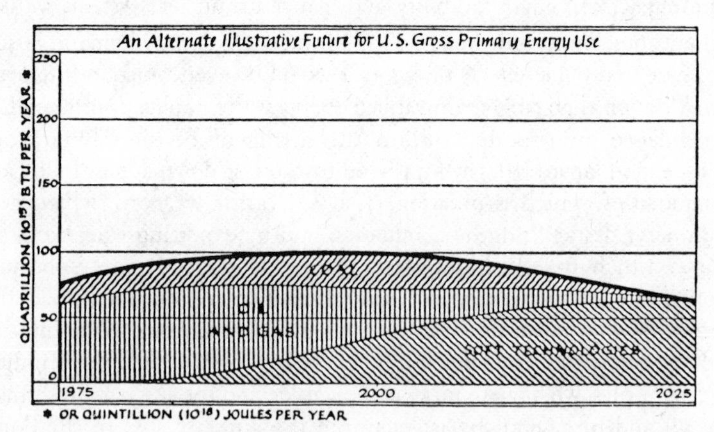

its parts: a coherent policy distinguished from the "hard" path of Figure 1 not by how much energy we use or by our choices of energy equipment, but primarily by the technical and sociopolitical *structure* of the energy system. This distinction will become clearer in a few minutes.

The hard energy path depicted in Figure 1 rests on the belief that the

FIGURE 4

AVERAGE PER CAPITA PRIMARY ENERGY CONSUMPTION IN DENMARK FOR HEATING AND COOKING[a]

	year	Gcal/yr
ca	1500	7–15
	1800	7
	1900	3
	1950	7
	1975	17

[a]*Source:* "*Energy in Denmark 1990–2005: A Case Study,*" *Report #7, Survey by The Work Group of The International Federation of Institutes for Advanced Study, c/o Sven Bjørnholm, The Niels Bohr Institute, The University of Copenhagen, September 1976.*

more energy we use, the better off we are. Energy is elevated from a means to an end in itself. In the soft path of Figure 3, on the contrary, how much energy we use to accomplish our social goals is considered a measure less of our success than of our failure—just as the amount of traffic we must endure to gain access to places we want to get to is a measure not of well-being but rather of our failure to establish a rational settlement pattern. The cornerstone of Figure 3, therefore, is seeking to attain our goals with an elegant frugality of energy and trouble, using our best technologies to wring as much social function as possible from each unit of energy we use and supplying that energy in the most effective way for each task.

Many people who have finally learned that energy efficiency does not mean curtailment of functions (that is, that insulating your roof does not mean freezing in the dark) still cling to the bizarre notion that not to use more energy nevertheless means somehow a loss of prosperity. This idea cannot survive inspection of Figure 4, which shows how much energy an average person in Denmark used at various times for heating and cooking (over half of all end use). Looking only at the values for 1900 through 1975, one might be tempted to identify increasing energy use with increasing well-being. But if that were true, the statistics for the years 1800 and 1500 would imply that Danes have only just regained the prosperity they enjoyed in the Middle Ages. Deeper analysis shows what is really happening. In 1500 and 1800 Denmark had a wood and peat economy, and most of the heat went up the chimney rather than into the room or cookingpot—just as it does in the Third World today. In 1900 Denmark ran on coal, burned efficiently in tight cast-iron stoves. In 1950 Denmark used mainly oil, incurring refinery losses to run inefficient furnaces. In 1975, the further losses of power stations were added as electrification expanded. This example shows that a facile identification of primary energy use with well-being telescopes several complex relationships that must be kept separate. How much primary energy we use—the fuel we take out of the ground—does not tell us how much energy is delivered at the point of end use, for that depends on how efficient our energy system is. End-use energy in turn says nothing about how much function we perform with the energy, for that depends on our end-use efficiency. And how much function we perform says nothing about how much better off we are, which depends on whether or not the thing we did was worth doing.

I shall suggest in a moment that a rational energy system can virtually eliminate conversion and distribution losses that rob us of delivered end-use energy. But let me focus first on the efficiency with which that end-use energy can do our tasks for us. There is ample technical evidence that Americans can double this efficiency by about the turn of the century by using only "technical fixes"—that is, measures which

- are now economic by orthodox criteria,
- use today's technologies (or, more often, 1920s technologies), and
- have no significant effect on lifestyles.

Such measures include thermal insulation, more efficient car engines, heat recovery in industrial processes, and cogeneration of electricity as a by-product of process heat.

Over the 50 years shown in Figure 3—long enough to turn over much of the stock of buildings and equipment—we can use technical fixes *alone* to treble or quadruple our end-use efficiency, yielding a convex curve [indicating a long-term decline in gross primary energy use] despite normally projected rapid growth in population, economic activity, comfort, and equity. People who consider today's values and institutions imperfect are free to obtain the same result by some mix of technical and social changes—perhaps substituting repair and recycling for the throwaway economy, or gradually shifting settlement patterns and concepts of work so as to keep voluntary mobility from being swamped by involuntary traffic. But my analysis assumes no such social changes. Our end-use efficiency can be improved to and beyond present European levels *entirely* through technical measures if we wish, saving both money and jobs in the process. Figure 2 shows that conservation is far cheaper than increasing supply. Both technical fixes (such as insulation programs) and shifts toward less energy-intensive consumption patterns produce far more long-term jobs, using existing skills, than alternative supply investments can. Recent input-output studies suggest that conservation programs and shifts of investment from energy-wasting to social programs create anywhere from tens of thousands to nearly a million net jobs per quad saved.

The top curve of Figure 3, showing total energy needs of, say, 95 quads in the year 2000, is by no means the lowest that can be realistically considered. [This 95-quad value was indeed exactly that projected by the U.S. Department of Energy in its September 1978 Status Report on the Domestic Policy Review of solar energy, assuming oil prices of $32 per barrel (1978$).] Projections are dropping rapidly. Dr. Alvin Weinberg's group at Oak Ridge projects for 2000 a primary energy demand of 101–126 quads—the lower value being more likely—even with a modest conservation program. But Dr. Weinberg's end-use energy projections are much *lower* than mine, since he assumes a half-electric economy. Further, a major National Academy of Sciences study is now taking seriously a year-2000 technical-fix projection in the range of 65-90 quads of primary energy. [On 14 April 1978, the Demand and Conservation Panel of this study (CONAES—Committee on Nuclear and Alternative Energy Systems) published in *Science* its projections for the year 2010 ("U.S. Energy Demand: Some Low Energy Futures"), including "pure technical fix" values of 77 and 96 quads and a further value of 63 quads that could have been a "pure techincal fix" had the panel not used such very conservative techincal assumptions.] One panel is even proposing a value around 40–50 quads by assuming some lifestyle changes that could arguably improve the quality of life far more than the lifestyle changes of a high-energy path.

The second major component of the soft path shown in Figure 3 is the rapid deployment of soft energy technologies, which I define by five characteristics:

- They are diverse. Just as a national budget is paid for by many small tax contributions, so the soft-technology component of energy supply is made up of small contributions by many diverse technologies, each doing what it does best and none trying to be a panacea.
- They operate on renewable energy flows—sun, wind, forestry or agricultural wastes, and the like—rather than on depletable fuels.
- They are relatively simple and understandable. Like a pocket calculator, they can be highly sophisticated, but are still technologies we can live with, not mysterious giants that are alien and arcane.
- They are matched in *scale* to end-use needs.
- They are matched in *energy quality* to end-use needs.

These last two points are very important, and I must amplify each in turn.

First, scale: we are used to hearing that energy facilities must be enormous to take advantage of economies of scale. But we are seldom told about the even greater *dis*economies of scale. There are at least five main kinds:

- Big systems cannot be mass-produced. If we could mass-produce power stations like cars, they would cost less than a tenth as much as they do now, but we can't.

- Big systems, being centralized, require very costly distribution systems. In 1972, out of every dollar that U.S. residential and commercial users of electricity paid to private utitlities, only about 31¢ paid for electricity; the other 69¢ paid for having it delivered. [The same was true of natural gas in 1977.] That is a major diseconomy of centralization.
- Associated with the distribution system is a pervasive web of energy losses in distribution.
- Big energy facilities tend to be much less reliable than small ones, and unreliablity is a graver fault in a big than in a small unit, requiring more and costlier reserve capacity. The unreliability of the distribution system compounds this problem, and can make a unit of new generating capacity some 2.5 times less reliable (from the user's point of view) if sited in a central power station than if locally sited in a small station.
- Big energy facilities take a long time to build, and are therefore especially exposed to interest costs, escalation, mistimed demand forecasts, and wage pressure by unions which know the high cost of delay.

Other diseconomies of large scale, less expressible in economic terms, are also important. For example, big systems entail the high political costs of centrism and vulnerability that I mentioned earlier. Big systems, which magnify the cost and likelihood of mistakes, are too costly and complex for technologists to play with, so an important well-spring of inventiveness and ingenuity is dried up. Big systems also complicate integration into total-energy systems. But even ignoring these qualitative effects, one can use orthodox economics, as I have done in the Oak Ridge paper, to reach a basic and perhaps surprising conclusion: that, in general, soft energy technologies are *cheaper* than the big "hard" energy technologies which one would otherwise have to use in the long run to do the same job. Thus, as can be calculated from Figure 2, a completely solar retrofitted space-heating system, with seasonal heat storage and no backup, has a lower capital cost than a nuclear-electric and heat-pump system to heat the same house; it also has a lower life-cycle cost than a coal-synthetics and furnace system to heat the same house. Making vehicle fuel from forestry and agricultural wastes is generally cheaper than making similar fuels from coal. And so it goes: even using today's rather cumbersome art, the soft technologies compete favorably with their long-run alternatives.

This is not to say that solar heat can always compete with unrealistically cheap gas; but that is irrelevant, for the days of cheap gas are numbered. Congress has already been asked for billions of dollars to subsidize the synthesis of gas from coal at perhaps $30 per barrel equivalent. Of course the soft technologies would have to be financed at the household, neighborhood, or town scale at which they would be built; but capital transfer schemes already in use can give even householders the same kind of access to capital markets that oil majors and utilities now enjoy. [Mr. Lovins expanded this proposal in "How to Finance the Energy Transition," *Not Man Apart* (FOE, San Francisco), mid-Sept./Oct. 1978, pp 8–10.]

FIGURE 5

END-USE ENERGY (HEAT-SUPPLIED BASIS) CLASSIFIED BY PHYSICAL TYPE, *ca.* 1973

type	United States		Canada[a]		United Kingdom[a]	
heat	58%		69%		65%	
below 100°C		35%		39%		55%
100°C and above		23%		30%		10%
mechanical work	38%		27%		30%	
vehicles		31%		24%		27%
pipelines		3%				
industrial electric drive		4%		3%		3%
other electrical	4%		4%		5%	

[a]*Preliminary and approximate data; probably within 5% for Canada and 10% for UK.*

Source: US data calculated in A. B. Lovins, "Scale, Centralization, and Electrification in Energy Systems," ORAU symposium "Future Strategies for Energy Development," Oak Ridge, 20–1 October 1976. Canadian data based on sources cited in A. B. Lovins, "Exploring Energy-Efficient Futures for Canada," Conserver Society Notes 1, 4, 5–16 *(May/June 1976), Science Council of Canada (150 Kent St, Ottawa). British data estimated by A. B. Lovins from official statistics, including Census of Production data, and consistent with estimates by the Energy Research Group of The Open University (see* e.g. Energy Policy, *September 1976).*

Soft energy technologies are matched to end-use needs not only in scale but also in energy quality. Figure 5 classifies our end-use needs by physical type. About 58% is in the form of heat, of which 35% is below and 23% above the boiling point of water. A further 38% is mechanical work—31% to move vehicles, 3% to pump fluids through pipelines, and 4% to drive industrial electric motors. The remaining 4% represents *all* lighting, electronics, telecommunications, smelting, electroplating, arc-welding, electric railways, electric drive for home appliances, and all other uses of electricity other than low-grade heating and cooling.

Electricity is a very expensive form of energy: Americans already pay typically from $40 to $120 per barrel equivalent for it. The premium applications in which we can get our money's worth out of this special kind of energy total only about 8% of all our end uses. With improved efficiency, that 8% would shrink to about 5% [later analyses suggest roughly half as much as this], which we could cover with our present hydroelectric capacity plus a modest amount of industrial cogeneration. [The reduced figure could readily be covered with present and small-scale hydroelectric capacity, a bit of windpower, and no cogeneration.] That is, we could advantageously be running this country with no central power stations at all—if we efficiently used electricity only for tasks that can use its high quality to advantage, so justifying its high cost in money and fuels. These limited premium tasks are already far oversupplied, so if we make more electricity we can only use it for inappropriate low-grade purposes. That is rather like cutting butter with a chainsaw—which is inelegant, expensive, messy, and dangerous.

This thermodynamic philosophy saves us energy, money, and trouble by supplying energy only in the quality needed for the task at hand: supplying low-temperature heat directly, not as electricity or as a flame temperature of thousands of degrees. Thus our task is not to find a substitute to produce thousand-megawatt blocks of electricity if we don't build reactors, but rather to perform directly the tasks we would have performed with the oil and gas for which the reactors were supposed to substitute in the first place. By thus matching energy quality to end use we can virtually eliminate conversion losses, just as matching scale virtually eliminates distribution losses. These two kinds of losses together make up *more than half* of the total energy used at the right-hand end of Figure 1 [representing the hard energy path in the year 2025], yet are all but absent at the same point in Figure 3. Delivered end-use energy is not very different in the two diagrams in the year 2025, but it performs several times as much social function in Figure 3 as in Figure 1, with a corresponding advantage in conventionally defined social welfare. We would literally be doing better with less energy.

Extremely rapid recent progress in developing soft technologies has produced a wide range of technically mature systems—ones that face no significant technical, economic, environmental, social, or ethical obstacles, and require only a modicum of sound product engineering. We already have enough mature, convenient, and reliable soft technologies now in or entering commercial service to meet essentially all our energy needs in about 50 years. This does not assume cheap photovoltaic systems (which will probably soon be, or may already be, available), nor indeed any other solar-electric technologies. Living within our energy income requires only the appropriate use of straightfoward solar heat technologies, organic conversion to clean liquid fuels for vehicles, modest amounts of wind collection (mainly non-electrical), currently installed hydroelectric capacity and the readily available small-scale hydroelectric capacity that uses existing dams.

I believe a *prima facie* case has been made that we already know enough to start planning an orderly transition to essentially complete reliance on soft energy technologies. But this transition, or indeed any other, will take a long time—perhaps 50 years—so we must build a bridge to our energy-income economy by briefly and sparingly using fossil fuels, including modest amounts of coal, to buy time. We know how to do this much more cleanly and efficiently than we are doing now, with technologies flexibly designed so that we can plug in the smaller soft technologies as they come along. Simple, flexibly scaled technologies are also rapidly emerging for cleanly extracting premium liquid and gaseous fuels from coal, so filling the real transitional gaps in our fluid-fuel economy with only a temporary (and less than twofold) expansion of coal mining. We can thus squeeze the "oil and gas" wedge in Figure 3 from both sides, making its middle section more slender than that of Figure 1 and thereby eliminating much medium-term importing and frontier extraction.

Figure 3, in summary, shows a different path along which our energy systems can evolve from now on. It does not suggest wiping the slate clean, but rather redirecting our effort at the margin, thus freeing disproportionate resources for other tasks. It does not abolish big technologies, but rather concedes that they have a limited place which they have long since saturated, and proposes that we can take advantage of the big systems we already have without multiplying them further. Because the long lead time of big systems means that many power stations are now under construction, we can use that backlog, current overcapacity, and the potential for rapid industrial cogeneration to ensure adequate supplies of electricity before improved effeciency starts to bear fruit.

Having these two energy paths before us in outline permits some illuminating comparisons. First, in rates: it appears that the soft and transitional technologies are much quicker to deploy than the big high technologies of Figure 1, because the former are so much smaller, simpler, easier to manage, and less dependent on elaborate infrastucture. For fundamental engineering reasons, their lead times are measured in months, not decades. For similar reasons, as I suggested earlier, they are also likely to be substantially cheaper than hard technologies. They are environmentally far more benign, and bypass the risk of major climatic change from burning fossil fuel. And they are more certain to work, for the risk of technical failure in the soft path is distributed among a large number of simple, diverse technologies that are in general already known to work and that require an especially forgiving kind of engineering. The hard energy path, in contrast, puts all its eggs in a few brittle baskets—fast breeder reactors and giant coal-gas plants, for example—which are not here and may or may not work.

Hard and soft technologies have very different implications for technologists. Hard technologies are demanding and frustrating. They are not much fun to do and are therefore unlikely to be done well. While they strain technology to (and beyond) its limits, the scope they offer for innovation is of a rather narrow, routine sort, and is buried within huge, anonymous research teams. The systems are beyond the developmental reach of all but a few giant corporations, liberally aided by public subsidies, subventions, and bailouts. The disproportionate talent and money devoted to hard technologies gives their proponents disproportionate influence, reinforcing the trend and discouraging good technologists from devoting their careers to soft technologies—which then cannot absorb funds effectively for lack of good people. And once hard technologies are developed, the enormous investments required to tool up to make them effectively exclude small business from the market, thus sacrificing rapid and sustained returns in money, energy, and jobs for all but a small segment of society.

Soft technologies have a completely different character. They are best developed by innovative small businesses and even individuals, for they offer immense scope for basically new ideas. Their challenge lies not in complexity but in simplicity. They permit but do not require mass production, thus encouraging local manufacture, by capital-saving and labor-intensive methods, of equipment adapted to local needs, materials, and skills. Soft technologies are multi-purpose and can be integrated with buildings and with transport and food systems, saving on infrastructure. Their diversity matches our own pluralism: there is a soft energy system to match any settlement pattern. Soft technologies do not distort political structures or priorities; they improve the quality of work by emphasizing personal ingenuity, responsibility, and craftsmanship; they are inherently non-violent, and are therefore a livelihood that technologists can have good dreams about.

Soft technologies, unlike hard ones, are compatible with the modern concept of indigenous eco-development in the Third World. They directly meet basic human needs—heating, cooking, lighting, pumping—rather than supplying a costly, high-technology form of energy with imported, capital-intensive, high-technology systems that can only enrich urban elites at the expense of rural villagers. Soft technologies would thus contribute promptly and dramatically to world equity and order. A soft energy path would also do more than a hard path to reduce pressures on world oil and coal markets and to avoid global capital shortages.

An even more important geopolitical side-effect of U.S. leadership in a soft energy path is that it promotes a world psychological climate of denuclearization, in which it comes to be viewed as a mark of national immaturity to have or desire either reactors or bombs. Nuclear power programs and their web of knowledge, expectation, and threats are now the main driving force behind proliferation of nuclear weapons. Yet foreign nuclear programs require a domestic political base, both public and private, that does not yet exist and must be borrowed from the U.S. Ten years' residence in Europe has led me to the firm political judgment that unilateral U.S. action to encourage soft energy paths at home and abroad, in tandem with new initiatives for nonproliferation and strategic arms reduction, could turn off virtually all foreign nuclear power programs and could thus go very far to put the nuclear genie back into the bottle from which we first coaxed it to emerge. Three decades after we secretly chose a fateful path which we then tried to force on the world, we can again lead the world by openly choosing, and freely helping others to choose, a path of prudence. But this shift of policy is urgent: we must stop passing the buck before our clients start passing the bombs.

If nuclear power were phased out, here or abroad, but no other policy were changed, the economic, environmental, and sociopolitical costs of centralized electrification would still be intolerable—even with energy conservation. As I suggested earlier, the key distinction between the soft and hard energy paths is in their architecture, the structure of the energy system, rather than the amount of energy used. Both paths entail difficult social problems of very different kinds: for the hard path, autarchy, centrism, technocracy, and vulnerability; for the soft path, the need to adapt our thinking to use pluralistic consumer choice, participatory local democracy, and resilient design to substitute a myriad of relatively small devices and refinements for large, difficult projects under central management. We must choose which of the two kinds of social problems we want. But the problems of the hard path can only be addressed much less pleasantly, less plausibly, and less consistently with traditional values than the more tractable problems of the soft path. The problems of the hard path, too, become steadily harder, while those of the soft path become gradually easier.

I believe that to a large extent the soft path would implement itself through ordinary market and political processes once we had taken a few important initial steps, including:

- correcting institutional barriers to conservation and soft technologies, such as obsolete building codes and mortgage regulations, lack of capital-transfer programs, restrictive utility practices, and lack of sound information;
- removing subsidies to conventional fuel and power industries, now estimated at well over $10 billion per year, and vigorously enforcing antitrust laws;

- pricing fuels at a level consistent with their long-run replacement cost. I believe this can be done in a way that is both equitable and positively beneficial to the economy, since unrealistically cheap energy is an illusion for which we pay dearly everywhere else in the economy.

[U.S. Department of Energy estimates in 1978 suggest that rolled-in pricing—whereby incremental consumers see average prices rather than, for example, the real present cost of Alaskan oil, Canadian gas, and recently commissioned power stations—currently gives an effective subsidy of more than $67 billion per year to hard technologies. This value would be even higher if calculated at true *marginal* costs—the costs of the *next* unit (such as power stations ordered now)—and higher still for long-run marginal costs (for systems ordered decades from now).]

I do not pretend that these steps will be easy; only easier than not taking them. Properly handled, though, they can have enormous political appeal, for the soft energy path offers advantages for every constituency: jobs for the unemployed, capital for businesspeople, environmental protection for conservationists, enhanced national security for the military, savings for consumers, opportunity for small business to innovate and for big business to recycle itself, exciting technologies for the secular, a rebirth of spiritual values for the religious, world order and equity for globalists, energy independence for isolationists, radical reforms for the young, traditional virtues for the old, civil rights for liberals, states' rights for conservatives. Though a hard energy path is consistent with the interests of a few powerful American institutions, a soft path is consistent with far more strands of convergent social change at the grassroots. It goes with, not across, our political grain.

The choice between the soft and hard paths is urgent. Though each path is only illustrative and embraces an infinite spectrum of variations on a theme, there is a deep structural and conceptual dichotomy between them. Soft and hard *technologies* are not *technically* incompatible: in principle, nuclear power stations and solar collectors can coexist. But soft and hard *paths* are *culturally* incompatible: each path entails a certain evolution of social values and perceptions that makes the other kind of world harder to imagine. The two paths are *institutionally* antagonistic: the policy actions, institutions, and political commitments required for each (especially for the hard path) would seriously inhibit the other. And they are *logistically* competitive: every dollar, every bit of sweat and technical talent, every barrel of irreplaceable oil, every year that we devote to the very demanding high technologies is a resource that we cannot use to pursue the elements of a soft path urgently enough to make them work together properly. In this sense, technologies like nuclear power are not only unnecessary but a positive encumbrance, for their resource commitments foreclose other and more attractive options, delaying soft technologies until the fossil-fuel bridge has been burned. Thus we must, with due deliberate speed, choose one path or the other, before one has foreclosed the other or before nuclear proliferation has foreclosed both. We should use fossil fuels—thriftily—to capitalize a transition as nearly as possible straight to our ultimate energy-income sources, because we won't have another chance to get there.

To fix these ideas more firmly, I should like now to sketch an example from one of the approximately ten other countries in which soft-path studies are under way: Canada, whose energy system is strikingly similar to that of the U.S. This example comes from a study which I did last March in the Canadian Ministry of Energy, Mines, and Resources under the auspices of the Science Council of Canada. The data are theirs, the conclusions my own.

FIGURE 6

CANADIAN ENERGY USE IN 1973 (POP. 22M)

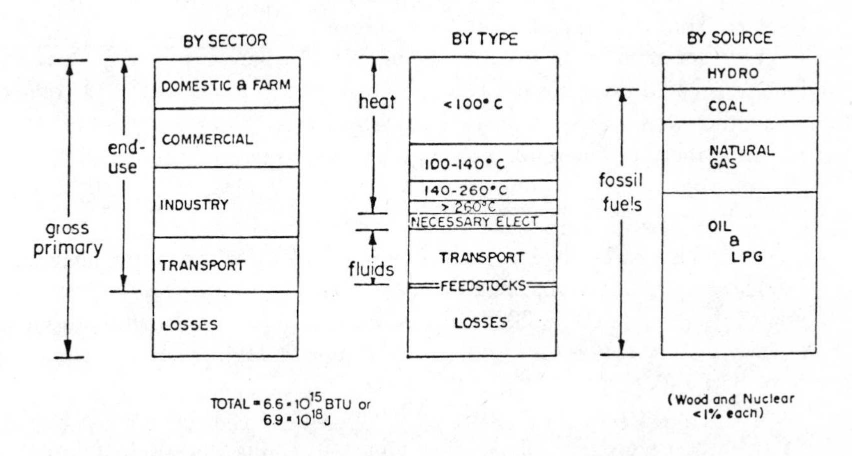

Figure 6 shows the approximate structure of energy use in Canada. Use by economic sector and by source broadly resembles our own, except that Canada has proportionally more hydroelectricity. The middle bar, showing the thermodynamic structure of end use, reveals that, as in the U.S., end-use needs are mostly heat (especially at modest temperatures) and liquids. Indeed, the electrical uses other than low-grade heating and cooling are such a small term that, on an aggregated basis, they *and* all high-temperature heat could be supplied by present hydroelectricity.

The Canadian Cabinet has approved technical fixes to improve end-use efficiency. The Ministry calculates that these rather modest measures, now being implemented, should hold primary energy use for the commercial and transport sectors and for heating houses roughly constant over the next 15 years despite normally projected growth in population and in sectoral economic activity. However, 15 years is really too short for these measures to bear much fruit, since it takes a much longer time to turn over the stock of buildings and equipment. I therefore looked ahead 50 years—about the end of the lifetime of a power station ordered today. I assumed population and economic growth similar to official projections (with minor and unimportant exceptions), plus technical fixes of the moderate, straightforward type already being implemented. The result over 50 years was a shrinkage of per capita primary energy to half today's level, or about the present Western European level. [This result was confirmed by a more detailed government study (Brooks et al, reprinted in Volume 2 of the Senate hearing record) and by later studies at both the federal and the provincial level.]

I also applied the same technical fixes not to a standard growth scenario but instead to the per capita activity levels of 1960—which might be a very rough surrogate for a luxurious form of "conserver society." This calculation yielded a further factor-of-two shrinkage, to a quarter of today's level or about the present New Zealand level.

I then returned to my original, higher set of estimates of energy needs for 2025, based on approximately a normal growth projection, and constructed a conservative estimate (exaggerating high grade needs) of its end-use structure. The result is the middle bar of Figure 7, with the current structure on the left for comparison. On the right side of Figure 7 I have drawn to the same scale some building-blocks of supply.

The block marked "hydro" is the minimum hydroelectric output already firmly committed for 1985, not counting James Bay. It exceeds the appropriate electrical uses in 2025. Below it is a minor contribution from roughly the present level of rural wood-burning and from anaerobic digestion, burning energy studies, etc.

At the lower right in Figure 7, I have assumed that 50 years is long enough to meet essentially all low-temperature heat needs with solar technologies. These are already technically and economically attractive in Canada, and the real question is not viability but deployment rate. If you think 50 years is not long enough to deploy all those devices, you are free to choose a target date later than 2025 and stretch out the transition. The 50 years is merely a first guess.

On the far right side of Figure 7 is a box labeled "liqwood." This is the name and number given by the Canadian Forestry Service to the net amount of fuel alcohols that they think they could sustainably produce from forestry. I suspect that closer ecological study will reduce this estimate, but it is still likely to be larger than the liquid fuel requirement for the transport sector (shown in the middle bar), even today (as shown in the left-hand bar). [This was confirmed in May 1978 by a very detailed study done for the Canadian government by InterGroup Consulting Economists, Ltd. (Winnipeg).]

FIGURE 7

CANADIAN ENERGY USE

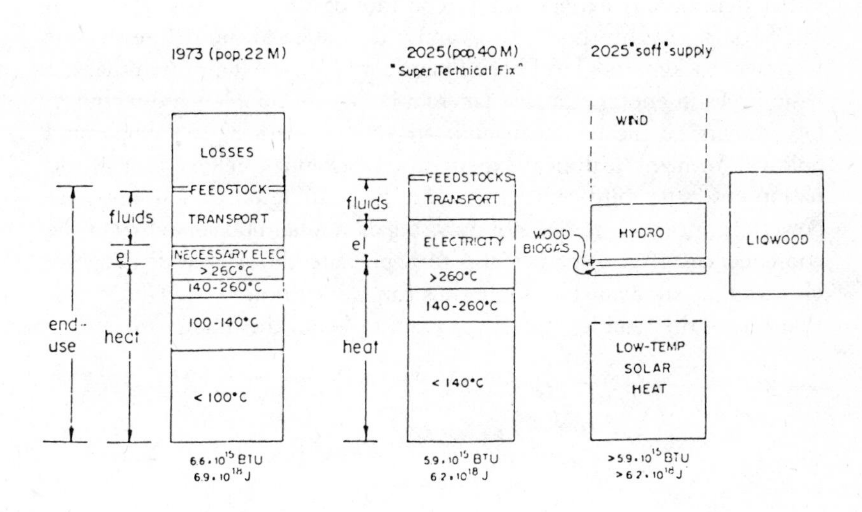

At the upper right in Figure 7 is an open-ended section labeled "wind." It is a very large number, far off-scale. The versatility of wind lets it fit anywhere into the end-use structure—pumping heat, for example, or compressing air to drive industrial machines.

Figure 7 suggests on its face that the "hydro" block already committed, plus any two of the other three, will yield *a surplus that matches the structure* of Canadian energy needs in 2025. In practice one would use a combination of all these sources. The next step in this exercise, which has not yet been done in Canada, is to refine the data, disaggregate by regions, and work backwards from 2025 toward now to see how to get there. Two things make me expect attractive results. First, such a study has already been done in countries less well placed than Canada, such as Denmark and Japan, and the transition looks quicker and cheaper than present policy. Second, if one pairs off the soft energy supply technologies with the hard technologies that one would otherwise have to use in the long run to do the same job, the soft ones are cheaper, and likewise for the conservation and transitional technologies; so when one adds them all up they should still be cheaper.

The value of thinking backwards in this way from where one wants to be in the long run—the right-hand end of Figure 3—is that it reveals the potential for radically different paths that would be completely invisible to anyone who merely worked forwards in time through incremental adhocracy. Such a person could only discover, perhaps, in 2010 that we might have gone in a very different direction if we had thought of it in 1980 before it was too late. To avoid such a trap we must be unabashedly normative in exploring our goals, then figure out how to achieve them, rather than blindly extrapolating trend into destiny.

Of course, in this short talk I cannot do justice to the richness of the technical background; but I hope I have conveyed the impression that the basic issues in energy strategy, far from being too complex and technical for ordinary people to understand, are on the contrary too simple and political for many technical experts to understand. I believe that as this nation enters its third century we must concentrate on these simple, yet powerful concepts if we are not only to gain a fuller understanding of the consequences of choice, but also to appreciate the very wide range of choices that are available. Only thus can we learn, as Robert Frost did, that taking the road less traveled by can make all the difference.

BLACKMAN vs. LOVINS

ERDA Official Denies that Hard and Soft Paths Are Mutually Exclusive

Almost immediately after the publication of Amory Lovins's Article "Energy Strategy: The Road Not Taken?" in *Foreign Affairs*, Dr. Robert Seamans, Administrator of the Energy Research and Development Administration (ERDA), requested his staff to analyze it. A memorandum written by Dr. Wade Blackman and dated November 1, 1976, was the result. Excerpts from the Blackman memo and Lovins's response to it (addressed to Dr. Seamans) follow:

Blackman: The subject article is well-written, is provocative, and has received a wide circulation and evoked a variety of responses. It is desirable that ERDA critically analyze the content of the article and be prepared to respond to the points that have been raised.

FIGURE 1

ENERGY SYSTEM CATEGORIZATION

ENERGY SYSTEM:	**Energy Production**		**End-Use Devices**	**Service Production**
Energy Categories:		Supply		Demand
Applicable Disciplines:		Engineering		Economics
Efficiency Measures:		2nd Law		Utilization Efficiency
Means for Effecting System Changes:		Technology Fixes		Life style Change

Figure 1 presents a categorization of the energy system considered by Lovins. Energy is produced in primary, secondary or tertiary forms which are utilized in end-use devices to supply an economic service. For example, primary crude oil is refined to become a secondary energy source which is transformed in an automotive end-use device to produce passenger transportation. The energy production and end-use devices make up the supply part of the system; and the service production effected by the end-use device constitutes the demand part of the system, a demand whose attributes can be considered to be slowly changing over time and which can be supplied in different ways by technological substitution in the supply part of the system. Energy supply is primarily the domain of the engineering disciplines, while demand is in the domain of the economists. The second-law efficiency measures provide a yardstick for determining the effectiveness of the supply part of the system and a utilization efficiency (such as load factors, etc.) determines the effectiveness with which the supply system produces the service it is intended to perform. Changes in the system may be effected through technological substitutions (technical fixes) in the supply part of the system and/or through life-style change on the demand side.

Lovins: Though I do not have a serious quarrel with Figure 1, which purports to summarize the structure I ascribe to the energy system, I would have disaggregated its center portion to distinguish primary from end-use energy, i.e. to show more clearly the role of conversion and distribution losses, and I would not have confined "Applicable Disciplines" to engineering and economics. (Indeed, a central component of my criticism of the ERDA Plan before the 1975 and 1976 CEQ hearings has been ERDA's tendency to ignore other and perhaps more important disciplines and perspectives.)

Blackman: Lovins's major points will be examined sequentially.

What are "Soft" Technologies? The alternate path proposed by Lovins utilizes a technological mix which is termed "soft," and it is asserted that this mix differs significantly from the mix of technologies currently emphasized by ERDA. The "soft" technology path would emphasize in the near-term solar heating and cooling and technologies which would improve the efficiency of energy production through technical fixes. In the mid-term, the "soft" path would employ coal technologies and biomass to effect a transition to a solar economy in the long term.

Table 1 presents a comparison of the relative capital intensities of selected technologies which are grouped into the "hard," "soft," and "transitional" categories. It can be seen that some of the "soft" technologies are as highly capital-intensive as some of the "hard" technologies. The low capital-intensity of the technical-fix technologies is noteworthy.

Lovins: Dr. Blackman should make clearer that I distinguish transitional fossil-fuel technologies (such as the use of fluidized beds for district heating and industrial boiler backfits) from soft technologies. Bioconversion of agricultural and forestry wastes, too, would begin in the short term, not medium, and technical fixes would be aimed primarily at end-use efficiency, not "the efficiency of energy production." The discussion correctly stresses that technical fixes are cheaper than marginal supply, but does not point out that in general, soft technologies cost less than the hard technologies which one would otherwise have to use in the long run to perform the same end-use tasks. This important argument is supported in detail by my paper for the October 1976 Oak Ridge symposium. It rests on the assumption, which Dr. Blackman should perhaps have emphasized in the caption to his Table 1, that one is speaking of the costs of *complete* energy systems at the margin, not just components of those systems.

TABLE 1

Approximate Investment Required (in 1976 $) to Increase a Consumer's Delivered Energy Supply by the Equivalent of One Barrel of Oil Per Day

Hard Technologies:	**Investment**
Petroleum in 1960s	a few thousand
North Sea Oil	10,000
U.S. Frontier (Arctic and Off-Shore)	10,000–25,000
Synthetic Oil and Gas from Coal	20,000–50,000
Coal-Electric	150,000
Nuclear-Electric	200,000–300,000
Soft Technologies:	
Conservation (Technical Fix)	
Often	0–3,500
Generally	8,000
At Most	30,000
Solar Space Heating in Mid-60s	
In U.S.	50,000–70,000
In Scandinavia	100,000
Biomass Conversion	10,000–25,000
Currently Available Wind-Electric	200,000
Transitional Technologies:	
Stal-Laval Fluidized-bed gas-turbine System with District Heating and Heat Pumps	30,000

[*These numbers are changing so rapidly that they are certain to be outdated by this book's publication date; already, the investment required for wind-electric is* much *lower than given here*—ABL.]

Blackman: *How Much Energy Will We Need in the Future?* The U.S. energy demand goal of the path Lovins proposes is about 95 quads in the year 2000, with a diminution to about 75 quads in 2025. Is this reasonable?

The U.S. with only 6 percent of the world's people accounts for approximately 32 percent of the world's energy consumption but it produces about 31 percent of the world's gross national product (GNP). Western Europe and Japan taken together account for the same proportion of the world GNP as does the U.S., and their combined energy-use almost equals that of the U.S. Energy-use and GNP are closely related in all parts of the world. For the past 20 years, there has been virtually no change in the energy use per unit of GNP, ranging around 11-12 million barrels of oil equivalent per billion of 1970 dollars of GNP in the U.S. Although Western Europe has averaged slightly less energy-use per unit of GNP, their curve shows a similar invariant behavior over the past 20 years. The constancy of these curves does not mean that the efficiency of energy-use has not increased during the past 20 years—much progress has been made in many areas. The constant relationship simply means that efficiency improvements have been offset by new uses for energy which have resulted from technological progress. New product innovation will be necessary in the future if the nation is to achieve its economic goals, and it would be illogical to assume that new uses for energy will not occur in the future and imprudent for planning purposes to assume that drastic changes will rapidly occur in the relationship between GNP and energy use over the next 25 years. [In fact, in the U.S. it fell by half during 1974-8.]

Because of long-term improvement in labor productivity, it takes progressively fewer workers to produce a unit of GNP. To produce today's GNP takes about half the number of workers required 20 years ago. About 27 million people were added to the U.S. work force during the past 20 years, and a further increase of 13 million more is expected during the next 10 years. Thus if reasonable full employment is to be maintained (as required by the Full Employment Act), it is mandatory that GNP growth be maintained.

If labor productivity were to grow at 2.5 percent annually, between now and the year 2000, real GNP would have to grow at an equal rate just to keep unemployment at existing levels if the work force were held constant (an extremely conservative assumption because of the expected increase in the labor force of about 13 million in the next decade). If the relationship between GNP and energy-use remained essentially constant, as it has for the past 20 years, an energy demand of about 137 quads in the year 2000 would be required. To have a demand of 100 quads in the year 2000 would require that the energy used per unit of GNP decrease at an average annual rate of 1.3 percent between now and the year 2000.

Lovins: Dr. Blackman considers my "proposal" for 95 quads of U.S. primary energy conversion in 2000. It is not a proposal, forecast, or recommendation, but rather a hypothetical and illustrative future used only as a qualitative vehicle for ideas. The text makes this clear. Page 76 of the *Foreign Affairs* paper also points out that 95 q is by no means the lowest value that can be realistically considered. Alvin Weinberg's group projects primary energy slightly higher, but end-use energy substantially lower, than I consider, and the 65-90 q CONAES scenario is also considerably below mine (to say nothing of the 40-50 q scenario being developed by the lifestyles panel of CONAES).

I take strong exception to Dr. Blackman's argument based on an alleged energy-GNP relationship. It is simply not true that "Energy use and GNP are closely related in all parts of the world": indeed, the statistics are notable for 3-5x scatter in nearby and similar countries. (See for example Schipper and Lichtenberg in *Science*, 3 December 1976.) Indeed, I am told that the group of eminent energy modellers in CONAES have concluded, to their surprise, that energy and GNP need have little to do with each other and that behavior of each can be varied within extremely wide limits without affecting the other. [See *Science*, 14 April 1978.]

It is equally misleading to suggest that past increases in labor productivity need continue into the future. In a country that cannot productively employ much of its labor force, it is even plausible to suggest that the disemployment of people by black boxes is due for a reversal. Thus, though I have a broader problem with any argument that relies on as meaningless an indicator as GNP, I consider both the premises and the logic of Dr. Blackman's argument faulty.

Blackman: *How Much Can We Rely on Conservation?* It is ludicrous to assume that human energy can be substituted for mechanical or electrical energy on any large scale through the use of labor-intensive technologies rather than capital-intensive technologies whose use correlates highly with energy use. Human beings are among the least efficient sources of energy. Human productivity is only increased by technological progress which provides tools and equipment powered by external forms of energy which leverages human capabilities.

Lovins: Dr. Blackman seems to have missed the point of my argument about conservation. I believe that end-use efficiency in the U.S. can be doubled by about 2000 and trebled or quadrupled by about 2025 *entirely* through technical fixes. People who consider today's values and institutions imperfect are free, if they wish, to consider some mix of technical and social changes instead, but lifestyle and value changes are not necessary to my argument. I do not assume a drastic behavioral change between now and the year 2000, even though I believe such changes are now occurring—partly in response to the structural and political side-effects of the hard path.

I have not suggested coolie-work or rickshaws; the argument, as Denis Hayes remarks, is rather about whether to propel two tons of metal half a mile to buy a dollar's worth of beer in throw-away aluminum cans, or whether to toast the feet of sparrows alighting on one's uninsulated roof.

Blackman: The preliminary barrier to the implementation of technical fixes which would serve to improve the energy efficiency of the economy is institutional rigidity which historically has changed very slowly. We differ significantly with Lovins on the extent to which it is possible to implement new technical fixes between now and the year 2000. Lovins assumes that it is possible to double end-use efficiency by the turn of the century through the implementation of technical fixes. ERDA's best estimates are that the implementation of conservation technologies will save about 500 quads between now and 2000.

Even if it is possible to achieve energy savings greater than 500 quads through technical fixes implemented between now and the year 2000, it is clear that the U.S. cannot risk reliance on conservation to the extent advocated by Lovins if it is to avoid the severe economic consequences which would result if an adequate supply of energy were unavailable in the future. Maximum efforts must also be made to expand energy supply.

Lovins: Evidently Dr. Blackman thinks that the institutional barriers represented by building codes, mortgage regulations, and restrictive utility practices are much more difficult to overcome than the economic and political obstacles to stripping coal, building breeders, shipping plutonium, piping Arctic gas, drilling for off-shore oil, and all the other things that cannot currently be done and show no sign of becoming easier. My own guess is that such constraints will prevent annual U.S. primary energy use from ever exceeding some number around 85 q. Of course overcoming institutional barriers isn't easy; but it's easier than not doing it.

Dr. Blackman asserts *ex cathedra* that even if we can conserve a lot of energy in theory, we need to pretend we won't, in case somehow conservation doesn't work: that is, having just gone out and bought an expensive raincoat, we now need to buy an even more expensive umbrella too. What is never clear to me is *why* one would want to do that. *Why* must "maximum efforts . . . also be made to expand energy supply"? I would have thought that anyone from an agency [ERDA] responsible for the projections so thoroughly demolished by independent review (e.g. by von Hippel and Williams, *Bulletin of the Atomic Scientists*, December 1976) would have learned more diffidence by now in proposing that conservation programs need not affect supply plans.

Blackman: *Are the Paths Mutually Exclusive?* Lovins argues that commitments to the energy strategy currently being pursued may foreclose future commitments to the alternate path he proposes. The validity of this argument rests on (1) the future availability of resources to implement various energy options, and (2) whether or not the current RD&D base is giving adequate attention to the technologies required to implement a "soft" path in the future.

Lovins asserts that the capital required during the next ten years to implement the production goals of [President Ford's] 1975 State of the Union Message following a "hard" technology path will be about three-fourths of the net private domestic investment (NPDI) over the next decade, and historically, the energy sector has required only about one-fourth of NPDI. In arriving at this conclusion, Lovins neglects to consider the extent to which the energy sector will internally generate capital flows to supply part of their future capital needs. When these effects are considered, studies by Bankers Trust, ERDA's Office of Planning and Analysis, and others indicate that capital requirements for a "hard" technology path will probably not exceed the energy sector's historical share of NPDI. If "soft" technologies which are less capital-intensive can be substituted for some of the hard path technologies, adequate capital should be available for the implementation of whatever mix of "hard" and "soft" technologies is deemed to be optimum in the future. The extent to which this capital will actually flow into the energy sector, however, will depend on the economic attractiveness of new energy ventures ("hard" or "soft") vis-a-vis competing investment opportunities.

Lovins: Dr. Blackman's allegation of an oversight in my capital calculations is wrong. Net private domestic investment is exactly what it sounds like—the "net" refers to its being net of capital consumption (depreciation allowance, i.e. replacement of old plant), and the "private" refers to all private investment, whether paid for by retained earnings, sale of debt, or whatever. Whether retained earnings will suffice, or whether debt-to-equity ratios and interest coverage ratios will become unsupportable, is a different argument. Mine merely suggests that the best data we have (the updated Bechtel data base) strongly indicate future problems of the gravest kind in trying to divert adequate investment into the energy sector. I am familiar with the Bankers Trust data base and do not consider it as thorough or reliable as the Bechtel one—though the conclusions of the BT study can be rather well reconciled with the Bechtel numbers if due allowance is made for differences in system boundary, inflation, treatment of cost versus price, etc.

Blackman: To develop the technologies for achieving the nation's energy objectives, ERDA has adopted a pluralistic approach toward the development of a wide range of technical options which encompasses RD&D related to the "soft" technology strategy as well as the "hard" technology strategy. Thus the two-path distinction drawn by Lovins is more apparent than real.

Benefit-cost studies conducted for APAE indicate that energy RD&D

budgets could be significantly increased and the expected benefits would justify the increased costs. Thus, there is a reasonable expectation that future energy RD&D expenditures will be adequate to support the development of a technology base which would be broad enough not to preclude either a "soft" path option or a "hard" path option.

How well does ERDA's current RD&D program support a "soft" path? A research agenda consisting of 71 discrete research opportunities was proposed by the landmark American Physical Society study on efficient use of energy. This study represents the state-of-the-art in terms of a technical fix type of conservation analysis, is referred to by Lovins, and has been the focus of questions during the recent hearings on planning conducted by Senator Bumpers. APAE staff members have estimated the extent to which the proposed research agenda is being pursued. Of the 71 research opportunities which make up the agenda, it is estimated that over three-quarters of the proposed programs are currently being pursued by ERDA, commercial interests, and/or other federal agencies. Thus it would appear that much of the RD&D support required for a "soft" path is being developed.

It is concluded that neither the resources available for implementation of future technological options nor the resources available for, or current commitments to, developing a technological base are likely to make the "hard" and "soft" paths mutually exclusive as is asserted by Lovins.

Lovins: Dr. Blackman's argument—that the two paths are not exclusive and that, anyhow, ERDA's program embraces both—is clearly wrong. The exclusivity to which I referred is not technical incompatibility of hardware but rather *institutional and cultural incompatibility and logistical competitiveness of evolutionary paths within a sociopolitical context.* Moreover, ERDA's program does not embrace most of the elements of a soft path (e.g. non-electric wind systems, most bioconversion, simple solar hardware, many transitional technologies); and my conversation with APAE staff on 10 December made it clear to all of us, I think, that ERDA simply lacks the staff to know what soft technologies are available. That knowledge will have to be injected from the outside. This will be difficult, since most ERDA staff wouldn't recognize a soft technology if it came up and bit them. It is at best disingenuous to pretend that studies of the proper role of soft technologies have been under way at ERDA before now.

Dr. Blackman's inventory of APS opportunities is irrelevant to soft supply technologies, and does not mention that many important opportunities are being pursued as a low priority, or in distorted or primitive form, or by people who do not appreciate the merits of technical simplicity. The shape of ERDA's budget and institutional commitments is an

Blackman: *Some of the Good Points.* Although there are many points made by Lovins with which we disagree, he makes many points which are valid. First, the point that there is currently an overemphasis of electricity-producing technologies in ERDA RD&D strategies, appears valid.

To use high-quality electrical energy for tasks which could be performed with lower quality energy forms is wasteful. Considering the fact that about 75 percent of ERDA's resources go into electricity-producing technologies and electricity currently accounts for only about 11 percent of end-use demand, makes the question of balance an appropriate question to raise.

Second, the question raised concerning the extent to which the scale and quality of energy systems under consideration is properly matched to end-use requirements appears to be worthy of future attention. To properly address these issues requires a strong systems orientation and a systems design capability which currently is believed to be lacking in ERDA. It is recommended that this must be given consideration in future resource allocation and organizational development decisions.

Third, the point raised concerning our relative lack of emphasis on cogeneration and district heating technologies is believed to be valid. It is believed that inadequate attention is currently being given to cogeneration and district heating in ERDA's program mix and that these technologies should be given greater emphasis in the future.

Finally, it is proposed by Lovins that the utilities be induced to enter the solar heating and cooling business. In the proposed scheme, utilities would finance the installation of solar heating and cooling units and would receive payments based on the fuel savings which the installed units would produce. When the cost of an installation was repaid, the fuel savings would then flow to the owner of the unit for the remainder of its life. Because a primary barrier to greater use of solar heating and cooling units is their high first cost, Lovins' proposal would provide a means for eliminating this obstacle. Although there are many other institutional and perhaps legal barriers which make the implementation of this proposal difficult, it is believed that this proposal should receive further study by ERDA.

Lovins: It is perhaps inevitable that my comments should have a mainly critical tone, particularly since I believe Dr. Blackman's main points (a 95-q scenario is implausible, the paths are not exclusive, ERDA is already working on most of the elements of a soft path, capital problems are not so serious, supply must go full steam ahead despite conservation) are misconceived and unsupportable. But I welcome Dr. Blackman's and ERDA's interest in the subject, believe we have much common ground on technical points, and look forward to working with Fred Weinhold's [ERDA] staff in their studies of decentralized energy systems.

3

ERDA vs. LOVINS

Adoption of a Soft Path Would Be Irresponsible, ERDA Claims

Senator Henry Jackson, Chairman of the Committee on Interior and Insular Affairs, requested "ERDA's views with respect to the technical analysis presented" by Amory Lovins at the joint hearing before his committee and Senator Nelson's Committee on Small Business. The response was an anonymously authored report entitled "ERDA Preliminary Views on Amory Lovins' Testimony." The report, with Mr. Lovins's comments interpolated, follows.

ERDA: ERDA is familiar with the views of Mr. Lovins and many others who advocate a "soft" energy policy. This view is essentially one that would de-emphasize large-scale central station electric power (either nuclear or coal) and instead rely heavily on decentralized solar technology coupled with large improvements in end-use efficiency and supplemented (for an interim period) by small scale, decentralized coal-fired systems. In particular, we have reviewed Mr. Lovins' testimony [before the joint hearing] on December 9, 1976, as well as two other recent papers by Mr. Lovins:

1. "Scale, Centralization, and Electrification in Energy Systems," presented at symposium held at Oak Ridge, Tennessee, October 20 to 21, 1976.
2. "Energy Strategy: The Road Not Taken?" *Foreign Affairs*, October 1976.

These papers, especially the Oak Ridge paper upon which the other two are based, contain a large number of assertions about the technical and economic prospects for both the centralized and decentralized technologies. The assertions are based partly on some rudimentary calculations shown in the paper, partly on citations from selected analyses published by others, and (apparently) partly on the author's intuition. The technical and economic arguments are interwoven with the author's judgments on the moral, ethical, and social merits of the alternate energy paths.

Lovins: I accept this characterization with three slight amendments: "decentralized solar technology" and "coal-fired" in the first paragraph are both somewhat too restrictive; the Oak Ridge paper is a technical companion-piece to, rather than the basis of, the *Foreign Affairs* paper (which was written months before it); and the "rudimentary" calculations appear to be adequate to support the broad conclusions drawn from them, though of course they can and should be refined.

ERDA: This memorandum contains ERDA's views on some of the major points raised by Mr. Lovins. Other points, as noted, will require analysis and experimentation before the extent of agreement or disagreement on the options advocated by Mr. Lovins can be established. Pending the outcome of such efforts, there cannot be an "official" ERDA position on every issue raised by Mr. Lovins.

Points of Agreement

We agree that some of the technologies included in Mr. Lovins' "soft" path are potentially very attractive. We are pursuing many of them now, although some may well be worthy of more attention. These include:

1. A *wide variety of end-use conservation technologies,* especially for buildings. Many of these are already available and are economically attractive, but their rapid implementation is inhibited by high initial costs, artificially low prices for fuels (e.g. natural gas), as well as institutional barriers. Others require some further development and demonstration. The development and commercialization of conservation technologies has the highest priority in ERDA's most recent National Plan.

2. *Solar heating of buildings,* and possibly other decentralized solar thermal applications (e.g., cooling, crop drying). ERDA is supporting development and demonstration of a wide variety of such concepts, and is continually evaluating concepts. Solar heating is likely to become increasingly attractive as the industry matures and as alternative space heating options become more costly. However, the extent to which solar heating will take over the nation's heating burden, and the rate at which this will occur, will depend on a number of institutional considerations and market factors that Mr. Lovins does not address. In this regard, ERDA does not foresee that solar heating will be as widespread or will meet a given need as completely as suggested by Mr. Lovins.

3. *Cogeneration and district heating.* More extensive use of these technologies could result in major savings in both energy and capital. But these require greater institutional cooperation than has been traditional and some loss of flexibility.

Lovins: It would be interesting to learn what "institutional considerations and market factors" I did not address. ERDA's doubt that solar heating "will be as widespread or will meet a given need as completely as suggested" in my papers reflects, I suspect, failure to address seriously the issue of symmetry of cost comparisons; ERDA's calculations currently compare solar costs with those of heating with cheap fossil fuels, not with those of heating with marginal hard-technology systems for which ERDA seeks billions of dollars in subsidies precisely because they cannot compete with the same cheap fossil fuels. I also suspect that ERDA has not gone thoroughly enough into the seasonal storage numbers, and has not properly examined the synergistic effect of solar design and conservation measures applied to the same buildings.

The barriers referred to under all three headings are real and must be dealt with more vigorously and imaginatively than ERDA or other agencies have done in the past. I do not see what "loss of flexibility" would arise or that the rewards are not great enough to justify an unprecedented effort.

In general I do not accept that ERDA is pursuing "many" soft technologies now in a way likely to speed their wide deployment. ERDA has an over-elaborate approach in every respect—administration (both inhouse and of contractors), design, conceptualization, analysis, hardware development—and it is an open question whether this is stifling rather than advancing soft technologies. I find also in ERDA a serious information problem, and efforts to overcome it are so far slow, sporadic, and not very successful.

ERDA: We agree with Lovins that the use of high-quality energy for tasks requiring only small temperature differences (e.g. electric resistance heating of buildings) is wasteful in many applications. As energy costs rise, there will be a growing economic incentive for many users to choose alternatives that are more consistent with the Second Law of Thermodynamics (e.g., solar heating, district heating, etc.). However, we feel that the economic trade-offs as determined from the users' viewpoint, rather than the Second Law itself, should pace this development.

It should be noted, however, that electricity can, in most fixed applications, be used more flexibly than other energy sources. This flexibility has economic value (e.g., ability to rapidly reconfigure a portion of a production process) that could offset a premium in direct energy cost.

Consistent with the above areas of agreement, ERDA's current program includes a large number of RD&D efforts in support of "soft" technologies. A research agenda consisting of 71 discrete research opportunities was proposed by the American Physical Society's landmark study on efficient uses of energy. It has been estimated by ERDA that over three quarters of the proposed programs are currently being pursued by ERDA, commercial interests, and/or other federal agencies. Thus it would appear that much of the RD&D support for a "soft" path is being developed.

Lovins: I agree that the Second Law is no substitute for economics, but ask that users be given correct economic signals—a goal ERDA has so far done less to promote than to retard, since correct price signals would be fatal to many favorite ERDA technologies. I agree that electricity is a premium energy carrier, but suggest that the value of its flexibility is (1) hard to quantify, and (2) unlikely to justify its high marginal cost in applications that do not really require such flexibility. From ERDA's and other agencies' involvement at some level (often not a very serious level) in most of the APS research tasks for end-use efficiency, it does not follow that ERDA has a good RD&D base for a soft path. Knowledge and understanding of the most basic soft-path concepts is still rudimentary in most of ERDA and, I suspect, will remain so until many staff are recycled.

ERDA: *Points of Disagreement.* While ERDA might argue about a number of specific numbers employed by Mr. Lovins in his papers, we focus our divergent judgment on two issues central to the energy RD&D strategy adopted in the National Plan. First, we do *not* agree that the "hard" and "soft" paths are mutually exclusive. Basically, we believe that those "soft" technologies that are really attractive economically will penetrate the market without extensive federal RD&D support (although this might be expedited by government help in RD&D as well as in alleviating institutional barriers).

Even with major successes in applying the soft technologies, there will be many applications where centrally supplied electric power is the most practical alternative because of the flexibility it provides rather than because it is required as an energy form. Thus, a "hybrid" energy economy, with both hard and soft technologies playing major roles, would evolve and would probably persist for many decades. Analyses by ERDA of (a) maximum energy savings that could result from a heavy emphasis on conservation ("technical fixes") and of (b) maximum likely impact of soft supply technologies, indicate that energy demand would still increase between now and 2000. But even if demand stays constant or declines slightly, new plants based on "hard" technology (coal, nuclear, geothermal) will be needed to sharply reduce our dependence on oil and gas, which currently supply three quarters of our energy needs.

Lovins: I suspect ERDA's view of exclusivity is that it is purely technological; the rejection of exclusivity here shows no appreciation of cultural or institutional exclusivity (nor indeed of resource competition, which is what all those fights with the Office of Management and Budget are about). I agree that soft technologies should propagate rapidly if institutional barriers are cleared away and markets allowed to work, but I doubt ERDA has its heart in the latter condition, which would be a disaster for most of the ERDA budget's main items.

The last paragraph starts with a view consistent with mine—that central electric stations have a useful role "for many decades"—but does not specify whether that means building *more* stations or just using those we already have. The relevance of soft-technology projections to demand projections is obscure.

ERDA does not explain—and I find it an exceedingly puzzling point—why, if energy demand stays constant or declines, new power stations must be built. Efficient direct use of the fuels (and fluxes) we have seems more sensible.

ERDA: Whether or not this "hybrid" energy economy would eventually evolve (on economic grounds) into a soft energy economy would depend on a number of technical developments (involving both "hard" and "soft" technologies) whose outcome is as yet highly uncertain. If a purely "soft" path does not evolve on the basis of economic advantages (and the odds now seem somewhat against it) then the nation might ultimately have to decide whether or not the noneconomic advantages of the soft path are worth the extra cost. Mr. Lovins argues that the soft technologies will gain from socio-political opposition to resource exploitation and plant siting that will ultimately foreclose the hard options. The inevitability of such a future is not as apparent to ERDA.

Lovins: This paragraph implicitly rejects, without details, my economic arguments for the advantage enjoyed by soft technologies. With regard to political opposition to hard technologies, the question is not so much whether it will inevitably foreclose those technologies, but how far society could or should be distorted in an effort to prevent this outcome.

ERDA: We have not seen any evidence that the development of new "hard" technologies—e.g., large coal-fired power stations, synthetic fuel plants, or breeder reactors—will hinder the development of any of the "soft" technologies, except—possibly—by offering a lower cost alternative. We agree with Lovins that the market prices of all forms of energy—"hard" or "soft"—should not be distorted by subsidies or regulation and should include the environmental costs and other external costs (to the extent that an agreement can be reached on how to quantify them). But subject to this condition, we see no reason why a competition among all the "hard" and "soft" alternatives—with the choices to be made by end-users—should not be in the best interest of the nation.

Lovins: This flies in the face of the history of U.S. and international energy RD&D since 1945—a classic case-study in resource competition. The reference to a "lower cost alternative" suggests that the anonymous author has not done a serious comparative analysis such as I have attempted, or is not looking at the cheapest ways to perform given end-use tasks, or both. I agree that a soft-hard economic competition, with complete internalization and no subsidy, would be a good idea, and am confident it would promptly mean the demise of hard technologies at the margin—provided, and it is a big proviso, that the alternatives all have to compete with each other, not some with each other and some with the cheap historic fuels they are all meant to replace.

ERDA: The second area of major disagreement encompasses Mr. Lovins' estimates of the difficulty of raising the *capital investment* necessary to implement the "hard" path. Mr. Lovins estimates that the capital investment necessary to implement the "hard" path would be over one trillion dollars. Studies by ERDA, Federal Energy Administration, and The Bankers Trust Company of New York indicate that capital requirements for a "hard" path probably would be about half that and would not exceed the energy sector's historical share of total fixed business investment—22 to 30 percent. For example, Bankers Trust estimated that the 1976 to 1985 capital requirements for energy industries at approximately $420 billion.

ERDA, using the complete Bechtel Energy Supply Planning Model, estimates the same 1976 to 1985 capital requirements at $551 billion. These estimates were calculated as part of a complete system analysis. (While Mr. Lovins cites isolated numbers from the Bechtel study, his final results differ from those of that study.)

Similarly, in its "1976 National Energy Outlook," FEA estimated that approximately $600 billion, or about 30 percent of total plant and equipment expenditures during the 1975 to 1984 period, would go to energy sector development.

Lovins: ERDA does not seriously address the straightforward capital-cost calculation I presented. I have compared the Bechtel analysis with that of Bankers Trust, for example, and find them not so dissimilar: in billions of 1974 $, the cumulative capital costs for 1976-85 are:

item	BT	Bechtel
oil and gas exploration/production	128.4	32.0
refineries	22.1	22.7
pipelines	16.4	8.9
other	13.3	11.3
coal mining	10.5	9.4
transport	10.3	9.0
synthetics from coal	6.0	2.2
from shale	6.7	5.8
(BT also had other minor terms)		
nuclear fuel cycle	6.6	14.0
flue-gas desulfurization	7.5	13.8
electricity: coal-fired		47.8
nuclear	not	116.1
hydro & pumped storage	specified	24.6
oil & gas		6.1

TOTAL ELEC. GENERATION	129.1	194.6
electric transmission	22.3	51.1
electric distribution	50.5	151.8
TOTAL ELEC. UTIL. EXCEPT NUCLEAR FUEL CYCLE	219.5	411.3
GRAND TOTAL	439.7	559.3

This comparison is subject to a few methodological queries: the two analyses did not consider quite the same program; BT nowhere defines capital cost, which I take to be the same as for Bechtel (direct construction cost exclusive of all user costs such as interest, escalation beyond the GNP inflation rate, land, design, and administration); BT does not state installed electrical capacities; I assume BT, like Bechtel, includes the cost of facilities under construction but not yet commissioned in 1985, though this is not stated.

On the whole, and given the large uncertainties inherent in such estimates, the agreement between the studies is excellent outside the electric sector, where BT assumes higher unit costs for power stations but much lower ones for transmission and distribution. I am familiar with the Bechtel data sources and consider them much more complete and authoritative than the thin BT documentation. (The same is true for the nuclear fuel cycle.)

The main point, though, is that all the figures are in 1974 $ for ordering in early 1974. Correcting to present [1976] dollars and ordering (along the lines of my *FA* note 8) yields quite similar results for both studies. This is broadly consistent with ERDA's data showing that the Bechtel model yielded 1974-$ results some 31% above those of the BT analysis. (My Bechtel data were also from a complete system analysis and were taken directly from the final Bechtel report, so it is not clear what sort of disagreement ERDA alludes to.) My impression of the FEA model was that it was considerably weaker than the Bechtel model but yielded results not very different. My comparison of total capital flow in to the energy sector with cumulative net private domestic investment is elementary, and if ERDA has found some mistake, I should be glad to have it pointed out.

ERDA: As far as ERDA can tell, Mr. Lovins' estimates of the *capital costs of delivered electric power* appear to be exaggerated as a result of his computational procedure, even though most of his basic inputs come from credible sources.

First, he tends to use the highest estimates and worst case examples (e.g., Con Edison experience in New York for electricity distribution costs). More importantly, he assumes a 7 percent capital cost escalation in *real terms*. We believe such an escalation is unrealistic over any extended period (though it did occur during recent years). Furthermore, no such escalation is apparently used for estimating the costs of "soft" technologies.

The combined effect of these is roughly a doubling of the cost per delivered kilowatt.

Lovins: The allegation of exaggerated electrical-system capital costs appears to be wrong because

- I do not use highest estimates and worst cases, but rather Bechtel's national average estimates, and in particular I nowhere rely on ConEd experience or data; I cite ConEd, after doing my basic calculations, as the only example of whole-system electric costs I could find in the literature, and note that it is indeed unrepresentative.
- I do not assume 7%/y real capital cost escalation; I convert 1974 base costs from the 1974 $ in which the Bechtel data base gives them to 1976 $ by using the best indices available, and then assume *zero* real cost escalation from the assumed 1976 ordering date onwards.
- Accordingly, it was not necessary to assume cost escalation for soft technologies, whose costs, as for hard technologies, were computed in constant dollars.

The factor-of-two discrepancy cited therefore rests on misunderstandings. These must have arisen from a very careless reading by someone in a hurry, as the Oak Ridge paper is very clear on these points.

ERDA: It is difficult with current knowledge to estimate the costs of the "soft path," which Mr. Lovins asserts will be cheaper. Many of the technologies he cites are not on the commercial market.

Lovins: I agree it is difficult to estimate the soft-path costs, though no more so than for the hard path, but have done the best I could, explicitly and with full documentation from *empirical* data. Taking care to incorporate many conservatisms that favored hard technologies, I still calculated a robust capital-cost advantage for soft technologies.

ERDA: Moreover, related system costs have not been considered in his estimates.

Lovins: I considered "related system costs" as far as the data allowed, and took care not to omit any that could significantly affect the results.

ERDA: The most significant example of this problem is in solar heating, which is the critical central technology of his "soft path." He claims this technology can be employed without other energy backup (e.g., electrical heating in extended periods of cloudiness) and seems to base his economics on average rather than most demanding situations.

Studies by ERDA indicate that the capital costs of solar heating systems depend strongly on the amount of storage capacity installed to meet needs when solar insolation is not available or inadequate (this obviously varies geographically and seasonally). These studies generally show that economically competitive solar heating would meet about 40 to 60 percent of space heating needs. Sizing a system to meet 100 percent of needs would greatly escalate capital costs several-fold or even an order of magnitude—making solar heating much more expensive than the "hard" technologies cited by Mr. Lovins.

Lovins: The paragraphs dealing with seasonal storage are answered in my exchange with Professor Bethe [in a later chapter]. I think the conclusion that seasonal storage "would greatly escalate capital costs several-fold or even an order of magnitude" is indefensible, especially in view of empirical data from season-storage buildings in Ontario, Prince Edward Island, Denmark, etc. Of course seasonal-storage designs should be, and generally are, based on abnormal rather than reference conditions, and the extra cost of doing so is moderate.

ERDA: *Issues Requiring More Extensive Analysis.* The principal areas requiring further analysis before a more complete definitive response can be made are twofold: (a) claims for performance of new "soft" technologies, and (b) performance of these technologies in an integrated energy system.

Mr. Lovins derives his estimates of "soft" technology performance and costs from a variety of sources, including a few solar and wind systems that he claims are, or soon will be, commercially available. Generally, his estimates for photovoltaic, solar heating, and wind are lower than the corresponding estimates currently used by ERDA's Division of Solar Energy. However, a considerable amount of work will be necessary to analyze Lovins' cost estimates (the bases for which are not included in any of the three papers), to verify the technical performances and lifetimes of the systems he uses as examples, and to compare these on a consistent basis to corresponding ERDA estimates.

Lovins: I am puzzled that the last paragraph says the bases of my soft-technology cost estimates "are not included in any of the three papers," as the Oak Ridge paper gives many references on this point (*Soft Energy Paths,* of course, gives even more). I agree someone should try to reconcile my numbers with ERDA's: the sources of disagreement should be illuminating.

ERDA: Lovins claims that biomass conversion could meet the entire liquid fuel requirement of the transportation sector at a cost that would be attractive compared to all alternative fuels except those derived from petroleum. In addition to the large current technical uncertainties, there are questions about land use, capital costs, environmental effects, and end-user cost penalties that would be associated with a very large biomass fuel industry. A considerable amount of research—plus a thorough analysis comparing biomass fuels with other options (synfuels from coal, or electric vehicles)—must be done before biomass could credibly be established as the preferred energy source for the transportation sector.

Lovins: The last paragraph seems (wrongly) to envisage a biomass-plantation approach, for which the comments given are indeed appropriate. It is generally accepted that bioconversion to fuel alcohols—starting with current residues, not special crops—should yield a product costing very roughly the same as the present price of gasoline, and therefore presumably costing less than long-run marginal oil.

I am astonished that ERDA suspects electric cars or synfuelled cars might be cheaper than biomass-alcohol-fuelled cars: this suggests a serious lack of information on biomass conversion and feedstocks. After all, Robert Seamans had been heard to remark in early 1976 that the main reason ERDA didn't do much about biomass was that it was already economically competitive and thus didn't need much RD&D support. (The previous year's argument was that methanol didn't fit in the Brookhaven model used for ERDA planning. Whatever the reason, ERDA's biomass program is still weak.) It appears that pending better information, ERDA is *assuming* that electricity and synfuels are the cheapest long-run transport fuels [a situation partly corrected by 1978], and thus continuing the comparative neglect of the biomass alternative.

ERDA: The second major area of required analysis involves examining the technologies in a systems context. Mr. Lovins compares technologies on the basis of capital costs. While this can be a useful indicator for early analysis, it cannot provide the definitive answer one would desire. For example, the lowest capital cost electrical generating equipment of those generally in use are gas turbines. But, their high fuel and operating costs result in expensive electricity compared to more capital intensive options. (Hence gas turbines are used only to meet peak-load demand.)

Lovins: This paragraph misses the point that if, as I claim, soft technologies have lower capital cost than hard technologies, they will *automatically* yield delivered energy at lower life-cycle cost because they have no fuel. (In theory, biomass conversion might be an exception, but in practice it doesn't seem to be.) No such case can be made, as ERDA points out, in comparing different kinds of power stations that trade off fuel cost against capital cost. But in the absence of the need to study both kinds of costs in order to decide which technology is cheapest, capital cost is an excellent first-order signal. [For calculations of delivered energy prices from hard and soft technologies, see Table 2 in Lovins's "Soft Energy Technologies," *Annual Review of Energy* 3:477–517 (1978).]

ERDA: A key consideration in analyzing the overall economics of a technology is determining the system capacity factor. This factor depends on (a) demand profile (e.g., peak to average requirements), (b) the degree of reliability/availability required of the system, and (c) routine shutdown requirements. This type of analysis has not been [done] by Mr. Lovins. (This is not leveled as a criticism—for he has done a great deal with limited resources—but as a statement of needed work.)

Lovins: I agree that a more detailed systems study would be useful. The parameters relevant to my conclusions, however, have already been considered in my analysis, so I do not expect the conclusions to be at risk. [This was broadly confirmed by research at Lawrence Berkeley Laboratory in 1978.]

ERDA: To illustrate this point, Mr. Lovins' claim about the cost advantages of [decentralized] electric power production—elimination of the transmissions costs, improvement in reliability, major reduction of "spinning reserve" requirements, etc.—must be subjected to an analysis in the context of a realistic utility system. It should be noted that reliability considerations were in fact the principal motivation for the utilities to form the power pools that interconnected a large number of generating plants and thus reduced the total requirement for reserve capacity. It is hard to see how a single, small generator located near the user could meet the current utility reliability standard unless it were interconnected with other generators. By and large, some aggregation of energy supplies systems helps improve reliability, decreases peak to average demand, and, thereby minimizes total required capital costs.

Lovins: The paragraph mixes up interconnection with centralization and unit scale. Clearly "a single, small generator located near the user" would require built-in backup units or substantial storage to be reliable—the former approach is in fact used successfully in commercial diesel total-energy systems—but I do not propose such systems, so the question hardly arises. Much of the systems analysis I call for on utility unit scale has already begun, and the results support my "diseconomics of large scale" argument much more strongly than I expected: see *e.g. SEP*, pp. 91–3, and Harding's memo cited in my response to the Council on Energy Independence.

ERDA: Many of the environmental aspects of the "soft path" are now in the realm of opinion and require analysis. Issues deserving attention include occupational safety, environmental controls, and health and labor effects.

Lovins: I agree these are good issues to study—though I also feel enough information is already available to support the case I make.

ERDA: Besides claiming technical and economic advantages for the "soft path," Lovins presents many non-economic arguments relating to the societal advantages of avoiding dependence on large-scale, remotely located energy facilities based on "high" technology. We have not tried to judge the merits of such arguments here but we recognize that many of them have broad appeal and might eventually motivate a national decision to accept certain penalties in cost or convenience in order to accelerate a "soft path" evolution.

Lovins: ERDA has not shown the existence of the "penalties in cost or convenience" referred to.

ERDA: *Conclusions.* ERDA recognizes that the technologies comprising Mr. Lovins' "soft" energy path offer an important means of reducing our dependence on scarce fuels. Aside from the ethical merits of energy alternatives, we recognize that an energy future like that articulated by Mr. Lovins has in a number of its aspects economic, strategic and social appeal. The fact that conservation programs have the highest priority in the most recent ERDA Plan *(ERDA 76–1)* reflects this understanding.

Lovins: The first conclusion tries, wrongly, to extend the high priority on efficiency-improvement technologies to imply an equal priority for the soft supply technologies, whose priority is in fact much lower. This paragraph also states that conservation has "the highest priority" in *ERDA 76–1.* It would be more precise to say, as *ERDA 76–1* says on p. 7, that conservation technologies "are now ranked with several supply technologies as being of the highest priority"; those supply technologies are "coal—direct utilization in utility/industry; nuclear—converter reactors; oil and gas enhanced recovery; gaseous and liquid fuels from coal; oil shale; breeder reactors; fusion; solar electric." Indeed, everything seems to be "highest priority" *except* soft supply technologies, hydrogen, geothermal, and electrical-system support technologies. (See Table II, *ERDA 76–1*, p. 8.) In this light, presentation of conservation as "the highest priority" seems a bit disingenuous. As usual, the ERDA conservation program remains weak—though it may be rejuvenated in the new [Carter] Administration.

ERDA: Nevertheless, it is far from established that the complete "soft" energy path advocated by Mr. Lovins is economically or technically sound;

nor is it established that the "hard" path is as grim as he would have us believe. Many of the difficulties currently confronting the large scale coal and nuclear technologies appear to be amenable to technical solutions. Only time will tell whether or not these problems are less tractable than those that currently hinder the widespread application of the "soft" technologies.

Lovins: The first sentence is judgmental, the second of doubtful relevance, and the third question-begging.

ERDA: Thus, we believe that the adoption of the energy policy advocated by Mr. Lovins—i.e., an immediate and exclusive commitment to a "soft" path, with total abandonment of nuclear and large-scale coal technologies, would be irresponsible. ERDA sees no reason not to develop both sets of technologies and let them compete in the market place for specific applications.

Lovins: The first sentence should refer not to foreclosing or abandoning hard-path options but to the cost of heroic measures needed to resuscitate and maintain them, so foreclosing other options that could have been pursued with the same resources. The last sentence would be more persuasive if a market place existed and if ERDA were making vigorous and symmetrical efforts to develop both sets of technologies.

4

BEHRENS vs. LOVINS

Analyst sees Growth/No-Growth As Key Energy Policy Issue

Dr. Carl E. Behrens, an analyst with the Environment and Natural Resources Division, Congressional Research Service, Library of Congress, wrote "an analysis and interpretation" of the testimony presented by Amory Lovins. Excerpts from his report follow.

Behrens: The publication of "Energy Strategy: The Road Not Taken?" in the prestigious publication *Foreign Affairs* in October 1976 crystallized into a simple construct a decade-long investigation of the effects of growth on the human environment and the earth's ecosystem. In his article, Amory Lovins describes two "paths" to the future, one employing "hard technology," the other "soft technology." The two paths, he finds, are mutually exclusive; pursuit of the hard path gradually makes the soft path less and less attainable. Over a period of 50 years, pursuit of the soft path would lead to substantial reductions in energy consumption; perhaps more important, it would, in Lovins' judgment, match energy inputs closely to their end uses, thus eliminating the tendency of hard-technology systems to use "chain saws to cut butter."

Lovins does not discuss in detail the interconnections between his soft-versus-hard construction and the general concept of "limits to growth" that has been developing as part of the emergence of the environmental movement since the 1960s. The concept is implicit in his work, however, as is indicated by his acknowledgment of the contributions of more than 80 persons to the formation of his ideas. He describes the soft path as "radically different," and a central difference is its compatibility with limited or no-growth concepts.

Contrast with the testimony of the other witness at the December 9 hearing, Linn Draper, makes this clear. The two witnesses disagreed substantially on the potential of many alternative energy sources, but the major difference, as Draper emphasized in his testimony, was over the idea that the two paths are mutually exclusive. The incompatibility of the two paths derives from the fact that the soft path is designed to reach a particular goal, whereas the hard path involves meeting future energy needs as they develop. Lovins' methodology consists of "working backwards" from 50 years ahead by "highlight[ing] what can be done in the long term by recommitting resources in the short term." Draper, on the other hand, works from "past and present energy consumption patterns" to determine "the changes that may be required, both technical and institutional, to meet future needs."

Lovins is dealing with a long-range goal, tailored to the concepts developed by the no-growth movement. Adoption of his soft path would have strong government policy implications, and it is in the context of policy decisionmaking that this analysis will discuss what is involved in adoption of the soft path. A central factor is the question of the role of long-range planning in the formation of policy.

Lovins: Dr. Behrens's section 1 characterizes my *Foreign Affairs* article as an encapsulation or a culmination of "a decade-long investigation" (presumably someone else's) "of the effects of growth on the human environment and the earth's ecosystem"—an approach distinguished by "its compatibility with limited or no-growth concepts" and its being "tailored to the concepts developed by the no-growth movement." Dr. Behrens also notes that the article "does not discuss in detail the interconnections between his soft-versus-hard construction and the general concept of 'limits to growth' . . . ."

The reason for this omission is that limits-to-growth concepts, and for that matter "limited or no-growth concepts," are not part of my thesis in the article, are neither necessary nor important to its argument, and were not uppermost in my mind when I wrote it. Of course these concepts are an important part of the intellectual life of my time, and any thinker is inevitably a product of his or her time; moreover, in the specific sense set out at pp. 12–14 of *Soft Energy Paths*, I am personally sympathetic to certain concepts that Dr. Behrens might include in "limited or no-growth concepts." For the purpose of my *FA* article, however, I have deliberately set aside these personal preferences in order to frame an argument equally compatible with orthodox growth economics—an orientation made quite clear in the article.

I also take issue with Dr. Behrens, or fail to follow his argument, in paragraph 1–3, where, he contrasts the soft path—"designed to reach a particular goal"—with the hard path, which "involves meeting future energy needs as they develop." The difference is essentially that between choosing destiny and extrapolating trend, by short-term increments, into *ad hoc* destiny. Both approaches entail planning (though the soft path generally works within a longer time horizon), and the most confirmed advocates of the hard path justify its enterprises by alleged needs decades from now; but the needs they foresee are derived by more or less straightforward extrapolation of past trends, without seriously questioning whether those trends, so extrapolated, remain desirable or feasible.

Behrens: *Long-Range Policy and Short-Range Goals.* "The seemingly urgent," Rufus Miles has written, "is usually the successful enemy of the truly important." As a long-time high official of the Department of Health, Education and Welfare, and later as president of the private Population Reference Bureau, Miles has had ample opportunity to observe this aphorism operate in the making of policy. The problems that must be faced by Congress and the Executive are generally urgent and critical, with short-term fuses, he says; this pressure tends to drive out long-range planning.

Two severe energy shortages within four years have certainly made the problems in this area both critical and urgent. Energy policy is the subject of much attention, controversy and activity. In forming this policy, the long term is frequently given lip service, or perhaps a bit more; but the pressures of near-term crisis must be countered first. Successful long-range planning is rarely rewarded in the political present, nor is lack of it likely to have negative effects for which leaders now in office will be criticized while they are there. In crisis—and energy seems doomed to recurrent crisis—the urgent becomes not merely important, but overwhelming.

The urgencies are complicated by the multiplicity of goals which must be met by an energy policy. The mixed signals that have been coming from Congress and the Executive Branch since the Arab oil embargo of 1973 are in large part due to the fact that these goals are sometimes incompatible or even mutually exclusive.

A primary short-term goal is alleviating the economic impact of sharp increases in energy prices, and of shortages of energy at any price. If the cost of energy relative to other goods and services has been too low in the past, as many contend, the process of changing this relative position is no easy task. At a time of many inflationary pressures, raising energy costs tends merely to be reflected in higher prices in other areas, with the relative cost of energy remaining essentially the same. The quadrupling of crude oil prices by the OPEC cartel, which accompanied the Arab embargo, was one of the major inflationary forces of 1974 and 1975, although other factors were also important. The "stagflation" of this period was particularly damaging, both economically and politically.

Lovins: I agree with the sentiments in these paragraphs—save that I think Dr. Behrens is doing himself an injustice by concentrating in 2-4 on the first-order inflationary effect of higher energy prices. In the testimony on which he is commenting, I suggested that countervailing second-order effects may be collectively larger. It is also possible that stagflation has much deeper causes (e.g. the Kondratiev-cycle mechanism suggested by Jay Forrester and others); and the Arab embargo does not explain the tendency of marginal prices of most sorts of energy to exceed historic prices since about 1970.

Behrens: Even if policies aimed at changing the relative price of energy, or of a particular energy source such as oil, are successful, the economic impact of such success on those who depend on them must be taken into account. In a market economy changes in price and availability of a necessary resource may take place, but policies deliberately undertaken to change the course of events carry with them the responsibility to minimize and alleviate the side effects created by their success. Nor is it easy to do so. Government compensation programs tend to be expensive, politically volatile, and frequently inequitable.

Lovins: The method I have suggested for approaching long-run marginal-cost pricing—a federal severance royalty—would avoid the difficulties of compensation programs to which Dr. Behrens alludes. [See "How to Finance the Energy Transition," *Not Man Apart*, mid-Sept./Oct. 1978.]

* * *

Behrens: The most direct path to the goal of reducing negative economic effects is to keep energy prices as low as possible, and assure the greatest possible supply of energy. A corollary is that any policy that implies increased prices or curtailed supply will be seen as a possible threat to the achievement of the goal. Such a policy might have compensating features: it may be designed to achieve other equally important goals. But these will have to be strong enough to overcome its disadvantages, or it will have to be demonstrated that the proposed policy has price and supply advantages for the long run. In Miles' terminology, the "truly important" must succeed over the "seemingly urgent."

Lovins: In this paragraph Dr. Behrens has explicitly claimed that cheap energy is a legitimate national goal. He has thereby failed to address my contention that "cheap" energy, such as we have enjoyed and still enjoy, is actually very expensive; we pay for it (by structural distortions) everywhere else in the economy.

Behrens: A second major policy goal consists of protecting the environment: of easing the environmental impact of increasing energy production and use. In many ways this goal is incompatible with the first-order goal of keeping prices low and supply high. Natural gas is the cleanest conventional fuel, but is in shortest supply. Oil is next least environmentally offensive, but it is also becoming scarce or expensive. Coal is environmentally least acceptable, expensive to clean up, even more expensive to convert to uses which now are filled by the first two fuels. Nuclear power can perhaps—an explosively arguable "perhaps"—be described as environmentally acceptable, since its immediate environmental problems are largely limited to the disposal of relatively small amounts of radioactive wastes, and the production of a certain amount of waste heat. But nuclear power has problems in the areas of safety and implications for the proliferation of atomic weapons—problems that have become highly politicized. Furthermore, nuclear power is applicable almost exclusively to the production of electricity, which is either too expensive or not yet adaptable to many energy uses. Solar power's near-term potential, though substantial, appears limited to the residential/commercial sector, and can be expected—once again, arguably—to carry with it price penalties which are antithetical to the goal of easing the effect of increased energy costs.

Lovins: The last sentence mentions only obliquely (by the "arguably") the argument that *marginal* costs are lower for solar systems than for hard-technology systems to do the same jobs, not only in houses and offices and shops but also in making transport fuel and much industrial process heat. If this argument is correct, then cheap energy—assuming it as a goal, which is far from self-evident—is more nearly achievable with, than without, solar energy in the short and medium term.

* * *

Behrens: There are two modes of conservation in a growing economy. One is to increase the energy efficiency of devices, or to change to devices that use other fuels that accomplish the same tasks more efficiently. The second is to limit the number of tasks performed. The first method is probably more compatible with economic goals, although some price penalties and economic dislocations are clearly involved. From an environmental standpoint, however, it is much less satisfactory than the second. The improvement in the environment of increasing the mileage of automobiles from 15 to 30 miles per gallon would be marginal, for example, if increasing numbers of more efficient vehicles were to jam the streets. Even switching to less polluting electric automobiles would lead to the necessity of large increases in electric generating capacity unless the number was also limited.

Lovins: It is not clear that "the energy efficiency of devices" takes into account the scope for improvement in the efficiency of the energy system through reducing or eliminating losses in conversion and distribution. The "price penalties" of improved end-use efficiency or energy-system efficiency are often negative. Of course second-order effects, such as Dr. Behrens's example of more cars, can occur, but they are often suppressed by other signals (such as the saturation of the car market or the capital cost of cars), yielding a substantial environmental improvement from improved end-use efficiency. In such cases as district heating or industrial cogeneration, the net benefit is almost always large.

* * *

Behrens: The energy policy goal concerning petroleum is clearly to cut back on consumption of oil. Substitution of other fuels is one technique to this end; increasing the price of imported oil, or all oil, is another. Limitation of consumption in certain sectors, such as electric generation and other industrial uses, would have some effect, as would the general energy consumption policies discussed with regard to environmental protection. The difficulty is that once again these policies tend to clash with one or the other of the first two energy policy goals. Limitation of general energy supplies would be consonant with environmental goals but contradict economic ones; stimulation of substitute fuels would be just the opposite. Pricing policies would run into the same type of difficulties.

Lovins: Here as elsewhere Dr. Behrens tends to use "energy" and "fuel" interchangeably. The difference is crucial. Substituting natural energy fluxes for depletable fuels can achieve his economic and environmental goals simultaneously.

* * *

Behrens: These are some of the "seemingly urgent" factors that complicate the formation of long-range policy. Nevertheless, there is much current discussion about what these long-term goals should be. As indicated above, much of the discussion revolves around the question of growth and its implications.

Lovins: Although the question of growth (and the much more interesting questions in which one considers *what* is to grow how fast with what costs and to whose benefit for how long) is much in the air, I have already made it clear that my article bypasses that question by assuming, for purposes of argument, whatever conventional growth assumptions are assumed by advocates of hard energy paths. I do not happen to consider those assumptions attractive or plausible, but it is interesting and useful that my soft-path analysis can cope with them anyhow. Thus, while I am sorry to disappoint Dr. Behrens in his desire to write a limits-to-growth essay somehow tied to these hearings, I consider the connection between the two to be tenuous.

* * *

Behrens: The concept of an end to affluence has by no means been universally accepted. One reason why opponents in the debate over energy policy seem to talk right past each other is the fact that one side refuses to recognize the limits-to-growth argument as a basis for policy, while the other insists that the limits are imminent and urgent and demand immediate action.

Lovins: My own *prediction*, not contained in my article, would be that supply constraints will make it hard to achieve primary energy supply of

75 q in 2000; my own *preference*, which I consider plausible, is for U.S. primary energy of the order 10–20 q/y in the very long run (long enough to approach practical limits on Second Law efficiency). My assumption, however, is consistent with average growth of 3%/y in delivered functions over the next 50 years, equivalent to even higher growth in gross economic activity than is assumed by most hard-path advocates. This assumption, embodied in my article and testimony, is what Dr. Behrens should be considering. Going beyond it, he does not really address the arguments I made.

Behrens: *Policy Formation: Options and Evidence.* It should be clear from the foregoing discussion that decision-making in the area of long-range energy policy is no easier than choosing among the various short-term goals. Not only must a policy-maker try to balance the conflicting goals of easing economic burdens, protecting the environment, and avoiding excessive dependence on foreign oil; he must also analyze the various claims regarding future trends and the limits to growth in a dozen social and technological areas, to determine how much leeway and time he has before committing himself and the society to one course or another.

Even the normal practice of keeping open as many options as possible until the future becomes clearer may be sharply limited. The lead time in many areas, from population growth to commercialization of new technologies, seems to have lengthened at the same time that major events are occurring at progressively shorter intervals. The main message of those who warn about limits to growth appears to be that, by the time the evidence is clear, it will be too late to take action.

Lovins: While Dr. Behrens's "policy-makers" are no doubt important people, I wonder whether they have as much influence on events as they might want. As I state in this hearing record, "I am not even sure that our attempts to influence the future are far more likely to have the effect we intend than the opposite one—though that doesn't mean we shouldn't try." According to Kenneth Boulding's Law of Political Irony, everything one does to try to help people hurts them and vice versa. I think we should try to shape the future—especially by avoiding plainly deleterious courses—but am wary of overinflating expectations by anyone.

Behrens: Thus there is mounting pressure to make decisions on the basis of evidence as it exists. Unfortunately, that evidence is not only unclear, it is widely inconsistent. Harman's model of "gestalts" was inspired by the observation that "one observer will perceive in a situation a particular meaning while a second, with essentially the same data at his disposal, perceives the situation quite differently." In fact, the data available to the policy-maker from opposing advocates are rarely "essentially the same." When one witness claims that "exciting developments in the conversion of agricultural, forestry, and urban wastes . . . now offer practical, economically interesting technologies sufficient to run an efficient U.S. transport sector," as Lovins does, and another rejects biomass conversion because it would compete with agriculture and estimates that solid waste utilization would produce a maximum of 3% of total energy consumption, and Draper does, then the difference is more profound than can be explained by gestalt psychology. It lies instead in the realm of engineering and arithmetic.

Lovins: Dr. Draper evidently misunderstood my bioconversion proposals, and his municipal-waste estimate—3% of a presumably high projection of U.S. primary energy—may be correct and is not inconsistent with my data. In my experience, numerical differences between analysts can always be traced to differences in assumptions, provided both analysts are arithmetically competent, and one can then choose whichever assumptions seem best documented and most defensible, taking due account of the consequences of being wrong in either direction. Differences in perception, however, determine what questions one asks: whether, for example, one asks what is the cheapest source of gigawatt blocks of busbar electricity, or what is the best way to meet a particular end-use need.

Behrens: Because of such discrepancies, of which there are many examples, the difficulties in forming policy are amplified. There is evidence, hazy but substantial, that we are approaching the end of the growth cycle in a number of areas. The effects of acting on such evidence are so large, however, that a paramount need is for responsible amplification and analysis of what data exist. Information selected and presented solely on the basis of its support of a particular "gestalt"—what might be called advocacy science—is of little use in balancing the conflicting demands of a changing society.

Lovins: "Responsible amplification and analysis of what data exist" is what I attempted, within the context of both a conventional and a different view of what the energy problem is. I hope this effort is useful for policy, am encouraged that it seems to be stimulating policy discussions, and am disappointed that Dr. Behrens did not discuss it more directly.

5

BETHE vs. LOVINS

Nobel Laureate Chages "Soft" Arithmetic, Wishful Thinking

Dr. Hans A. Bethe is a Nobel Laureate and Professor of Physics Emeritus, Cornell University. An exchange of letters between him and Amory Lovins was published by *Foreign Affairs* in April 1977. That exchange, rearranged in dialogue form with Professor Bethe's explicit approval, is reprinted below.

To the Editor [of Foreign Affairs]

In his article, "Energy Strategy: The Road Not Taken?" in the October issue of *Foreign Affairs*, Amory Lovins tries to show that providing sufficient energy by a "soft technology" requires less investment than by a hard one. Unfortunately, the figures say otherwise.

The mainstay of Lovins' soft technology appears to be space heating by solar power. The energy which can come from this source has been examined very carefully by J. D. Balcomb of the Los Alamos Scientific Laboratory. He studied in detail the sunshine hours in various parts of the country, as well as the "degree days" of heating required. He found that in the most favorable areas (excluding Southern California) one square foot of solar collector will provide about 170,000 Btu per year; while for the average of the country, the figure is nearer 110,000 for space heating. If hot water is also provided by solar heat, this may add 50,000, for a total of 160,000 Btu per year for the average. One barrel of oil per day (which Lovins chooses as his unit) means 2.1 billion Btu per year, and is therefore equivalent to 13,000 square feet of solar collector.

Lovins: Professor Bethe's letter illustrates how sensitively conclusions depend on a complex chain of many assumptions which rest in turn on one's access to up-to-date information. My own assumptions, deleted from my article as too long and technical, can only be sketched below, but are fully explained and documented in my widely circulated Oak Ridge paper, presented October 21, 1976. I hope that Professor Bethe and others will kindly consult that paper, the updated substance of which will appear in my spring 1977 book, *Soft Energy Paths: Toward a Durable Peace,* [published in May 1977 by Friends of the Earth/Ballinger].

I am not familiar with Dr. Balcomb's calculations in detail. His value for the heat supplied by an average U.S. flat-plate solar collector (160,000 Btu per square foot per year on average) is 24 percent below the roughly 210,000 Btu I assume, but above the 146,000 Btu assumed in the main calculations of my Oak Ridge paper, which were concerned with the less favorable climate of Denmark. Dr. Balcomb's figure suggests some combination of a moderately inefficient collector, high losses, suboptimal orientation or working temperatures, or a poorly insulated house. These are consistent with his [assumed] design philosophy, which seeks only 50 percent solar heat and thus is presumably optimized to compete with fuels priced well below their long-run marginal cost of replacement. I, however, assume long-run marginal-cost pricing, which Professor Bethe advocates. This leads to 100 percent solar heat with seasonal storage, and to a quite different optimization of collector area, operating parameters, house insulation, etc. My approach entails much more storage (currently costing about $21–$34 per cubic meter of water—less than for backup capacity, which requires steel, copper, and concrete), but not a "greatly increased" collector area. Storage and collectors can be traded off, and the former enables the latter to work more of the time [for a higher annual-average efficiency]. I assume collectors designed for high efficiency in cloudy northern European conditions.

Bethe: The cost of solar collectors alone is anywhere from $12–$18 per square foot. If we include heat storage, circulation system, etc., about $20 should be a reasonable average price. Thus to replace one barrel of oil per day by solar heat requires an investment of about $250,000. However, Lovins in his footnote 29 [of the *FA* article] gives an investment cost of $50,000–$60,000, too low by a factor of four to five.

Dr. Balcomb's numbers are for an installation providing 50 percent of the heat for a building; Lovins' number supposedly is for 100 percent. If 100 percent solar heating is desired, the cost will be much higher, partly because the area of solar collectors must be greatly increased to provide for cloudy days, and partly because much more heat storage will be required. In Europe, conditions for solar heating are much worse because of the short winter days north of the Alps, and the much greater cloud cover compared to the United States.

It should be mentioned that Dr. Balcomb himself is working actively on solar heating and wants very much to make it successful, which includes quantitative studies of the implications of his favorite technology.

Lovins: Our disagreement about price is clearer. Professor Bethe's quoted cost of $12–$18 per square foot is far too high. Many good collectors were available even in mid-1974 at a price (not a cost) of about $1–$5 per square foot f.o.b. factory, or about $6–$11 per square foot installed. I assume whole-system installed prices—in 1976 dollars and with seasonal heat storage more than ample for northern Europe—of about $14 per square foot in the late 1970s and $9 per square foot in the mid-1980s. These estimates are consistent with others by A. D. Little, the Office of Technology Assessment, and industrial specialists in mass production of solar hardware. The Energy Research and Development Administration seeks a $10 per square foot average installed price (without seasonal storage at about $1 per square foot) by 1980—a common price in California today. [Lovins assumed "field-erected" collectors assembled at the building site, whereas Professor Bethe assumed packaged collectors whose cost (including assembly labor) is subject to markup by the manufacturer, wholesaler, retailer, and installer. This accounts for most of the price discrepancy. Installed system prices for good flat-plate collector systems in the U.S. in 1978 were typically about $25–40/ft^2 with packaged collectors and $10–15/ft^2 with field-erected collectors. See also the exchange between Lovins and Alvin Weinberg in the Appendix.]

Such relatively low prices show why active solar heat at the margin (let alone the much cheaper passive approach) competes nicely with the $82-per-barrel-equivalent nuclear electricity and the $33-per-barrel-equivalent synthetic gas which Professor Bethe conservatively projects. Moreover, far simpler kinds of plastic and flat-plate collectors are commercially available today for about $3–$4 per square foot installed, and even cheaper solar ponds and paper collectors show promise, though I assume no such innovations. Likewise I do not assume neighborhood-scale solar systems, which are much cheaper than single-house systems, nor collectors integrated with roofing.

Bethe: In the case of nuclear energy, Lovins' mistake is in the opposite direction. The cost of nuclear power plants ordered now is about $500–$600 per kilowatt electric (kWe), excluding interest during construction and inflation. The interest has been estimated at about 30 percent. Since Lovins' numbers are in 1976 dollars, one must not include inflation during construction. Lovins wishes to consider delivered electric energy rather than installed capacity. Taking a capacity factor of 60 percent (present experience) gives an investment cost of $1,000–$1,300 per deliverable kilowatt. A barrel of oil per day is equivalent to 67 kilowatts, and thus requires an investment of $75,000–$90,000. Lovins gives a figure of $200,000–$300,000, this time too high by a factor of three.

Lovins: Solar costs *are* uncertain—nearly as uncertain as nuclear costs. (U.S. reactors have been built consistently at twice the expected real capital cost and, lately, at thrice the expected real fuel-cycle cost.) But there are structural reasons to expect solar costs to fall and nuclear costs to rise.

I need not invoke that observation, however, to explain our disagreement over the capital cost of nuclear systems. Professor Bethe assumes a power station can be built for $650–$780 per kilowatt electric (kWe) of installed capacity (including interest charges), using 1976 dollars. For my estimate I have taken the conservative Bechtel model (cited in footnote 2 of the *FA* article) and converted its figure of $585 per kWe, in 1974 dollars, to 1976 dollars by use of the index obtained by Professor Bupp (cited in footnote 3) through simple or multiple regressions of light-water reactor costs. This conversion yields respectively $761 or $929 per kWe of installed capacity. [Better 1978 analyses cited in Lovins's exchange with Jay James (Appendix) suggest that even the higher figure is too low.] I also assume a capacity factor of 55 percent, rather than 60 percent. So far there is no major difference.

However, since I am computing the cost of a *complete* nuclear *system* per kWe *delivered* (not installed or sent out), I included some straightforward costs that Professor Bethe omits. These are a surcharge for 10.7 percent transmission and distribution losses (the Bechtel analysis assumes 16 percent); an initial core priced at $100 per kWe installed (in 1976 dollars); and marginal investment, all in 1976 dollars per kWe installed, of $76 for fuel-cycle facilities, $86 for transmission equipment, and—a big term—$525 for distribution equipment. (These three terms use Bechtel data and the 1974-1976 cost index used by Brookhaven.)

Combining all these figures yields $3,152 or $3,496 per kWe delivered, on the alternative two assumptions about how to convert power station costs from 1974 dollars to 1976 dollars. The lower of these two figures corresponds to $211,400 per delivered daily barrel equivalent, 2.4 to 2.8 times Professor Bethe's estimate.

Even my higher figure is conservative because both Professor Bethe and I omit many costs: marginal capital investment in reserve margin and spinning reserve; federal regulation, security services, and research and development; marginal investment in future waste management and decommissioning; the 6–8.5 percent of the station's output required to operate the reactor's fuel cycle; cost escalation, after 1976 ordering, at any rate exceeding the GNP inflation rate (constant-dollar cost escalation, which I assume to be zero, is now running at about 20 percent per year for LWRs, 7 percent per year for coal-electric stations); and end-use devices. A realistic calculation, including all but the last of these terms, would yield a nuclear capital cost nearer $300,000 per delivered daily barrel equivalent.

Bethe: This comparison does not yet do justice to electric energy. It has been assumed so far that electricity is used in the form of resistance heaters for space heating, a very wasteful procedure. It is possible instead to use heat pumps which Lovins himself recommends in a different part of his article. With heat pumps, electricity can give two to three times the heat value of resistance heating, at an investment cost for the pumps of a small fraction of solar heating.

Lovins: Professor Bethe is correct to assume heat pumps. My Oak Ridge calculations on space heating do so, assuming a coefficient of performance of 2.5 (the present U.S. average is 1.9, and Professor Bethe suggests 2–3). [Lovins assumed this 250% efficiency not only under average conditions but also on the coldest winter day. In practice this is a large conservatism, because actual heat pumps save little or no generating capacity under these winter-peak conditions—not the severalfold saving implied by Professor Bethe.] Completely solar U.S. active space heating nonetheless turns out to have a substantially lower marginal capital cost than a nuclear and heat-pump system to heat the same house. Indeed, the 100 percent solar system is cheaper than the nuclear-and-heat-pump system in life-cycle cost—where solar also beats synthetic gas—even in cloudy Denmark at the latitude of southern Hudson's Bay, and even there solar may win on capital cost too. [Later calculations show it does.] (If it cost more, there would still be good reasons to use it anyway; but its being cheaper makes a neater argument for people who value narrow economic rationality above unemployment, inflation, centrism, vulnerability, proliferation, and other concomitants of the nuclear option.)

Bethe: I submit that these considerations invalidate Lovins' arguments for a soft technology from the point of view of investment.

But also for other reasons a decentralized production of electricity does not seem to me attractive. I can only admire the reliability of the electricity supply from our utility; apart from the famous blackout in November 1965, service has hardly ever been interrupted for as much as a minute. My argument is not that "small is ugly," but that some technologies do not lend themselves to smallness if they are to be reliable. Small, community power stations cannot afford sufficient trained manpower (engineers, operators, linemen) to cope with breakdowns, nor would there be possible the very effective interconnection of grids which helps the big utilities' reliability.

Lovins: Professor Bethe's remarks about the unreliability of decentralized electricity generation will be news to the American Public Power Association's 1,400-odd utilities, which reliably supply electricity to more than 30 million Americans at least as cheaply as big utilities could. The latter buy their excellent reliability, however, at extremely high marginal cost, and big stations tend to be so much less reliable than smaller ones that the smaller ones have lately been costing less per kWe sent out. Centralized big stations also entail transmission and distribution grids so costly that they accounted for 69 percent of the electricity bill paid by average nonindustrial customers of U.S. private utilities in 1972. The pervasive diseconomies of large scale discussed in my Oak Ridge paper make the virtues of bigness an open issue that awaits detailed analysis.

But Professor Bethe's strictures on decentralized electricity generation are a non sequitur, since I nowhere advocate this approach. (Radical decentralization of electrical generation—beyond the level of industrial cogeneration—would probably make sense only if cheap photovoltaic cells are developed. This may happen very soon—indeed, some analysts believe it has already happened in U.K. and U.S. projects—but my article nowhere assumes this or any other solar-electric technology.) My reference to diesel generators was not a proposal to deploy millions of them, but rather an analogy to make the point that if we could make power stations the way we make cars, they would cost less than a tenth as much as they do, but we can't make them that way because they're too big.

What I did advocate is matching scale and energy quality between end-use needs and marginal energy supply, thus virtually eliminating the costs and losses of conversion and distribution. This implies, first, taking advantage of the big systems we already have, which will be with us for decades, without multiplying them further; and second, redirecting our effort at the margin away from centralized electrification. This will free disproportionate resources that can yield more energy and more jobs faster if invested in conservation and in soft and transitional supply.

Bethe: The above calculations do not necessarily make solar heating impractical financially. Under favorable conditions on interest and taxation, solar energy can provide heat at about $7 per million Btu (MMBtu) at the most favorable locations in the United States, $9 at average locations. Natural gas presently costs about $2 per MMBtu; the price may rise to $4. Synthetic gas may cost $4 at the factory and perhaps $6 delivered. Electricity, priced to correspond to the investment cost for future nuclear and coal power plants, may deliver heat at $15 per MMBtu by resistance heaters and, compared to this, solar heating will indeed be economical. But the relative economics of electric heat pumps versus solar heating will differ from place to place, and needs to be investigated. Solar heating is almost certainly economical for the heating of water.

Parenthetically, these calculations show the economic importance of developing and building coal gasification plants.

There are a number of points where I am in agreement with Lovins, such as: (1) we need more energy conservation; (2) we should make maximum use of technical fixes, although they will not save as much energy as Lovins believes; (3) energy should be priced according to the replacement cost of energy installations, not according to the (much lower) original investment; (4) agricultural wastes should be exploited for energy; (5) large-scale solar electricity generators are probably not economical; (6) passive heating by the sun is a promising technique. And there are a few more.

The energy problem of the United States and other industrial countries is extremely serious. We need to combine many different techniques to solve it. But it cannot be solved by combining "soft" arithmetic with wishful thinking.

Lovins: The real challenge to Professor Bethe is to resolve the inconsistencies in his own analysis. We both advocate improved efficiencies (induced both by proper pricing and by specific policies), solar heating, bioconversion of wastes, and (presumably) cogeneration, district heating, and transitional coal technologies. I contend that these, with quite modest amounts of frontier oil and gas, will make his proposed reactors, coal-electric stations, and coal gasification plants unnecessary. We simply do not need maximum feasible expansion of all forms of supply—especially the most costly, nasty, speculative, and inflexible kinds.

In short, if we buy an excellent raincoat, as I propose, then we shall not need an umbrella, a tent, and weather modification too. Nor can we afford them all. Of course we need diversity, but we cannot do everything, and if we try, practical constraints will force us to choose priorities. Priorities inevitably advance some options while retarding others. Some options are logistically competitive or are institutionally or culturally incompatible with others—so much as to foreclose them. My article suggests that some *combinations* of options can mesh to form a coherent policy consistent with our needs, while other combinations produce merely a hash. Until Professor Bethe has squarely faced this choice, he will not have begun to apply his physical insight to a public policy problem whose solution would make many other problems fall into place.

[Dr. Bethe raised no objection to publication of the foregoing exchange or to "interleaving his and my arguments because this makes the exchange much clearer." He denied permission, however, to republish subsequent letters from him to Lovins which Lovins had already inserted in the Senate hearing record. Had he intended that correspondence to be published, says Dr. Bethe, he "would have had to work on it much longer and more carefully." We respect his decision; but we regret it, since the further exchange of correspondence clarifies a number of matters (notably the feasibility of seasonal heat storage, upon which 100 percent solar heating depends). So that the value of this exchange will not be totally lost to readers, we paraphrase arguments that Dr. Bethe has denied us permission to quote verbatim.]

Paraphrase of Bethe: Your estimate of seasonal storage requirements is wrong by a factor of ten or more.

Solar energy is not needed for heating in the summer, so it must all be stored for later use. This is perhaps two-thirds of the yearly total, and thus amounts to 380 megacalories per square meter of solar collector.

Water in a heat-storage tank might at best range from about 40°C in the spring to 80°C at the end of summer. With a 40°C "swing" between maximum and minimum storage temperatures, one would need to heat 9.5 million grams of water per square meter of collector, neglecting heat losses from storage and pipes. This is ten times the figure you give.

Lovins: Thank you for your letter of 7 March. One remaining difference of opinion concerns the seasonal storage of solar heat. This note will be in three parts: first, sharpening quantitative assumptions; second, conceptualizing seasonal storage; third, conclusions.

Quantitative assumptions

1) The 180 W/m^2 for a rather bad U.S. site, or 125 W/m^2 for Denmark, which I assume as the total direct-plus-diffuse solar flux on an optimally oriented fixed flat plate is, as you suppose, the flux averaged over all states of the earth's rotation and orbit. I agree that my assumed 0.42 First Law collector efficiency is conservatively low; Dr. Balcomb is correct that one can obtain 0.50 or more with good design. If, as you seem to imply, he uses his collectors only in the winter, that is his bad luck for not having a seasonal storage load to work into; but I doubt that is what he does.

2) I do not agree that a 40°C storage swing is implausibly high; most seasonal storage designs without heat-pump boosting (which allows ~90°C) assume at least 50°C. The Hunn et al. paper [an example discussed in Lovins's letter but omitted from this abridgment as too technical and detailed] considers 68°C as its base case. I assume 50°C, which is probably conservative. The engineering for heating with 40°C (or even somewhat cooler) water is not as hard as it may seem, nor is it difficult to heat unpressurized water in a solar collector to 90–95°C. Indeed, one can often thermally stratify the storage tank by a variety of simple methods so as to feed water to the collector at the lowest available temperature, thus improving its First Law efficiency.

3) To simplify the formalism, I include collector heat losses in the First Law efficiency of the collector, and other heat losses (chiefly from the storage tank) in the m^3/m^2 ratio, i.e. storage/collector area. I assume normal losses from a well-insulated single-building storage tank, even though proper design can often make the tank losses go into the building, and losses from neighborhood-scale tanks would be much lower owing to their larger ratio of volume to surface area.

4) For the large tanks we are discussing—nearer 10^2 than 10^1 m^3—I assume, following Fischer, a 1976 installed price nearer $21/$m^3$ than

$34/m^3 (1976 $ throughout). The latter figure is a maximum, characteristic of tanks of only about 10 m^3. Both figures of course include all excavation and other work. Your figure of $30/m^3 is too high, as you will find if you study the ACES experience and consult the Midwest Bunker Silo Co. (Charlotte, Michigan) cited by Fischer. Indeed, even lower prices than $21/m^3 could realistically be assumed: uninsulated ferrocement tanks, which could easily be sprayed with insulating foam, are now being built in many countries for only $5–10/m^3. I do not assume this cheaper technology.

5) I do assume that domestic hot water as well as space heat is taken from the solar system; the Oak Ridge paper does not say so because this is the usual practice. It increases the effective duty factor of the collectors and can improve their First Law efficiency.

6) The $100/m^2 professionally installed solar heat systems I mentioned are generally of high quality and look durable. (You might ask Jerry Weingart of Caltech/IIASA about this.) No doubt Dr. Balcomb can find people who will charge 2–3x as much, but he doesn't have to pay that for a good product. Parenthetically, I note that the lower range of the system price for his demonstration house, namely $129/m^2, plus 1 m^3 of storage at my $21/m^3, exactly equals my Oak Ridge estimate of $150/(m^2+m^3) for a complete solar system in the late 1970s. I am glad we agree about that.

7) You are not correct in assuming that a 4.3-kW average heat load for a 125-m^2 house cannot easily be reduced further. This error has important implications for solar design, as my *Foreign Affairs* response points out: it affects chiefly the winter space heating load and hence the relative and absolute amounts of collector area and storage volume required for what is essentially a peak load. An average, decently insulated, 125-m^2 Danish house today has an annual average space heating load of about 2.3 kW. If better insulated, to a level that pays back in a few years at today's OPEC prices, it needs only about 2kW for all net low-temperature heat: typically 1.2 kW space heat plus 0.8 kW hot water. The next step is heat recuperators on exhaust air and water; their heat recovery (83–90% for air and 50% for water), together with night shutters, allows the 116-m^2 Lyngby house to be comfortable with an annual average of only 0.52 net kW—0.26kW space heat plus 0.26 kW hot water. (The former figure [0.26 kW for space heating] could have been reduced to zero with some further refinements, chiefly in window gain.) The marginal capital cost of all the Lyngby insulation, recuperators, etc. is zero or negative—that is, the Lyngby house, neglecting its solar system, need cost no more to build than the corresponding fabric of a conventional, well insulated Danish house the same size. But in general such arbitrarily low-energy performance can also be achieved by retrofit at capital investments less than about k$25–30/(bbl·day) enthalpy saved, a level well below the marginal

cost of any long-term heating system. Even a very straightforward commercial insulation, weatherstripping, and glazing retrofit in UK dwellings, at prices that compete nicely with cheap North Sea gas, reduces the approximate net annual-average space heating load from 2.06 to 1.11 kW for an average 2-story cavity-wall detached house (100 m^2 floor), and from 0.22 to 0.086 kW for an average 1-story cavity-wall intermediate flat (60 m^2 floor). A very thorough, but economically attractive, retrofit of a 1.6-kW postwar semidetached UK house can reduce its net space heating load to 0.3 or even 0.15 kW. There is an extensive literature on these subjects, a little of which is cited in note 15 of my *Foreign Affairs* paper. [Lovins later described an extremely efficient no-backup solar house in Saskatchewan as empirical proof of his thesis: technical details are in *Soft Energy Notes* 6 (March 1979), Friends of the Earth, San Francisco.]

8) Taking account of points 2 and 4, and assuming my 1985 solar capital costs, will yield a cumulative difference of a factor of two from your handwritten calculation. That is, wrongly assuming (as I shall show below) that "*all*" the summer heat—assumed to be 2/3 of the total heat—must be stored for winter, the delivered price of 1 MBtu (10^6 Btu) would be not $22 but $10.5. A heat price very similar to $10.5/M Btu can be obtained by taking your *Foreign Affairs* estimate of $6/M Btu for delivered synthetic gas, dividing it by a First Law furnace efficiency of 0.7 (which you neglected to do in your *Foreign Affairs* comparisons with $7/M Btu solar heat), and, if you aren't feeling charitable, adding a further $1.75/M Btu for a gas heating system at the margin, a total of $8.6/M Btu without or $10.3/M Btu with the gas heating system. Within the margin of error, both these figures are indistinguishable from the solar heat price. That $10.5/M Btu price, however, is inflated 2.4-fold from about $4.4/M Btu by your incorrect assumption about how much storage volume is required. I shall consider this point next.

Paraphrase of Bethe: A rather low cost for collection and storage of solar energy is $500 per square meter of collector, implying an investment of $220 per MBtu per year. This is eight times the figure you give. Seasonal heat storage seems to make matters worse instead of better; this isn't surprising because an awful lot of storage capacity is required.

Balcomb cites a cost of $20–30/$ft^2$ for a professional solar heating installation. One wonders how well the much cheaper installations you mention will perform and how long they will last.

Heat-loss calculations based on ERDA's data for a typical house (8 Btu lost per ft^2 of floorspace per Fahrenheit degree-day) indicate that in a climate like that of the Lyngby house in Denmark (or the Bethe house in Ithaca, New York), a heat storage tank would need to be roughly equivalent in volume to the house itself. Houses probably can't be made much more heat-conserving than an average rate heat loss of 4.3 kW for the Lyngby house.

Your *Foreign Affairs* article, and your letter in answer to mine, may lead politicians to spend a lot of money on solar research and development that isn't as promising as you suggest.

Lovins: *How seasonal heat storage works.* Your basic assumption is that all the summer heat, which you say means at least 2/3 of all the annual heat captured, must be stored for winter. From that assumption you compute, logically enough, a storage requirement of 9.5 m^3/m^2 for a 40°C storage swing (7.6 with 50°). In contrast, the Zero Energy House in Lyngby, Denmark, with 42 m^2 of collector, exists (I have visited it and a real family is living in it) and, unfortunately for your theory, it got through the 1975-6 winter very nicely with zero backup heat and only 30 m^3 of water storage, a ratio of 0.71 m^3/m^2 or 7½% of your ratio. [The Lyngby house later had technical problems irrelevant to this argument and not shared by successful 100% solar houses elsewhere in Denmark, in Norway, Canada, and the U.S.]

The largest ratio of which I am aware is in the 100% solar Provident house now operating near Toronto, a 325-m^2 dwelling with an average heat load of a few kW. Its 67 m^2 of Miramit collectors feed 273 m^3 of water storage (plus 0.47 m^3 of storage and 7.43 m^2 of collectors for the separate hot water supply), a ratio of 4.1 m^3/m^2 for the space heating system alone or 3.7 for both together. The marginal capital cost of the entire solar system is said to be about k$10 (~$134/m^2 collector). Even higher m^3/m^2 ratios occur in two other 100% solar designs proposed for Southern Ontario, also by Frank Hooper of the University of Toronto: the 1860-m^2 Aylmer old people's apartments (15 kW average heat-plus-hot-water load), now at tender stage, is to use 910 m^3 of storage for 186 m^2 of collectors (a 4.9 ratio), and—as an extreme case—a 93,000-m^2 warehouse now being proposed would use 45,425 m^3 of storage and 6041 m^2 of collectors, a 7.5 ratio. Both these buildings, however, are considered by their very competent designers to be microeconomically attractive. The simulations used to analyze them (including analyzing the internal heat flows in the storage tanks) have been empirically verified in Provident.

Most designs for northern climates use somewhat lower m^3/m^2 ratios than Hooper: the high-technology Aachen house, for example, uses only 46 m^3 for 20 m^2 of selective collector, a 2.3 ratio, but achieves a 90°C swing by boosting with a heat pump, so that the annual storage is ~40GJ. (The 116-m^2 house requires an average of 0.95 net kW of space heat and, with a 75% efficient recuperator on waste water, 0.11 net kW of water heat.) The 80–90% solar Lorriman house in Toronto, with 142 m^2 of floor and 57 m^2 of collector, also boosts with a heat pump, but uses only 19 m^3

of storage (0.33 m^3/m^2), split between two tanks each of which is heated by the collector while the other is cooled by the heat pump; total marginal capital cost is about k$6 ($105/m^2). An intermediate design, the 330-m^2 Ark on Prince Edward Island, is heated almost entirely with 78 m^2 of collector, and combines 83 m^3 of water storage with 85 m^3 of greenhouse rock storage (it is both a house and an indoor farm). [Lovins's 1978 *Annual Review of Energy* article more clearly distinguishes two types of "seasonal" storage: one (discussed below) with buffer storage for a few weeks, the other with many months' (i.e. true seasonal) storage, and both yielding 100% solar heating.]

More diverse examples could be adduced, as the freedom of choice in m^3/m^2 ratio is considerable and the choice often rests as much on architectural convenience as on cost (the cost optimization curves, in other words, are often fairly flat over a wide range, so the higher m^3/m^2 ratios of e.g. the Hooper designs are not necessary in any technical or strongly economic sense). But the Lyngby house alone would make the point: according to Boulding's First Law ("Anything that exists is possible"), something must be wrong with your assumption.

What is wrong is that I am not talking about houses above the Arctic Circle—for which zero-backup, purely solar houses would indeed be subject to your $10.5/MBtu logic if anyone cared to build them. (Other designs, including wind/solar/heat-pump hybrids, would be more attractive.) Below the Arctic Circle, a substantial amount of insolation is available *during the winter*, and the object is to store only enough heat to meet the time series of intermittent deficits between that winter input and the winter heat needs. Conversely, and contrary to your statement, heat *is* required for space and water heating at other times of the year.

Let us consider for a moment the unfavorable case of Denmark. The insolation on a Danish wall averages above 100 W/m^2 from January until early October, and above 150 W/m^2 from mid-February to about August. It hits a minimum of about 65 W/m^2—more than half the annual average of 125 W/m^2—around November, and is below about 85 W/m^2 only from mid-October to mid-December. Now, Lyngby has about 1700 h/y sunshine and about 3000 Celsius degree-days (you estimated 3333); the monthly mean temperature is below 0°C only in January and early February, and below 5°C only November through March. The standard of insulation and recuperation in the Lyngby house corresponds therefore to a total net heat load (space heat *plus* hot water) of about 47 $kJ(°C_{<20}-d)^{-1}m^{-2}$ floor, or 29% of the value you assume to be all but irreducible for space heat *alone*.

These meterological and architectural conditions have an important result: *at no time* does the monthly average rate of solar heat supply to the storage tank at Lyngby fall below about half of the total low-temperature heat load (space heating plus hot water) of the same house for the same month (not even in November: remember that the debit columns below also show storage losses). In particular, a typical year at Lyngby might go somewhat like this:

mo.	captured heat to storage kW-h/mo	captured heat to storage monthly av kW	heat drawn & lost from storage kW-h/mo	heat drawn & lost from storage monthly av kW	heat stored kW-h
Jan	860	1.18	840	1.15	20
Feb	930	1.27	800	1.10	150
Mar	1260	1.72	690	0.94	720
Apr	670	0.92	575	0.79	815
May	400	0.55	500	0.68	715
Jun	415	0.57	460	0.63	670
Jul	535	0.73	485	0.66	720
Aug	470	0.64	475	0.65	725
Sep	585	0.80	475	0.65	835
Oct	425	0.58	480	0.66	780
Nov	250	0.34	640	0.88	390
Dec	475	0.65	850	1.16	25
Total	7275		7270		6565
Mean	606	0.83	606	0.83	547

The variation in the rate of heat capture reflects variation both in insolation and in First Law collector efficiency: with the unsophisticated collectors used, this drops from about 0.30–0.35 with winter inlet temperatures of about 45–50°C to about 0.10–0.15 with summer inlet temperatures of about 75–90°C, so that even the summer insolation, though strong, is not captured much—and not needed much either. (As mentioned earlier, Hooper's systems maintain high collector efficiency throughout the year through careful design.) The variation in heat load reflects the hot-water base load plus winter space-heating peaks. The stored heat runs through an annual cycle, but it is far from a simple cycle, for the stored heat is continuously being both drawn down and topped up—notably by winter insolation captured at high First Law efficiency.

It can readily be calculated that in a typical year the 30 m^3 storage tank at Lyngby stores about 6 Gg-cal. If it were all stored in one accumulation and discharged (in the dark) in one drain, as you assume, it would require a temperature swing of 200°C. In fact, however, the continuous winter insolation makes the tank operate less as an in-once, out-once seasonal store than as a heat buffer on a time-scale of days and weeks, so its actual temperature swing of about 45°C suffices.

I hope you will now appreciate the fallacy of your assumption that storage and discharge are both independent, monotonic processes forming a single huge cycle and hence requiring huge storage volumes. In fact both processes occur simultaneously—the exact analysis being quite complex—and the balance between the time-series of charge and discharge is more favorable than it looks. Discharge, by decreasing the collector inlet temperature at Lyngby, increases the collector efficiency and so uses the short winter days to greater advantage. I might add that the classical calculation that seasonal storage at northern European latitudes requires a water volume comparable to the volume of the house is a well known, almost a textbook-example, fallacy perpetrated by people who don't understand solar systems. Proper analysis invariably shows a storage volume of the order of 1 m^3/m^2 (less for U.S. latitudes, possibly excluding Alaska). Of course, the collector area and the storage volume, as I make clear in my Oak Ridge paper and *Foreign Affairs* response, can be traded off over a very wide margin, from the Hooper high values to the Lyngby low value (or lower). But it should br obvious, as it evidently is to Hooper, that if, as I argue, a collector system costs about four times as much per m^2 as a storage tank costs per m^3, there is little incentive to substitute the former for the latter, and every incentive to substitute the latter for capital-intensive backup capacity used at low capacity factor.

I am not aware of any conventioal solar systems analysis that has tried to test my argument with realistic data. There are, however, several designers who have proved their (and my) seasonal storage theories in real buildings. Probably foremost among them are Civil Engineer Esbensen and Professor Vagn Korsgaard at the Technical University of Denmark, 2800 Lyngby, and Professor Hooper at 790 Bay St., Suite 1022, Toronto M5G 1N8. I am sending copies of this letter to them both in the hope that they can let me (or both of us) know if they spot any egregious blunders or wish to supplement or refine my arguments. They are both skilled in the use of daily or hourly simulations, which are the only way (aside from their buildings and similar ones) of assessing how various seasonal storage configurations could work.

Conclusions

Because of your misunderstanding of how seasonal heat storage works, your estimated storage requirements are about an order of magnitude too high. Your estimates of delivered solar heat cost, both for that reason and for the reasons of quantitative detail noted earlier in this letter, are about 5 times too high, as stated in my response to you in the April *Foreign Affairs*.

Paraphrase of Bethe (in a later letter): Thank you for your detailed discussion of seasonal storage with solar heating. I am very much surprised that it works, the figures seem based on solid statistics.

An independent assessment of the potentialities of solar heating in various parts of the United States would be a good idea. If such an assessment supports the concept of seasonal storage, and if solar heating is given favorable tax treatment, it could compete for new houses in many parts of the U.S.

They Lyngby house works, it is said, largely because of its excellent insulation. This is the best way to save heating energy in all houses.

Lovins: I was very glad to get your letter of 12 April today and to see that you are prepared to accept my seasonal heat storage numbers. I agree that a U.S. assessment is needed for various regions—though Denmark and Ontario should be harder cases than practically anything in the U.S. outside Alaska. I believe that mating the more sophisticated U.S. collectors with insulation of the level of refinement shown at Lyngby should further improve performance.

The Swedes are doing good work on seasonal storage and I hope to examine it in the next few weeks. I also note from the spring 1977 *Co-Evolution Quarterly* that the Ark in Prince Edward Island didn't start charging its heat store until September (much too late), then had the coldest winter in a hundred years, but stayed toasty anyway, and the backup wood furnace was never even lit.

Paraphrase of Bethe: Old houses are generally unsuitable for (active) solar heating. Passive solar heating (designed so the structure of the house itself collects and stores heat) may be the best partial solution. Neither old nor new houses seem very adaptable to solar heating in densely populated areas. Synthetic gas and electric heating by heat pumps still seem necessary.

Lovins: I am not as pessimistic as you about solar retrofits. If, as you say, south walls in existing houses have many windows—and I think that is correct—then one can go a very long way toward a passive design just by retrofitting a very high standard of insulation and heat recovery. [A recent study by Balcomb shows that an attached greenhouse, such as many people have successfully added onto existing houses, works at least as well as a flat-plate collector system. (It is also cheaper.) During 1977-78, therefore, passive solar retrofits came to be viewed as an attractive option.] But you will have noticed from my *Foreign Affairs* paper that I envisage solar retrofits, especially in densely populated areas, as being mainly on a neighborhood rather than a single-house scale, and using the heat distribution systems left over from transitional district heating. [A later Canadian study by Hollands and Orgill, cited in Lovins's *Annual Review of Energy* article, found seasonal-storage neighborhood-scale solar heating, with collectors at roughly the price assumed by Professor Bethe, to compete with $8/barrel oil.] This approach greatly simplifies the collector siting problem, as buildings that can take excess collector area (schools, shopping centers, hospitals, etc.) can be married to others that can take little or none. If you look down on Manhattan from a tall building, you will observe that practically any notional neighborhood contains at least one building of the former kind, roughly adequate to heat the whole neighborhood. (One would often have to strengthen the roof to support the weight of the collectors, but that does not seem an insurmountable problem.) Even in single-building retrofits, a roof, at U.S. latitudes, is as good as a south wall, and in fact one can do very well indeed with a non-south wall at any lower-49 latitude. The key to the whole problem, as you say, is very effective insulation and heat recovery, and that will be cheaper than any sort of supply at the margin.

6

LAPP vs. LOVINS

Consultant Accuses Lovins of Irresponsibility

One of the most vociferous critics of Amory Lovins's *Foreign Affairs* article is Dr. Ralph Lapp, whose letterhead describes him as an "Energy/Nuclear Consultant." In the familiar format, we interrupt Dr. Lapp's critique in order to bring Mr. Lovins's rebuttals into closer juxtaposition to the points rebutted. Departing from our usual practice, however, we give the first word this time to Mr. Lovins because the introduction to his comments on the Lapp critique summarizes the basic differences that underlie disagreements between the two men.

Lovins: Dr. Lapp and I evidently have fundamental differences about the nature of the energy problem. In his view, appparently, the problem is mainly a technical and economic one of how to expand energy supplies (especially domestic supplies) to meet the extrapolated homogenous demands of a dynamic economy. This is the view broadly characterized in section II of my *Foreign Affairs* article. In my view, on the other hand, the energy problem is chiefly a social, political, and ethical one of how to accomplish our society's diverse goals by meeting its heterogeneous end-use needs with an elegant economy of energy (and other resources) supplied at appropriate scale and quality. Our disagreement about the nature of the energy problem and the subject matter of energy policy is not a technical but a trans-scientific disagreement and cannot be resolved by technical arguments.

I suspect that Dr. Lapp and I also have different ideas not only about the means of production but about the ends of production (of energy and other things). He seems to have in mind a future that is like the present only more so: more factories, consumer ephemerals, Los Angelization, big cars, and hassle. It is not clear to me why people would want that. Of course the poorer people in our society want, and are entitled, to live as well as anyone else, but that is quite a different matter. In my view of the future, people who want to live as Dr. Lapp envisages can do so, but those who don't needn't: there is room for diverse lifestyles, including those representing post-industrial values. I emphasize again, as I did in my testimony, that my soft path can be construed entirely as a technical fix: it can, if Dr. Lapp wishes, support just the same pattern and level of industrial production that he assumes. I myself would not choose that, but my own preferences are only of academic interest, since I have deliberately set out to sketch the technical and social basis for a pluralistic future that would be as appealing to the Dr. Lapps of this world (if he understood it properly) as to me. I seek a world that can give us both the leeway to find what we want—not to force my own (or Dr. Lapp's) personal preferences on other people.

Dr. Lapp repeatedly charges that I rely upon rhetorical devices rather than upon substantive discussion. He doth protest too much. I shall try in this response to treat his semisubstantive points while ignoring his *ad hominem* attacks—to the extent that the two are separable.

Lapp: The issues set forth in Mr. Lovins's article concern the magnitude and make-up of the U.S. energy supply as projected to the year 2025. Since projections beyond the year 2000 are highly conjectural, this review of the article will focus on his extrapolations to the year 2000. These appear as Figures 1 and 2 in the article and as Tables 1 and 2 in the *Congressional Record* reference. The two tables for 2000 project:

TABLE 1:		TABLE 2:	
Oil and gas	68q	Oil and gas	35q
Coal	58q	Coal	25q
Nuclear	34q	Soft Technologies	35q
Total	160q	Total	95q

Table 1 is taken by Mr. Lovins as a representative ERDA estimate, while Table 2 is set forth as Mr. Lovins's estimate.

Lovins: The issues set forth in my *FA* article are concerned not with the homogeneous aggregate of primary energy supply but with the heterogeneous structure of end-use energy needs and with the structural implications of a possible supply system appropriate to those needs. It is in failing to grasp this difference that Dr. Lapp gets derailed before he has even gotten up steam.

[In commenting earlier on a one-page summary of Lapp's critique, Lovins wrote as follows:] My *Foreign Affairs* article makes clear that its two figures show hypothetical and illustrative curves, not forecasts or precise recommendations. I have not projected any particular energy futures, but rather constructed two plausible examples as a heuristic device—a qualitative vehicle for ideas. It is not I but most official energy analysts who project—or have until recent months projected—U.S. primary energy use of the order of 230 q in 2025. That may be the trend, but it is certainly not destiny and will not happen.

The point I think Dr. Lapp misses when he compares two levels of per-capita primary energy use is that primary energy is not end-use energy, is not function performed, is not welfare obtained. It is only the last of these that people care about, and only the next-to-last that economists care about. It does not take an economic model to see or show that if conversion and distribution losses are virtually eliminated and if end-use efficiency is trebled or quadrupled, the ratio of function performed to primary energy used will increase roughly 4- to 5-fold from today's level. . . .

[Back, now, to Lovins's comments on the full-length Lapp critique:] Other people—not just ERDA, but others I mentioned—have extrapolated energy futures of which Figure 1 is a rough composite. I have sketched Figure 3, which is unabashedly normative (whereas the precursors of Figure 1 were tacitly normative). [See Lovins's prepared testimony.] Of course any estimates beyond 2000—or, for that matter, beyond 1977—are highly conjectural; but Dr. Lapp seems not to realize that it is the qualitative implications, not the precise quantitative details, of the energy future that are important for the type of analysis I attempt here. It is especially important not to assume that the world ends and then (presumably) starts afresh in 2000, but to trace out where we want to go over a period long enough to get somewhere, and then to ask whether the pre-2000 path is going to get us nearer to or farther from that goal.

For what it is worth, my year-2010 values (about 197 q for Figure 1 and 86 q for Figure 3 are nicely within the range of 70-210 q considered by the the National Academy of Sciences/National Research Council CONAES study, due to report mid-1977.

Lapp: Since almost all energy analyses project reductions in oil and gas supply, there is no need to comment on this energy sector. Similarly, there is general agreement that coal will have to increase markedly so the coal sector is not a critical issue as such.

Lovins: There is a vast disagreement on how "markedly" use of coal will increase and for how long, and on what we should do with the coal once we've mined it (where?). Again, it is essential to ask what we are trying to do with all that energy and how best to do it, not just what kinds of fuels we feed into the hopper.

Lapp: Three basic differences stand out:

1. Mr. Lovins' forecast slows total energy growth rate to slightly more than 1 percent per annum, approaching zero growth on a per capita basis.
2. Mr. Lovins recommends total abandonment of nuclear power.
3. Mr. Lovins urges reliance on "soft technology."

Lovins: (1) It is not a forecast. To adopt Dr. Lapp's approach for a moment, I show primary energy conversion rates changing at the following

5-year-averaged rates-of-change (equivalent compound interest):

1975–80	2.7%/y	1975–85	2.4%/y	2%/y in 1980–5 roughly consistent with President Carter's 20 April 1977 message
1980–85	2.0			
1985–90	0.8			
1990–95	−0.2	1975–2000	1.1%/y	(0.4%/y per capita on Series III projections of U.S. population growth from 213M in 1975 to 245M in 2000)
1995–2000	−0.6			
2000–05	−0.9			
2005–10	−1.1	2000–2025	−1.5%/y	
2010–15	−1.7			
2015–20	−2.1			
2020–25	−1.8	(slowing down as Second Law practical limits are approached)		

I would not want anyone to attach too much significance to these numbers, since they are not intended—as I have often pointed out—to be a forecast or a precise recommendation. What I would point out, however, is that while primary energy use is rising, leveling off, and falling again, end-use energy is steadily rising—to a value around 2000-05 perhaps about half again as great as it is now—then falls again, very slowly, to about 115% of its present value by 2025 (probably over 2/3 of its present value per capita). Moreover, since end-use efficiency throughout the 50 years is steadily and dramatically increasing, functions performed increase to some 4-5x their present level.

(2) I recommend that expansion of nuclear power be stopped—this has in fact happened in the U.S. since about 1974, and will continue in the absence of a federal bailout of hundreds of billions of dollars—and that nuclear power then be gradually phased out in an orderly way.

(3) I recommend a steadily increasing reliance, starting now, on a diverse array of soft technologies—not on some monolithic, homogeneous "soft technology."

Lapp: Three key points of controversy arise:

1. Can the U.S. economy continue to grow if the energy supply flattens out?

2. Do the facts warrant rejection of nuclear power?

3. Can "soft technology" supply 37% of the year 2000 postulated energy and 100% in the year 2025?

Lovins: (1) As noted above, real functions performed—which is what economists measure—increase under the illustrative soft-path assumptions by perhaps 4.4-fold over 50 years, equivalent to 3%/y compound. This is because (to put it in Dr. Lapp's terms) the rate of growth in end-use efficiency and in energy-system efficiency comes to equal and then surpass that of GNP.

(2) Whether nuclear power should be rejected is a question not of facts

but of values. Facts (to the limited extent that they can be disentangled from values) are relevant but not dispositive. Dr. Lapp and I have different values.

(3) Deployment rates are crucial, and are discussed in detail in Chapter 5 of *Soft Energy Paths*. Broadly speaking, the growth rates I assume for the total contribution from soft technologies is lower than that which most nuclear forecasters assume for their technology, which is technically much slower and more complex. Moreover, each joule of energy supplied by soft technologies is in a form fit for direct use without further conversion and distribution losses, so its effective primary contribution is considerably greater than that of 1 joule conventionally supplied and used.

The exact date when soft technologies can saturate all end-use needs is not important to my argument. I suspect that in practice it might be as early as about 2015 if we are skillful and lucky, perhaps 2040 if we are not.

Lapp: Seriatim comments follow on a paragraph-by-paragraph basis for each section in the article.

"The first part resembles present federal policy and is essentially an extrapolation of the recent past."

Comment. The recent past is not defined. If one tracks the 1969-76 period, annual growth rates range from −2.5 to +5 percent and average under 2 percent per annum. Mr. Lovins uses a 160q value for the year 2000 but this is surely too high. More likely, an annual 2.5 percent growth rate will form the average for the next 23 years. This results in a 135q total for the end of the century and means substantial reduction in the oil and gas requirement.

Lovins: In the minds of the forecasters on whose work I based Figure 1, "recent past" usually means the 1960s, as such people tend to assume that the faltering of energy growth in the early 1970s is just a lamentable but transient hiccup. U.S. primary energy use—which I have suggested above is not a very relevant number—grew at an average compound rate of 5.4%/y in the 1950s, 8.1%/y in the 1960s, and 3.9%/y in 1970-3. Dr. Lapp's "under 2% per annum" for 1969-76 fits nicely with my 2.4%/y for 1975-85.

I agree that 160q in 2000 is "surely too high." Out of the six ERDA-48 (June 1975) scenarios, though, only two fall significantly below this level—I. Improved Efficiencies in End-Use (122q) and V. Combination of All Technologies (137q). The rest are 0. No New Initiatives (165q), II. Synthetics from Coal and Shale (165q), III. Intensive Electrification (161q), and IV. Limited Nuclear Power (158q). So as far as one can tell, these scenarios were the basis for ERDA's planning at the time when my article went to press, and may have remained so through the first months of the new [Carter] Administration. One can, even today, find well-known analysts who think values above 160 q are desirable or even essential.

Dr. Lapp does not say where his 2.5%/y comes from; it merely falls within the range (roughly 45 to 190 q) proposed by other forecasters. More importantly, since my argument rests on the *qualitative* nature rather than the exact quantitative details of the energy system, in both the hard and soft path, changing Figure 1 from about 160 q to 135 q in 2000 would not significantly change my conclusions.

Lapp: "It relies on rapid expansion of centralized high technologies to increase supplies of energy, especially in the form of electricity."

Comment. Central station power, primarily electric generation, forms a "hard" technology that is the target for Mr. Lovins' discontent. The sentence quoted here is curious because coal-fired steam electric plants have been with us for so long that they would not appear to qualify as "high technology" except in scale. However, the real point which is central to this increased electrification is missed, namely, that the shift from fluid to solid fuels forces reliance on central station generation.

Lovins: Modern big pulverized-fuel supercritical steam stations, fighting hard (and not very successfully) for the last tenth of a percent of thermal efficiency, are indeed a high technology and a great tribute to their designers. Whether they are a good use of those designers' talents is another matter. Dr. Lapp *assumes*, rather than demonstrates, a need to shift from fluid to solid fuels (*i.e.*, coal and uranium). His unstated prior *assumptions* are that direct use of coal on a significant scale is not feasible, that large-scale electrification is, that energy use will grow as much as he thinks it will, and that only fuels—not natural energy fluxes—can play a significant role before 2000. He then further *assumes* that electrification must continue and must be highly centralized.

Lapp: "The first path is convincingly familiar, but the economic and sociopolitical problems lying ahead loom large, and, eventually, perhaps, insuperable."

Comment: Mr. Lovins makes a sweeping and monumental evaluation of known technology which must be reckoned as an extreme judgment. The perils of plunging ahead into a soft-technology future could be described in equivalent prose.

Lovins: I do not accept Dr. Lapp's implication that hard technologies are "known" and soft ones speculative.

Lapp: "The second path, though it represents a shift in direction, offers many social, economic, and geopolitical advantages, including virtual elimination of nuclear proliferation from the world."

Comment. It is naive, in the extreme, to state that giving up nuclear power would eliminate nuclear proliferation. One wonders not just at Mr. Lovins' naivete or ignorance of the existence of a nuclear weapons stockpile numbered in the tens of thousands, but at the lack of critical editorial review by the publication.

Lovins: I am acutely aware of the existence of nuclear arsenals—and of their connection, in both directions, with "peaceful" nuclear energy. Giving up nuclear power would not by itself eliminate proliferation (which is not the same thing as eliminating existing stocks of nuclear weapons); but as part of a package of synergistic policies (*FA*, section VIII; *Soft Energy Paths,* Chapter 11) it could, in my opinion, virtually eliminate proliferation. I have been gratified recently to find the argument in *SEP* Chapter 11 commending itself to some previously fatalistic analysts of the proliferation problem.

Lapp: "It is important to recognize that the two paths are mutually exclusive."

Comment. With this statement, Mr. Lovins undoes much of his argument. It becomes clear that he is so obsessed with nuclear proliferation, based on diversion of nuclear material from power cycles, that he is unable to deal with the realities of nuclear armaments. Thus Mr. Lovins begins his article with fallacious reasoning which is actually not essential to making a case for conservation or for exploitation of new technologies.

Lovins: Dr. Lapp's comment does not address the exclusivity issue nor explain why it is "fallacious" or "undoes" much of my argument. Proliferation is in my view a very important reason for proceeding with the soft path, but if that issue did not exist, there are plenty of other reasons for following the same course.

Dr. Lapp does not seem to realize that there are more relationships between nuclear power and nuclear weapons proliferation than just the diversion of strategic material from civilian fuel cycles (see *SEP* Chapter 11 for other connections, mainly political). Conversely, I do not see how Dr. Lapp's proposed energy future would make it any easier to cope with the growing military nuclear arsenals that we both worry about.

Lapp: . . . Mr. Lovins uses the blitz technique, condemning federal programs without treating them objectively or dispassionately. He fails to come to grips with the reality of energy demand by an industrial and highly mobile society whose billions of energy-consuming engines, ovens, and devices have special fuel requirements. Existing energy sources are deemed to be "bad" whereas non-existant energy sources are necessarily "good."

Lacking experience in large-scale systems, Mr. Lovins fails to appreciate the inertia that applies to energy systems. By the same token he suffers from the Aladdin Syndrome; he apparently believes that non-existant energy sources can spring forth from an R&D lamp, fully developed and economically viable. In a certain sense, nuclear energy could be called an Aladdin source but it must be recalled that it depended on the discovery of fission (1939), the success of the chain reaction (1942), and the engineering of large-scale economic reactors (1966). Yet even under growth patterns that are unmatched in modern times nuclear power contributes only 3 percent of total energy consumption in 1976.

Lovins: I am aware of U.S. end-use patterns, and of the problems of changing them, in considerable detail. I believe changes would be harder with the homogenizing electrification Dr. Lapp favors than with my more fine-grained and adaptive approach. I am especially interested in backfitting existing boilers and in converting surplus biomass to fuels appropriate for existing vehicles.

I have been engaged for some years in calling to public attention the rate-and-magnitude problems (limits on how fast one can do things) of the large energy systems, especially in the U.S. (See for example *World Energy Strategies: Facts, Issues, and Options,* FOE/Ballinger, 1975; available since September 1973 in a different form.

Dr. Lapp has confused the hard/soft dichotomy with an existing/nonexistent dichotomy of his own invention. If anything, I think he has it exactly backwards. *Every* technology I assumed in Figure 3 has already been shown to work, is being or can quickly be commercially deployed, is economically competitive or nearly so with today's cheap fuels, is cheaper than its hard-technology competitors at the margin, and faces no significant environmental, ethical, or structurally political obstacles—though some face obvious institutional barriers. I believe what we need with soft technologies now is not a long-term R&D program but deployment of what we already know how to do. Dr. Lapp's example of nuclear fission is a bad analogy because it is much more complex, demanding, and technically novel than any technology I have in mind.

Lapp: Lovins overstates the electrical supply as "more than doubling" in 10 years. This corresponds to a 7 percent per annum rate whereas a growth pattern of 5.5 percent is more appropriate. Such a growth is attainable, although by spouting sheer statistics on requirements Mr. Lovins does a disservice to his readers.

Lovins: . . . My text refers to a Bechtel analysis based on President Ford's January 1975 State of the Union Message program. (It is interesting that Chauncey Starr likewise proposed in summer 1975 a national goal of maintaining a 10-year doubling time of electricity generation through 2000, a 5.4-fold increase.) Reproducing the Bechtel facility requirements was intended not as a disservice but as an aid to appreciating the consequences of a policy such as President Ford's.

Lapp: . . . Lovins overstates the electrical supply as "more than doubling" in 10 years. This corresponds to a 7 percent per annum rate whereas a growth pattern of 5.5 percent is more appropriate. Such a growth is attainable, although by spouting sheer statistics on requirements Mr. Lovins does a disservice to his readers.

Lovins: The 450-800 was ERDA's figure when *FA* went to press. Again, reducing 800 reactors to 500 (or a later and much lower figure) does not alter my qualitative conclusions (e.g., 500 reactors produce a bomb's worth of plutonium about every half-hour rather than every 20 minutes).

Lapp: "Massive electrification . . . is largely responsible for the release of waste heat sufficient to warm the entire freshwater runoff of the contiguous 48 states by 34-49°F."

Comment. Surely, this is misleading to a reader. Mr. Lovins must be aware that all cooling water for power plants is subject to EPA regulations and that cooling towers are used for most inland plants.

Lovins: My cooling-water analogy is carefully phrased not to mislead people: obviously the stations will not be cooled in that way, but I thought it useful to give a geophysical illustration of how much waste heat is involved. This is rather similar to the common practice of representing, say, volumes of nuclear waste in terms of aspirin tablets, football fields, etc., without meaning to imply that they will be swallowed, played upon, etc.

Lapp: "The commitment to a long-term coal economy many times the scale of today's makes the doubling of atmospheric carbon dioxide concentrations in the next century virtually unavoidable, with the prospect then or soon thereafter of substantial and perhaps irreversible changes in the global climate. Only the exact date of such changes is in question."

Comment. Mr. Lovins should not intermix possibilities and certainties and he clearly, judging from his record, lacks the expertise for such certitude in this judgment.

Lovins: My conclusions about CO_2 are consistent with all authoritative climatological studies, the best of which I cited in my testimony, and with the findings of Dr. Weinberg's group at Oak Ridge and Dr. Häfele's group at IIASA.

I have been active in work at the interface between climatology and energy policy since 1970 and have read and written a good deal on the subject, including a paper in 1971 that was widely circulated and well received in the climatological community. (It dealt with technical and policy aspects of global heat balance.) In 1972 I was rapporteur of the climatic-change working group of the U.N. Environment Programme/International Federation of Institutes for Advanced Study meeting at Aspen on Outer Limits. In March 1977 I was one of eight members of the World Meterological Organization's ad hoc expert group on climatic change (Geneva).

Lapp: Energy waste in electric energy generation.

Comment. The thermal inefficiency in production of electric energy is a technological waste of energy, but this must be balanced against the productivity that is gained by the on-off switch of power and its ease of deployment. Furthermore, the shift to solid fuels makes it essential to employ the known technology of central station conversion to electricity. Mr. Lovins demonstrates here that he is a narrow channel commentator on technological issues and is singularly devoted to airing only the issue which fits his case. His comments on British primary and end use of energy demonstrate the superficiality of his prose. Instead of solid analysis of an energy system he makes side-swipe commentary.

Lovins: Of course power stations partition entropy between electricity and condenser-water heat, sending nearly all the negentropy to the former and most of the enthalpy, as the inevitable price to be paid, to the latter. Convenience is likewise partitioned: to the users, but not to the utility, for which the users' ease of switching on and off is a costly headache. The British example, which obviously has a good deal of solid analysis behind it, was first called to my notice around 1973 by Gerald Leach; I quoted it because it provides a graphic illustration of where the U.S. is heading.

Lapp: Intractable capital cost.

Comment. Admittedly, capital costs for energy production have escalated, but this indictment is meaningless unless Mr. Lovins is able to prove that his alternatives are less capital cost sensitive.

Lovins: I believe I have made a strong case that the soft and transitional technologies do indeed have substantially lower capital cost than the hard technologies otherwise required to do the same job (see the Oak Ridge paper, or, for a fuller treatment, *Soft Energy Paths*, Chapters 6-8).

Lapp: High capital cost of electric systems.

Comment. It is certainly true that electric systems involve higher capital costs than direct-fuel systems. However, Mr. Lovins should address himself to consumer costs and to consumer requirements. And to the environmental consequences of direct-fuel energy systems.

Lovins: Capital costs of power stations and grids *are* ultimately reflected in consumer costs. That is why electricity today—whose cost is based largely on stations built before the days of really rapid cost escalation—typically costs about $50-120 per delivered barrel equivalent, and will cost more when the much higher capital costs of new facilities have worked through the system. Whether consumers can conceivably use extra—marginal—electricity to sufficient advantage to justify this high marginal cost is up to them, but I doubt they can.

I have considered the environmental effects of direct-fuel systems, and find them in general, for the technologies I advocate, to be much less than those of equivalent central-electric systems to do the same jobs.

Lapp: Shell Group in London—"no major country outside the Persian Gulf can afford these centralized high technologies on a truly large scale, large enough to run a country."

Comment. No reference is given and it would be more scholarly to cite and quote the conclusions of the Group, including their assumptions. Clearly, underdeveloped countries with low per capita income will have a hard time electrifying their economies; that's obvious and one needs no Shell Group for such a conclusion. But what is the point of the paragraph?

Lovins: The Shell analysis, which referred chiefly to electrical systems, was done in PLE-2, Shell Centre, London SE 1, mainly by Dr. Gareth Price (now at Marven SA, Aptd. 809, Caracas 101, Venezuela), and was reported in October 1974 to the first meeting of the MIT Workshop on Alternative Energy Strategies. In late 1975 I asked Dr. Price for a written reference in which I could cite his and his colleagues' views, but he said the macroeconomic impossibility of electrification on a truly large scale was by then (in their view) so well known that he had stopped keeping track of the many professional papers in which it had been considered.

The point of the paragraph is to suggest that Dr. Lapp's numbers don't add up: the U.S. can't afford to build all those electrical facilities (or, though the constraints are not quite as severe, all those coal-synthetics facilities either), and if we tried, electricity prices would become astronomical.

Lapp: 1976-1985 capital outlays for energy systems.

Comment. A combination of the escalation of fuel prices, inflation, and excessive regulation has increased capital costs, and money problems are critical for deployment of new energy systems. But new plants require energy inputs and by his own projection Mr. Lovins demonstrates that his soft technology cannot provide an energy answer for the next decade.

Lovins: It is not clear to me how increased fuel prices have increased capital costs (except as a second-order effect). Nor is it clear what Dr.

Lapp is trying to argue. If he says power stations need fuel, obviously he is right, and I suggest the fuel can be used directly to greater effect and at lower cost. If he is saying soft technology cannot solve our energy problems by 1985, I agree, but suggest that a power station ordered today probably cannot do anything by 1985 either, except sit there and eat interest. I think there is good evidence that a dollar invested in a soft path—all its ingredients, including soft technologies—can yield more energy, jobs, and profits, even by 1985, than the same dollar could yield if invested in nuclear power (where, so far, the profits are almost universally negative among the principal world vendors). All these arguments, however, seem irrelevant to the paragraph of my article that Dr. Lapp sought to criticize.

Lapp: Utilities' failure to base historic prices on the long-run cost of new supply.

Comment. Surely, Mr. Lovins must be aware that utilities operate under strict regulation by State Public Service Commissions.

Lovins: As for utility regulation, I am well aware of it, and of how utilities have long sought and still seek to make it more "responsive" to their own desires. So far most utilities have been conspicuously unwilling to find out in practice whether price elasticities of demand are as low as they like to think. On the contrary, utilities have sought and received for all new power stations a taxpayer subsidy of 10% as investment tax credit, about another 10% as accelerated depreciation allowance, and lavish interest deductions, thus giving us all the illusion that electricity is cheaper than it really is—to say nothing of fuel-cycle subsidies, whether nuclear, coal, or oil, amounting to many billions of dollars per year.

I am not suggesting that utilities could instantly persuade regulators to let them price properly; rather that utilities have a great deal of influence, which so far they have generally exercised in the opposite direction, giving themselves strongly promotional and economically unsound rate structures.

Lapp: Patterson quote.

Comment: The quote is not referenced. It must be dated because the British energy situation has changed. It would be well to cite more recent U.K. publications.

Lovins: The reference is W. C. Patterson & C. Conroy, "Energy Alternatives and Energy Policy: A UK Viewpoint," memorandum to the Energy Resources Sub-Committee, Select Committee on Science and Technology, House of Commons, May 1976. This date is just a couple of months before the *FA* press date, so I could hardly have quoted something more recent. I did not cite it because no citation seemed necessary. For the purposes of Patterson's analysis, the British energy situation is unchanged from May 1976 through the present (May 1977) and does not look likely to change in a hurry.

Lapp: 1972 Interior Study.

Comment. Any studies prior to October 1973 need to be recognized as pre-crisis analyses. The 1972 study predicted a year 2000 192q energy consumption. Certainly, the U.S. has backed off from such a projection.

"Some analysts still predict economic calamity . . . ."

Comment. The analysts are simply recognizing that there is demonstrated parallelism between energy consumption and the GNP—and employment.

"But what have more careful studies taught us . . . ."

Comment. The studies are not identified, but this is a typical Lovins technique—of personal characterization of the quality of comparative analyses while not identifying the reports or displaying a basis for his evaluation. A good editor would simply strike this paragraph.

Lovins: Obviously the 1972 Interior study was pre-embargo; but Interior hasn't backed off its pre-embargo numbers nearly far enough. The Chase Manhattan Bank in 1976, too, sounded alarmingly like the Chase Manhattan Bank in 1973, and the FPC hasn't greatly changed its analysis either.

"Parallelism"—correlation—is not causality. Is Dr. Lapp seriously saying that insulating your roof will cause (a) freezing in the dark, (b) depression, or (c) unemployment?

The "more careful studies," as is obvious from context, are those cited in the remainder of this section of the *FA* article, and their citations in turn constitute quite a good basic bibliography on energy conservation up to mid-1976.

Lapp: Two ways to make a cheap, low-technology plug.

Comment. Apparently, the two ways are changes in life style (dress, living habits, car-pooling, smaller cars, walking, bicycling) and technical fixes.

Mr. Lovins fails to calibrate the energy effectiveness of these suggested life style changes, to appraise feasibility of style-changing and to estimate its economic impact. It's true that much energy would be saved by ordering Detroit to manufacture only Volkswagens. But the politics and socioeconomics of this conversion are formidable challenges to any society.

Would Mr. Lovins, for example, practice what he preaches? Or would he continue to live a high-energy life style with globe-circling jet plane travel?

Lovins: The reason I did not estimate the energy, political, and economic impacts of energy-conserving changes in lifestyles is that I assumed in my analysis that no such changes, hence no such impacts, would occur, *i.e.*, that my figure 3 is entirely a technical fix. In section X of the *FA* article I discuss some aspects of feasibility—which is obviously for the reader, and for our whole society, to judge. I have nowhere proposed the sort of ham-fisted direction of industry that Dr. Lapp tries to put into my mouth, and think it would be a bad idea. Being a pluralist, I happen to think that people who want to drive big gas-guzzling cars should be free to do so—provided they pay the full social costs of doing so.

My personal energy use is slightly below the U.K. average if I count my plane travel as a marginal energy cost (*i.e.* assume that the plane flies whether I'm in it or not), and above the U.K. average but well below the U.S. average if I count the aviation fuel as a pro rata share of the fuel burned, according to the ratio of my weight to the payload (not the total weight) of the plane.

I have only circled the globe once—in order to cross-examine on, and respond to, some evidence which Dr. Lapp presented to the Ranger Uranium Environmental Inquiry in Australia and which I believed to be incorrect. I minimize my flying by turning down most invitations, spending several thousand dollars a year on international telephoning, combining trips, and taking trains whenever possible. I should be delighted to stop flying altogether as soon as the Dr. Lapps of this world make it possible for me to do so. I wear sensible clothes, live in a small space, and have never owned a car. Several months a year spent in the mountains (a very low-energy lifestyle) bring down my annual average. I can heartily commend such a lifestyle to Dr. Lapp, but suspect he would find it rather alien.

Lapp: Technical fixes.

Comment. One cannot quarrel with heat-saving through better insulation and with more efficient furnaces, but one has to consider the dq/dt, i.e. the time scale on which savings can be effected and the cost involved.

Lovins: I do consider how quickly the savings can be made, and conclude they are faster than increasing supply. The cost is typically about an order of magnitude less than that of marginal supply, and in many cases is trivial, zero, or even negative. There is a very large professional literature on these subjects, and I suspect my familiarity with it will bear comparison with Dr. Lapp's.

Lapp: "Theoretical analysis suggests that in the long term, technical fixes alone in the United States could improve energy efficiency by a factor of at least three or four."

Comment. Lovins' quote does not reflect the realities of his own citation. Mr. Lovins should have cited Table 7, page 472 of the [1976] *Annual Review of Energy* in which potential savings from various energy conservation strategies [as tabulated by Lee Schipper] range from 1.2 to as high as 12 percent. Mr. Lovins goes on to assert that "we could have steadily increasing economic activity with approximately constant primary energy use for the next few decades . . ." but he fails to back up this contention with any economic evidence. Admittedly, energy-consuming devices can be made more efficient although many already approach their practical limits of efficiency. A community served by a steam-electric plant of 28 percent efficiency could replace it with a modern plant of 40 percent efficiency. But the community decision to make such a replacement involves raising of funds to build a new plant which may, in fact, cost ten times more per kilowatt installed than their present plant. Thus theory and practice diverge.

Lovins: Dr. Lapp neglects to mention that the *Annual Review of Energy* table he cites shows 1-12% energy savings *per end-use category*, or a total saving, over the whole economy, of 43%. It does not take much economic sophistication to see that if secular growth in economic activity and improvement in end-use efficiency take place at about the same rates, my quoted conclusions follow.

Replacing an existing power station with a more efficient one is not my idea of a technical fix to emprove *end-use* efficiency—though if it is Dr. Lapp's his conclusions become more understandable.

Lapp: The Swedish-American energy comparison is often put: "Why can't the United States be more like Sweden?"

Comment. Sweden does use less energy per capita than does the U.S., but there are striking differences between the two countries in geography, life style and GNP composition. Sweden, for example, is about the land area of two U.S. states (Oregon and Washington). U.S. devotes as much land to food production for export as all the land area of Sweden. Transportation of things and people in U.S.A. involves car, truck, train and plane travel of greatly different dimension than in Sweden. Life styles are different especially when you compare suburban homes in U.S. with Swedish counterparts. Living space is greater, refrigerators are larger and energy devices are more abundant. Now the United States could emulate Sweden in saving energy but one has to reckon with the capital cost of introducing such changes and with rather radical life-restyling.

Lovins: Dr. Lapp might have failed to consult the Schipper & Lichtenberg paper which I cite (since published in *Science* 194:1001-13, 3 December 1976; see also correspondence about this paper in *Science*, 8 April 1977). It shows that Sweden has 18% *lower* average population density that the U.S., a 67% *colder* climate, and per-capita production of steel, cement, fertilizer, and paper (a cross-section of the most energy-intensive industrial materials) respectively 110, 125, 64 and 241% that of the U.S. Swedes have a luxurious diet (slightly less so than average Americans), much higher social indicators, transport at least as ample and pleasant, more second homes, and comparable domestic conveniences.

Dr. Lapp's statements about living space and appliances, and his implication that imitating Swedish efficiency would require "rather radical life-restyling," are simply wrong, as the study bears out. (Further confirmation comes from the new Resources for the Future cross-sectional study of energy use in the U.S., Sweden, and elsewhere.) Schipper and Lichtenberg conclude that if Americans were as efficient as Swedes, other things being equal, we could produce the same output and live as well on about 30% less energy than we use now. But that is a moving target: Sweden has enormous room to improve her own efficiency and is developing major conservation programs too.

Lapp: "And there is overwhelming evidence that technical fixes are generally much cheaper than increasing energy supply, quicker, safer, of more lasting benefit."

Comment. This contention is not documented. Where is the overwhelming evidence?

Lovins: Here as in many other places, *FA* deleted extensive documentation for lack of space and the nature of its audience; I put most of the documentation back into the cited Oak Ridge paper. On this simple point, however, Dr. Lapp could consult p. 8 of ERDA 76-1, numerous speeches by Frank Zarb [then the federal "Energy Czar"], or virtually any competent study of the subject, including my *FA* notes 11, 12, 13, 16, etc.

Lapp: "Even making more energy-efficient home appliances is about twice as good for jobs as is building power stations."

Comment. Again, contention without evidence. Lovins' source (*Annual Review of Energy*) cites only 2 percent energy savings for improved home appliances.

Lovins: The 2% is about right, which is why I said "even"—it's only one of many small (but cumulatively important) terms. [A more complete Danish study in 1978 found a 3.5-fold improvement in the average efficiency of household electric appliances is now cost-effective. In the U.S. this would save about 0.75 quads of electricity or 3.2% of 1973 primary energy.] The jobs-per-quad figures are available from Bruce Hannon's input-output studies at the University of Illinois Center for Advanced Computation at Champaign Urbana [*SEP*, p 9, n 10].

Lapp: "The capital savings of conservation are particularly impressive."

Comment. Mr. Lovins does not document this, but goes on to maintain "Indeed, to use energy efficiently in new buildings, especially commercial ones, the additional capital cost is often negative; savings on heating and cooling equipment more than pay for the other modifications." Again, no documentation.

Lovins: The documentation is in *FA* note 15 [which cites three papers].

Lapp: "To take one major area of potential savings, technical fixes in new buildings can save 50 percent or more in office buildings and 80 percent or more in some new homes."

Comment. Table 7 in *Annual Review of Energy*, [1976] page 472, lists 5 to 8 percent savings in space heating due to improvements in insulation, heat pumps, and much less percentage in air conditioning due to insulation, window improvements and design. Mr. Lovins uses quotations for individual installations without putting the issue in realistic terms.

Lovins: It merely muddies the waters to mix up how much of a building's energy use can be saved [the 50-80%] with how much energy that action saves out of the national energy budget [the 5-8%]. Savings in the latter are made up of precisely such individual contributions. Thus, in the cited analysis of Ross and Williams, replacement of resistive residential heating with heat pumps (COP = 2.5), improved residential air-conditioner efficiency, residual heat losses cut in half by modestly improved insulation and weatherstripping, and reduced air-conditioning load from the same reduced infiltration contribute respectively only 0.6, 0.4, 3.3, and 0.4 q of the 6.6 q that could have been saved in the 1973 residential sector by simple technical fixes; but these and similar changes throughout the economy together add up to the dramatic 43% national energy savings that I cited. [By 1978 it had become clear that the space-heating saving could readily be 100% through heat-conserving and passive solar design, even in severe climates.]

Lapp: "A second major area lies in 'cogeneration' or the generating of electricity as a byproduct of the process steam normally produced in many industries. A Dow study chaired by Paul McCracken reports that by 1985 U.S. industry could meet approximately half its own electricity needs (compared to about a seventh today) by this means."

Comment. Industrial plants generate about 5 percent of total U.S. electric generation. It would be very helpful if Mr. Lovins would bring his gushers of statistics into some kind of summation which permits perspective.

Lovins: One such summation is that it would be economically justified to install, over the next decade, cogeneration equipment in the chemical, petroleum refining, and pulp and paper industries which would produce the equivalent of at least half of *all* U.S. electricity use in 1975. The ultimate potential is much larger. (Thermo-Electron Corp. report to FEA, "A Study of Inplant Electric Power Generation in the Chemical, Petroleum Refining, and Paper and Pulp Industries," PB-225 659, NTIS, June 1976.) Von Hippel and Williams (*Bull. Atom. Scient.*, December 1976) conclude on rather similar data that cogeneration, conservatively calculated, could by 2000 replace some 230 GW(e) of nuclear capacity. [The most recent and authoritative assessment, published in 1978 by R.H. Williams in *Annual Review of Energy* 3:315-356, similarly estimates a technical and economic scope for 208 GWe in 2000; the thermodynamic potential is 1000 GWe.]

Lapp: "So great is the scope for technical fixes now that we could spend several hundred billion dollars on them initially plus several hundred million dollars per day—and still have money compared with increasing the supply!"

Comment. Again, no documentation. Mere assertion. Several hundred million dollars per day equals 3 billion per year. [Webster's defines "several" as: "Consisting of an indefinite number more than two, but not very many." Hence, it is clear that "several hundred million dollars per day" is not a definite number and does not equal any definite amount per year. It is also clear that Dr. Lapp's $3-billion/year figure is wrong: taking "several" to mean three, several hundred million dollars per day would equal $3 billion in only ten days.] Without setting forth a technical basis for his estimate Mr. Lovins lays himself open to a charge of flamboyant rhetoric. Mr. Lovins' estimate must therefore be regarded as unsupported speculation.

Lovins: The capital costs of improving end-use efficiency—summarized at *FA* p. 73—are based on an extensive survey of the professional literature of energy-conservation economics. If, in the approximatory spirit of Dr. Lapp's and my discipline (physics), we assume that conservation requires an investment of the order of $10,000/(bbl·day) saved, then Ross and Williams's 43% saving corresponds to 16 million bbl/day or to an investment of order $160 billion. My comparative capital costs show this investment is substantially less than for any significant increments of supply. Moreover, 16 million barrel/day correspond to a daily fuel cost, at OPEC parity, of the order of $230 million/day—considerably more for synthetic fuels or electricity. These figures are consistent with my statement. The calculation is so elementary that I did not think it necessary to set it out in my *FA* article.

Lapp: Mr. Lovins defines many of the "real world" factors that rule against the easy implementation of his technical fixes. These do constitute practical constraints on changes in energy use in U.S.A. and they are not subject to magic wand dispersal.

Lovins: I agree; that is my point. I do think, though, that (1) it is easier to overcome these problems than not to, and (2) if we aren't clever enough to overcome them, we certainly won't be clever enough to overcome the problems that will arise from our failure to do so.

Lapp: The 2010-40 period is beyond comment.

Lovins: The period 2010-40 is what we must be thinking about if we want to arrive at a place worth getting to. As Dr. Lapp points out, the energy system changes only very slowly; with his technologies, 1987 is already cast in concrete.

Lapp: "Many analysts now regard modest, zero or negative growth in our rate of energy use as a realistic long-term goal."

Comment. This says nothing about the main body of opinion about zero and negative energy growth and about its disastrous economic implications.

The reference here to Dr. Weinberg's work is unfortunate since it is unpublished and has not been subject to critical external review. The same applies to the U.S. NRC study since the overall report has not been published and will not be for some time. This is not good science but borders on gossip.

Lovins: My statement is empirically verifiable; Dr. Lapp's is not (save by sufficiently restrictive definition of "main body of opinion"). The "disastrous economic implications" of zero or negative growth in energy use have been rebutted in many good studies, from the Ford Energy Policy Project through the eminent Economics Panel of the CONAES study.

The Oak Ridge study was available before my *FA* paper was published. It was subjected to critical external review: very critical in my case (as a member of the ad hoc review panel invited by Dr. Weinberg in June 1976), though my main comments were disregarded. Many of Dr. Lapp's colleagues were on the panel too, and I am sure he could have had a copy from mid-1976 onwards.

The NRC CONAES study is to report in late 1977, involves many more of Dr. Lapp's colleagues, published an interim report in January 1977 describing its 70-210q range for 2010, and is widely discussed in the energy policy community. [The main CONAES report was still not published by March 1979, but key results from the Demand and Conservation Panel were published in *Science*, 14 April 1978.] It was in the knowledge of these things that I made a factually correct statement about the CONAES 70-quad scenario, and in deference to the future publication of the final report that I did not identify the study. I checked the form of words in advance with CONAES staff to be sure there was no objection. Such treatment of important studies in progress is a normal practice in scientific literature.

Lapp: Here Mr. Lovins introduces his Figure 3 which shows his as yet undefined "soft technologies" supplying 1 q of energy by about 1981 and 10 q by 1987.

Comment. For a commentator who apparently abhors growth, Mr. Lovins emerges as a champion of a ten-fold growth of soft technology in 7 years. This is a growth rate of about 40 percent per annum!

Lovins: The definition and explication of the "as yet undefined" soft technologies begins on the same page as the graph. As for the growth rate, it is elementary that growth is rapid in the early stages of building up anything (whether soft technologies or reactors) from almost nothing: the growth rate of U.S. solar heating in mid-1977, for example, is probably more than 140%/y compound (doubling every 6 months or so). There are some kinds of growth that I like and advocate. [The U.S. solar market collapsed later as consumers waited for Congress to pass the promised solar tax credit. At the end of 1978, the market appeared to be picking up again—minus about a third of the vendors, mainly small businesses that went under meanwhile.]

Lapp: Mr. Lovins' description of "soft technology" is semi-mystical.

Lovins: The five-point definition seems to me clear and simple, as a definition should be. It is unfortunate that Dr. Lapp seems to find it much less comprehensible than many other people with less technical training have done. The definition is followed by an extensive discussion, augmented by references and a backup paper.

Lapp: Despite the obfuscation, it seems that Mr. Lovins is talking about solar energy in one form or another as his "soft" technology. But having assigned a quantitative growth curve to this technology in Figure 3, one would expect some quantitative interpretation of what this soft technology is. It's apparently some kind of idealistic match of energy to end use without regard to the practicality of available resources and the rate at which they can be exploited. Mr. Lovins' explanation of his soft technology is so diffuse that it defies comprehension.

[To refresh readers' memories, the five criteria that define soft energy technologies are (1) they rely on renewable sources, (2) they are diverse, each designed for maximum effectiveness in particular circumstances, (3) they are flexible and are relatively simple and understandable to the user (though they can still be technically very sophisticated), (4) they are matched to end-use needs in scale, and (5) they are matched to end-use needs in energy quality. An energy technology must satisfy *all five* criteria to be soft.]

Lovins: The "soft technology" curve shown in Figure 3 represents a plausible pattern for the sum of delivered end-use energy from all soft technologies, constrained by their practicable deployment rates but given vigorous support through the three methods suggested (plus appropriate engineering development). [The three methods referred to are (1) removing institutional barriers to conservation and soft technologies, such as obsolete building codes; (2) removing subsidies now given conventional fuel and power industries, and vigorously enforcing antitrust laws; and (3) pricing fuels at a level consistent with their long-range replacement cost.] No currently unproven or exotic technologies are assumed. As documented at p. 98 of *Soft Energy Paths,* my estimates of deployment rates fall roughly in the midrange of published estimates in the technical literature—even though apparently all other estimates are based on assumed soft-technology competition with cheap fuels rather than with other alternatives to those fuels at the margin.

Lapp: Mr. Lovins seems incapable of understanding that the trend to electrification in an industrial society does have its own logic and that the trend is now intensified by the necessity to shift from fluid to solid fuels.

Lovins: The logic has lost its force with changing circumstances.

Lapp: Mr. Lovins fails to recognize the efficiency and versatility that electricity has brought to industrial production and to homes. Granted there is a thermal conversion loss in electric generation, there are distinct advantages to electricity and these, in fact, account for the doubling every decade growth that has characterized the electric power industry.

Lovins: The marginal utility of electricity is not the same as the historic utility of electricity; moreover, convenience and versatility can be, and are, dearly bought. The efficiency of using electricity in current practice is often not dramatically greater than the efficiency of using fossil fuels (also very high-quality energy) and is sometimes less. The 7%/y past growth is, I suggest, related to promotional rate structures, advertising, and a real residential electricity price that fell fivefold from 1940 to 1970.

Lapp: Mr. Lovins becomes infatuated with his prose. Nuclear temperatures of millions (of degrees) cited by Mr. Lovins illustrate his naivete.

The highest temperatures reached in nuclear reactors are fuel pellet center-line temperatures of about 4000°F which reduce to 600°F at the pellet-clad surface.

Lovins: I am well aware of the kilodegree limits of reactor materials and of the even lower limits imposed by the chemical reactivity of Zircaloy cladding. My reference, by analogy to another primary heat source (flame temperature), was to the effective temperature of the nuclear reaction—that is, the energy of a fission event, about 194 MeV, divided by Boltzmann's constant. This temperature is indeed many millions of degrees—some 2.2 trillion degrees Kelvin. In an important sense, the elaborateness and high cost of a reactor arise from the severe mismatch between this high-energy event and the tolerance of materials and people. The point of my example is that a teradegree (10^{12} degrees K) temperature is restrained to heat fuel to kilodegrees (all it can stand) to make sub-kilodegree steam to drive a turbine, etc., partly to make dekadegree heat.

This is a good example of a correctly stated minor phrase which, like the one about the diesel generators, a few people have misunderstood. I put the *FA* article through many stages of adversarial review and about twelve drafts to try to anticipate and correct in advance all the possible ways in which people might misunderstand it, but I fear I was not quite imaginitive enough.

Lapp: Mr. Lovins uses a $50 to $120 figure for a per barrel-equivalent in discussing electric energy. As stated here this comparison is meaningless. Delivered electric energy at 1 cent for 10^3 Btus (common in U.S.A.) cannot be compared in terms of oil barrel Btus without specification of the end use and conversion efficiency.

Lovins: As the cited Oak Ridge paper and *Soft Energy Paths* (Chapter 8) discuss in detail, it is of course correct that different forms of energy—compared here on the standard "heat supplied" or enthalpic basis that is the basis of U.K. energy statistics—must be compared with due regard to their First Law efficiency of end use. I argue in the Oak Ridge paper, and more fully in *Soft Energy Paths*, Chapter 8.2, that marginal electricity cannot compete with present soft technologies in supplying heat (most of U.S. end-use needs), nor in other applications outside the roughly 8% of end use that is currently for premium, appropriately electrical, applications.

Lapp: "So limited are the U.S. end uses that really require electricity that by applying careful technical fixes to them we could reduce their 8 percent total to about 5 percent (mainly by reducing commercial overlighting), whereupon we could probably cover all those needs with present U.S. hydroelectric capacity plus the cogeneration capacity in the mid-to-late 1980s."

Comment. Mr. Lovins goes on to add: "Thus an affluent industrial economy could advantageously operate with no central power stations at all!" The exclamation point is really all that is needed as a comment.

It is exceedingly difficult to take Mr. Lovins seriously when he is so obviously out of touch with the real energy world. Imagine the plight of an American housewife doomed to live with 5 percent of her electric energy needs! She will scarcely believe Mr. Lovins' technology is "soft."

Lovins: My calculation is quite clear: technical fixes à la Ross and Williams, for example, can readily reduce the current appropriately electrical end-use needs of the U.S. from about 4 to about 2.5 q or less; present hydroelectric production averages about 0.8 q under adverse conditions; and the remaining 1.7 q can be supplied by cogeneration by the mid-1980s or in the longer run by microhydro and wind, roughly half and half. (The Thermo-Electron study mentioned above estimates an economic potential of 1.7-3.3 q of cogeneration in the three industries studied—by 1985.)

Dr. Lapp, uniquely, has contrived to misread "5% of all U.S. end-use energy, if we use energy efficiently, is for purposes that can use electricity to advantage" to mean "I expect people to get along with only 5% as much electricity as they need."

Lapp: "A feature of soft technologies . . . is their appropriate scale, which can achieve important types of economies not available to larger, more centralized systems."

Comment. This is pretty murky since Mr. Lovins consistently avoids coming to grips with giving his readers any picture of his "soft energy" system, but presumably he is relying on solar energy for household needs. Surely, he must be aware of the fact that solar heating is not applicable to many millions of households and in most cases is a supplemental option which requires reliance on central station energy systems. Mr. Lovins, to rephrase Jonathan Swift, kicks common sense outdoors.

Lovins: Dr. Lapp perhaps overlooked the catalogue of soft technologies on *FA* pp. 81-3. The households for which solar space heating, especially on a neighborhood scale, can never be retrofitted at a reasonable cost represent a tiny fraction of U.S. space heating requirements (I would guess a few percent), and many of them will not be here in 50 years. The solar technologies I assume in my technical and economic analysis are not supplemental but complete and hence require no backup.

Lapp: "Small systems with short lead times greatly reduce exposure to interest, escalation and mistimed demand forecasts—major indirect diseconomies of large scale."

Comment. Small may be beautiful in the eyes of the beholder, but not always as viewed by a Public Service Commission.

Lovins: A PSC or PUC should like this idea if it saves utility customers money. Some commissions are starting to think exactly that. For example, the California, New Jersey, Maryland, and Wisconsin grids have been the subject of excellent studies of how smaller power stations save money by reducing reserve-margin requirements.

Lapp: "Presumably a good engineer could build a generator and upgrade an automobile engine to a reliable, 35-percent efficient diesel at no greater total cost, yielding a mass-produced diesel generator unit costing less than $40 per kW."

Comment. Mr. Lovins here is playing with numbers, divorced from the reality of energy system requirements. Apart from the fact that he neglects to include cost of electric conversion and regulation, he seems to be heading down the wrong path. I thought he was in flight from a hydrocarbon economy and now we find he is recommending small-scale diesel-electric engines for supplying households with power. In small scale the peaking demands of a user impose very high resrictions on the system. While on an average a homeowner may have relatively low power, turning on an electric dryer hits the system with a heavy overload.

Would Mr. Lovins populate a cluster of ten or twenty houses with a diesel-electric source? He would have a distribution problem and a community environmental problem—noise, fumes, siting, to mention a few—and an economic impact in the form of high community capital cost plus very high operating costs. Diesel fuel costs alone run 4 to 5 cents/kW-hr.

Rather than speculate on $40/kW diesel-electric units, Mr. Lovins would be living up to the responsibilities of an author if he did his homework and investigated the realities of his proposal. For example, he could make a phone call to the Rural Electrification Administration where he would learn details of the Alaskan Village Electric Cooperatives whose 45 plants have a capacity to 15,500 kW. He would learn that total power costs for small 250-350 kW generation run: Alakunuk, 14.2 cents, and Anvik, 48.3 cents, per kilowatt/hour. I would think that there would be some consumer resistance to Alaskanizing U.S. suburban electric supply.

Lovins: I am not recommending diesels, but rather using them for a scale analogy. But Professor Paul Schweitzer, a pioneer of modern diesels, tells me the U.S. has 3 GW(e) of diesel generators in 280 stations spread over 2/3 of all counties in 44 states; the stations are typically 15-40 MW(e), use engines of a few MW capacity, and are competitive. It is worth noting also the advantages of smaller diesel total-energy systems for housing developments and apartment buildings—the reason HUD, Oak Ridge, and Fiat are working on internal-combustion-engine total-energy-system packages. Diesel fuel costs about 3.8 cents/kW(e)-h in the lower 49 states, not 4-5 cents.

I have, as Dr. Lapp suggests, done a bit of homework on the Alaskan Village Electric Cooperatives diesels, and have found out that they are not typical of conditions in "the U.S. suburban electric supply." At Alakunuk, for example, where Dr. Lapp cites 14.2 cents/kW-h electricity, diesel fuel is flown in by the barrel in twin-engined aircraft, probably 400 miles from Bethel (the nearest deepwater port), then stored in small tanks (of order 10,000 gallons). Mechanics and spare parts are flown in 2000 miles from Seattle. Everything is done in Arctic weather in an isolated fishing village without decent roads or port facilities, with bad communications, and with all the high overheads one might expect under those conditions.

This is all irrelevant, however, since I was not proposing diesel systems; I was illustrating the price advantage of mass producing small-scale units, which large-scale centralized systems cannot enjoy.

Lapp: Medium-scale solar technologies.

Comment. Mr. Lovins says that "the medium scale of urban neighborhoods and rural villages offers fine prospects for solar collectors" but we are left with his personal judgment and no analysis which involves numbers.

Lovins: The projects I mention, another in Sweden, and several California apartment-house solar projects provide useful data. The rough magnitude of expected savings can be calculated from first principles (such as the ratio of volume to area and of fixed to variable cost for storage tanks, the ratio of labor to materials cost for collectors, etc.). I believe much more quantitative analysis and experience in this field is urgently needed, but that the data available are sufficient to support my broad engineering and economic conclusions.

Lapp: ERDA's solar research budget. "The schemes that dominate ERDA's solar research budget—such as making electricity from huge collectors in the desert, or from temperature differences in the oceans, or from Brooklyn Bridge-like satellites in outer space—do not satisfy our criteria, for they are ingenious high-technology ways to supply energy in a form and at a scale inappropriate to most end-use needs. Not all solar technologies are soft."

Comment. Mr. Lovins fails to mention the $290 million solar budget Congress appropriated for ERDA. In rejecting large-scale solar projects Mr. Lovins severely limits the options available for his alternative energy sources. He is very confused about what is "soft" and "hard" technology. Large-scale solar-electric generation is "hard" but medium scale is "soft." Again, there is no quantification and no analysis is introduced. We are simply left with Mr. Lovins' personal evaluation of high technology. He is certainly entitled to his opinion as a layman, but he should be careful to qualify his judgments for the consequences of following the wrong energy paths can be disastrous.

Lovins: Most of ERDA's solar budget goes to hard technologies, and most of what goes to soft technologies is in my view misspent, so it does not much matter how big the budget is unless priorities change drastically. I limit the options I consider to those that we need.

It is not I who am confused about what is a soft technology. The soft/hard dichotomy is not the sort that depends on numerical hairsplitting, so quantitative specification of the definition is not necessary. It is also not appropriate, since the essence of the distinction is social, not technological: as Joe Nye recently remarked, one doesn't conclude that a man doesn't love his wife just because he can't say, "I love my wife 0.7."

Lapp: Solar heating and cooling are depicted as examples of "genuine soft technologies . . . now available and . . . economic."

Comment. Apparently, "soft" is "small."

Lovins: Soft is not necessarily small and the two are certainly not synonymous. Small devices can use nonrenewable energy, or be complex, or be mismatched thermodynamically to end-uses, or be mismatched in scale to big end-uses—all qualities that are excluded, by definition, from soft technologies.

Lapp: Commercialization of solar heating.

Comment. Given the seriousness of the U.S. energy situation solar heat sources should be tapped for residential application. But the systems lack the performance data and flexibility required to make them first options for new houses. What is required here is a realistic projection of the solar potential—not an overselling of possibilities. The closest Mr. Lovins comes to quantification is the following: "If we did this [solar heating through use of 'passive collectors' i.e. large south windows or glass-covered black south walls] to all new houses in the next 12 years, we would expect to save about as much energy as we would expect to recover from the Alaskan North Slope." Mr. Lovins fails to define the time span over which he would have this energy. Energy recoverable from the North Slope is about 10^{17} Btu. The average U.S. house heating requirement is 1.2×10^8 Btu per year. If we take a 30 year period this amounts to 4.2×10^9 Btu per house. Given the practical factors such as wall area, solar access, relative number of single houses, it would appear that only a few million houses could be built in the next 12 years and that solar-passive systems would account for about 2×10^9 Btu/house/30 yrs. That's 5×10^{15} Btu. Mr. Lovins errs by a factor of at least 20, assuming crash programs implemented his plan.

Lovins: New houses should use passive solar systems. Active solar systems [for retrofitting those existing houses whose shape or site make them unsuitable for passive retrofits such as attached greenhouses] have many currently commercial forms with ample data available on performance and cost. Deployment rates are discussed in *Soft Energy Paths*, mainly in Chapter 5.

Dr. Lapp disagrees with the R. W. Bliss calculation about the energy savings from building the house right, including taking advantage of high window gain. As should be obvious, the saving to which Dr. Bliss and I refer is over the lifetime of the houses, conservatively assumed to be 30 years. The differences of assumption, as set out in Dr. Bliss's article, are:

assumption	Bliss	Lapp
primary energy saved per house by building it right (10^6 Btu/y)	100	67
lifetime of each new house (y)	30	30
new housing starts (10^6/y) (actually 1.64 in 1960-7)	1.5	
new housing starts assumed, for the purposes of Bliss's example, to be built right (10^6 over next 12 y)	18	2.5
total primary energy saved by those houses over their lifetimes (10^{15} Btu)	54	5
Recoverable North Slope reserves (10^{15} Btu)	56	100
(10^9 bbl)	10.2	18

It is easy to see where Dr. Lapp's factor-of-20 discrepancy arises. He has assumed nearly twice as much North Slope oil as the Department of the Interior estimates (9.7 billion barrels), and has made his own assumption that only 14% of new houses would in fact be built so as to save 10^8 Btu/y (most of their space-heating load) through passive solar design and good insulation. But my text says "If we did this to all new houses in the next 12 years"—all, not 14%—because Dr. Bliss wanted to show the magnitude of the savings to be had if we did innovate in this way. [In some solar conscious parts of the U.S. such as the Santa Fe area, Pitkin County, Colorado, and parts of Northern California, at least a quarter of the new housing starts in 1978 were passive solar. It is also noteworthy that during a single year in 1977-78, *half* of the 250,000 houses in Nova Scotia were weatherstripped and insulated by their occupants.] Incidentally, neither Dr. Lapp nor I has included commercial buildings in the example (which would nearly double the potential savings).

Lapp: Organic conversion technology. Biomass energy utilization.

Comment. Opportunities should be seized to recover energy from agricultural wastes and these provide a potential for perhaps 10^{15} Btu annually. However, the orientation of U.S. farmland and woodlands to massive biomass exploitation raises profoundly difficult environmental and even moral problems. The world may not look kindly upon diversion of U.S. food producing lands to use as fuel sources for our automobiles, especially if it cuts back food exports.

Lovins: According to the careful study of Poole and Williams [*Bulletin of the Atomic Scientists,* May 1976, pp. 48–58], the potential yield of net fuels from U.S. agricultural wastes is 2.3 q/y, not 1 (the gross resource is 5.2 q/y of crop residues plus 0.7 of feedlot manure; Poole and Williams assume 80% recovery and 50% conversion efficiency). Urban waste at the 1973 level of generation and pulp waste, converted by pyrolysis at 70% efficiency, yield a further 1.4 q/y. And 4.5 q/y of logging wastes, collected at 80% efficiency and converted at 60%, yield 2.2 q/y. The total of these terms is 5.9 q/y. In contrast, today's entire liquid fuel requirements for transport, if it ran at the best European efficiency of today (three times better than the present average U.S. efficiency), would need only 5.7 q/y. Thus I am already in the right neighborhood without resorting to any biomass plantations whatever.

Lapp: Converting U.S. wine cellars and breweries into fuel for the transport sector.

Comment. Careful, Mr. Lovins, millions of Americans would rather drink than drive (and many do both).

Mr. Lovins' estimate that 10 to 14 times scale-up in the wine-beer industry could supply one-third of U.S. gasoline requirements, assuming a three-fold increase in automotive efficiency, is grossly in error. He is wrong by at least a factor of 10 in his estimate. Such shoddiness with numbers is inexcusable. Mr. Lovins does a trapeze act jumping from one topic to another obviously confusing methanol and ethanol production (certainly an absurd performance considering the pressure and temperature parameters for the two production processes).

Lovins: Dr. Lapp apparently thinks I proposed making fuel alcohols in a beer-and-wine industry scaled up 10-14 times. I did not, and my calculations are correct. The example makes clear through its opening sentence ["The required scale . . . can be estimated"] and two caveats—"(not all alcohol, of course)" and "(in gallons of fluid output per year)"—that I am comparing the physical scale of two chemical systems, *i.e.*, the required size of vats and plumbing. Fluid output per year is the appropriate basis for such an analogy.

Further, it is not I who has confused ethanol and methanol production.

It should be obvious to anyone acquainted with bioconversion technologies, and not unreasonably literally-minded, that I was not proposing to make fuel alcohol by the same processes, namely microbial fermentation, that we now use to make beverages: the aim would be to make as much alcohol as possible, not to make something that tastes good (and is only moderately toxic). With present technology—which current bacterial and enzymatic research promises to improve rapidly—destructive distillation and pyrolysis are among the methods of choice, and methanol the main prospective product.

Of course these processes differ markedly from beverage-making, but one can still fairly talk of the physical scale of the industry as being measured, to first order, by how many gallons of fluid output per year it has to handle. That was the clear object of my analogy, and I fear Dr. Lapp has picked the wrong nit again.

Lapp: Windpower is now introduced as a soft technology.

Comment. Again, Mr. Lovins fails to deal with one of his alternative sources in anything resembling a quantitative manner. Perhaps if he did it would not qualify as a soft technology.

Solar process heat, Mr. Lovins asserts "is coming along rapidly as we learn to use the 5800°C potential of sunlight." That is a curious statement since Mr. Lovins abhors nuclear technology which is even lower in temperature. His judgment on progress in this area is certainly not based on documented evidence. Again, we find the paragraph redundant with diffuse prose that defies comprehension.

Lovins: I am not envisaging zillions of multi-megawatt wind machines running a national grid; most of my wind machines would not make electricity at all because it would not be needed. The wind machines I have in mind are soft technologies (see the Oak Ridge paper and *Soft Energy Paths*, Chapter 7.2).

The thermodynamic potential of nuclear fission is treated above. That of sunlight, about 5800°C, is higher than the tolerance of known materials. Even completely scattered light on a cloudy winter day has a potential of more than 2000°K.

Documentation is in the cited Oak Ridge paper and in *Soft Energy Paths* (mainly Chapter 7.4). More is not hard to find. In one recent example, D. P. Grimmer and K. C. Herr of Los Alamos Scientific Laboratory (LA-6597-MS, December 1976) believe a Winston collector yielding 315°C would have a retail price of $11/ft^2 or less with today's technology, with concentration factors up to about 9 and an average efficiency (taking into account both collector performance and off-axis cosine factor) of about 0.44. They estimate that the output temperature is adequate to meet nearly 2/5 of U.S. primary energy needs today for the industrial, commercial, and utility sectors. Their price estimate implies delivered process heat, assuming a real fixed charge rate of 10%/y, at a price of about $5/GJ, probably competitive with 1980s OPEC oil, if annual average insolation is their assumed 181 W/m^2 (my own analysis assumes a U.S. average of 180). Storage would still leave the price well below that of hard-technology alternatives at the margin. Similar performance, and even higher temperatures, can readily be attained in cloudy northern latitudes (*e.g.* Scandinavia) by using evacuated tubular or flat-plate collectors with selectivities of order 50-60.

Lapp: Energy storage. "On the whole, therefore, energy storage is much less of a problem in a soft energy economy than in a hard one."

Comment. Paragraph fails to provide substantive basis for this contention. Mr. Lovins uses dams as illustrative of his thesis, but he must surely recognize that this form of energy storage is very limited and subject to stiff environmental opposition, i.e. witness ConEd's long struggle for stored energy through pumped power on the Hudson (i.e. Storm King).

Lovins: The cited paragraph is self-explanatory. It is precisely because the electricity on which Dr. Lapp would have us heavily rely is difficult to store in bulk, as at Storm King, that I would store the small proportion required for baseload as water behind existing conventional (not pumped) hydroelectric dams. See *Soft Energy Paths,* Chapter 7 and *passim,* for further discussion.

Lapp: "Recent research suggests that a largely or wholly solar economy can be constructed in the United States with straightforward soft technologies that are now demonstrated and now economic or nearly economic."

Comment. It now becomes clear that the sun is Mr. Lovins' solution, although he has blurred the issue with murky prose. But by excluding a number of solar options, he has narrowed down the possible energy sources to a precious few. He fails to recognize that he is proposing dangerous gambles as solutions to America's energy supply in the future.

Lovins: In the unlikely event that already-demonstrated soft technologies cannot be deployed as quickly as I think they can, the transitional sources can be stretched out a bit longer, still at higher efficiency, lower economic and environmental cost, and lower risk than Dr. Lapp's proposed hard technologies. This is inherently a low-risk strategy, technically and in every other way.

Lapp: "Figure 3 shows a plausible and realistic growth pattern, based on several detailed assessments, for soft technologies given aggressive support."

Comment. Alas, these "detailed assessments" are not referenced and the author has now clearly failed to address himself to the basic issue, namely, defense of his guestimates for soft technology. He has futured the U.S. energy situation in a purely conjectural manner not really susceptible to scientific analysis. Mr. Lovins concludes: "To a large extent, therefore, it is enough to ask yourself whether Figure 1 or 3 seems preferable in the 1975-2000 period." It may be enough for Mr. Lovins, but it will not satisfy a nation that demands energy and a continued high level of industrial activity. Mr. Lovins has failed to make any substantial case for his soft technology future—or for reduced energy consumption.

Lovins: See *Soft Energy Paths* Chapter 7 for technical details, Chapter 5 (especially at p. 98) for comparisons with other estimates. My numbers are broadly consistent with those of NSF-NASA (1972), AEC Subpanel IX (1973), and FEA PIB (1974), and my solar deployment rates are generally lower than those of Wolf, Morrow, and CONAES (see *SEP* p.98), none of whom assumes symmetrical cost comparisons as I do.

Lapp: Mr. Lovins projects total dependence on soft technology for the U.S. energy supply in the year 2025.

Comment. Given the present deployment of energy consuming devices, the demography and energy distribution systems, it is whimsical for a person to project a 100 percent revolution of the U.S. energy economy in 50 years. This is especially true when Mr. Lovins is incapable of providing any evidential basis for the contributions of solar energy on a rational time scale.

Lovins: I do not "project" and the dependence is not quite "total." The thrust of Dr. Lapp's comment, though, is that the U.S. energy economy cannot undergo drastic changes in 50 years. History suggests a different story. According to Bureau of Mines data, the proportion of U.S. primary energy supply met by coal increased 8-fold from 1850 to 1900, then halved to 1950; the market share of firewood decreased 4-fold from 1850 to 1900; from 1900 to 1950 the market share of oil increased 17-fold and that of natural gas 7-fold; in 1947-72 (only 25 years) the rate of U.S. primary energy use more than doubled.

In absolute (not market-share) terms the figures are even more striking. U.S. extraction of natural gas increased about 34-fold from 1921 to 1971; U.S. extraction of crude oil and natural gas liquids increased about 57-fold from 1900 to 1950; U.S. extraction of coal increased about 31-fold from 1850 to 1900. And if Dr. Lapp thinks these figures outdated he can try electricity generated by U.S. nuclear power, which increased roughly a millionfold from 1952 to 1971 (22-fold from 1961 to 1971)—and which, in WASH-1139 (74), was projected to increase a further 20-

to 27-fold in the 25 years from 1975-2000. In this light my 50-year estimates for much simpler, smaller devices with lead times one or two orders of magnitude shorter appear positively sedate. See also *SEP*, Chapter 5.4.

Lapp: "To fuse into a coherent strategy the benefits of energy efficiency and of soft technology, we need one further ingredient: transitional technologies that use fossil fuels briefly and sparingly to build a bridge to the energy-income economy of 2025 . . ."

Comment. Mr. Lovins states the problem inversely. What is required is to demonstrate that his soft technologies are capable in an economic, scale and practical sense of providing energy as illustrated in his Figure 3.

Lovins: I believe that the *FA* article and its citations (including its Oak Ridge backup paper), and in greater detail *SEP*, make a strong *prima facie* case for this proposition. If it is broadly correct, then the basic role of fossil fuels should be seen as transitional, not endlessly extrapolated off a cliff.

Lapp: The "coal bridge" as a transitional technology.

Comment. From a societal viewpoint the coal bridge should be compared with the nuclear bridge with careful analysis of societal costs and benefits. This Mr. Lovins fails to do.

Lovins: Such a direct comparison (though not favorable to nuclear power) is inappropriate because the coal technologies I have in mind are even more different in character from nuclear stations than big coal-fired power stations are. In the *George Washington Law Review*, August 1977, I also discuss in detail why there is no scientific basis for comparing the costs, risks, or benefits of even coal-fired and nuclear power stations, let alone nuclear stations and, say, fluidized-bed boilers. I do believe, however, that within the traditional framework of internal private costs, big coal-fired stations are generally cheaper at the margin in the U.S. than big nuclear stations—which is no reason to build either one.

Lapp: Coal technology . . ."is now experiencing a virtual revolution."

Comment. This must be characterized as an inability of Mr. Lovins to appreciate technical development in a realistic way. It bespeaks the Ralph Nader approach of seizing upon tidbits of reports on development and blowing them up into revolutionary solutions to energy problems.

Lovins: The references in *FA* and in the Oak Ridge paper fully justify my statement. I used "virtual revolution" advisedly, based on substantial firsthand knowledge of recent international developments.

Lapp: "Perhaps the most exciting current development is the so-called fluidized-bed system for burning coal."

Comment. This may be exciting to Mr. Lovins as an individual, but society would be better advised to depend on the mature judgment of the engineering community. This is especially true if coal is to be regarded as a transitional fuel. The whole history of societal use of fuel illustrates that many decades are required to move from a concept to full-scale deployment of energy systems. By the time Mr. Lovins would anticipate the erection of a few trusses in his coal bridge he would find time was his enemy. This failure to deal realistically with dq/dt and the reluctance to assess the number of years required to bring on line 1q of an energy source marks Mr. Lovins as an energy amateur.

Lovins: My conclusions about the scope for prompt and rapid commercial deployment of fluidized beds, ranging from domestic to industrial scale, have since been broadly confirmed by studies at Oak Ridge, Princeton, and the Science and Technology Committee of the House of Representatives, as well as by international contractual developments. [For a recent survey, see W. C. Patterson and R. Griffin, "Fluidized Bed Energy Technology: Coming to a Boil," INFORM (25 Broad Street, New York 10004), 1978.]

It is inappropriate to generalize from the development and deployment rate of our most difficult and unforgiving energy technologies, with inherently large scale, long lead times, and novel institutional and technical problems, to smaller, simpler devices that use existing infrastructure, are well understood, and can be retrofitted. (I have seen a good experimental fluidized-bed combustor put together in the laboratory in a few minutes; one can hardly do the same with a reactor.) Evidently Dr. Lapp is not familiar with the commercial and pilot experience with fluidized beds in Europe, where results are exactly contrary to his predictions.

Lapp: "Scaled down, a fluidized bed can be a tiny household device—clean, strikingly simple and flexible—that can replace an ordinary furnace or grate and can recover combustion heat with an efficiency of over 80 percent."

Comment. Assuming that this Aladdin-type home furnace can be made into a marketable item with consumer acceptance (very big assumptions if one proceeds from a flimsy footnote to such a stage) what is the time scale for such development? What is the distribution system for the coal-feed? (As a boy I used to shovel coal in the street where it was unceremoniously dumped because the space between houses did not permit truck entry and I cannot believe such terminal delivery will be acceptable again.) What is the environmental impact of burning sulfur-laden coal in tens of millions of home units? How do homeowners solve maintenance problems of the fluidized-bed slag block? What about the slag disposal problem?

Lovins: I have seen such devices operating in England. Field tests (gas-fired) in houses in the Netherlands have been successful. Commercial licensing for production is starting. The time-scale for development is already very nearly finished. The granular coal is fed by pneumatic hose from a tank-truck—just like the ones that now deliver oil—to a bunker, which then feeds by air or gravity to the furnace. This looks like a conventional furnace except that it is cheaper, quieter, and much smaller. The ash—there is no "slag-block"—overflows either into a bunker that is pneumatically pumped out by the delivery truck or into a snap-on receptacle like a vacuum-cleaner bag that is replaced when fuel is delivered. Thus both coal delivery and ash handling are sealed operations indistinguishable (save by their lower cost) from oil delivery.

Since not much fuel for such devices will be needed over the next decade or so, low-sulfur coal, perhaps Pennsylvania anthracite, might be a convenient fuel to start with; if desired, limestone or (preferably) dolomite can be added either with the coal or from a separate bunker and will remove, typically, upwards of 90% of the sulfur, forming calcium-magnesium-sulfate granules that make a good aggregate. In pressurized fluidized beds, such as might well be used for industrial cogeneration, sulfur removal can be as much as 98%. Small fluidized beds have successfully burned coal of all sulfur contents and of ash content up to 70%—essentially bituminous sandstone. Virtually all this information is in the cited Oak Ridge paper and its references (see also *Soft Energy Paths,* Chapter 7.1, especially n.3, p.118).

Lapp: Adaptability of transitional technology to soft technology in the future.

Comment. This paragraph is lacking in substance and does not merit comment.

Lovins: This adaptability is a key concept, and is especially important for the transition from fossil-fuelled urban district heating to urban neighborhood-scale solar heating. I am sorry Dr. Lapp does not think this concept worth discussing.

Lapp: "Both transitional and soft technologies are worthwhile industrial investments that can recycle moribund industrial capacity and underused skills, stimulate exports, and give engaging problems to innovative technologists."

Comment. Pure Lovins. Words designed to appeal to wishful thinkers.

Lovins: I think the statement quoted has the merit of being true—as many recent research and commercial developments confirm. I am sorry if Dr. Lapp does not like it or does not feel readily recyclable into a soft-technology enterprise.

Lapp: "Coal can fill the real gaps in our fuel economy with only a temporary and modest (less than twofold at peak) expansion of mining, not requiring the enormous infrastructure and social impacts implied by the scale of coal use in Figure 1."

Comment. There is only a modest difference for the year 2000 between Figures 1 and 3, i.e. the respective areas under the curves are not greatly different and do not justify Mr. Lovins' use of the adjective "enormous."

Lovins: There is a 2.3-fold difference in 2000 and an 8.5-fold difference in 2025. The ratio of cumulative coal mining in the two graphs is about 3.4-fold, rising sharply after 2025. These are substantial differences. Moreover, the hard path necessarily entails large-scale Western coalstripping, with all the dislocations of people, communities, water, railroads, etc. that that implies, while the soft path does not.

Lapp: Mr. Lovins stresses the incompatibility of "the two paths." He describes his "soft" path as "a radically different path" and since he so characterized it,. it behooves society to look before it leaps into such a radical future. Surely, what we must follow is not a hard or a soft path but a prudent path. This means examining all the possibilities and assessing the probability for each energy option to become an economically viable, socially and environmentally acceptable source for the future.

Lovins: This examination is what I have attempted—with special attention to the crucial sociopolitical implications discussed in *FA* section IX. The look before the leap is exactly what my article sought to encourage. If Dr. Lapp means to imply that given any dichotomy, the truth lies somewhere in between, I suggest that that is a rhetorical device with little relation to reality. I could easily have fitted into such a compromise scheme by choosing a Figure 3 of a much more radical character—indeed, I feel the energy use I showed there is unrealistically high in the light of more recent studies—but I chose instead to describe a prudent Figure 3 that is technically and socially rather conservative.

Lapp: "Soft technologies are ideally suited for rural villages and urban poor alike, directly helping the more than two billion people who have no electric outlet nor anything to plug into it but who need ways to heat, cook, light and pump."

Comment. Here, it appears, Mr. Lovins should have begun his article because it is clear that any realistic projection of world energy supply (see, for example, R. E. Lapp "America's Energy," 1976—page 88: "The World's energy-poor nations exist on very low per capita energy consumption ranging from 1 million Btus for Ethiopia to 5 million for India. The latter accounts for 15 percent of the world's population but uses less than

1.5 percent of its energy. As fuels become more scarce and increase in price, it seems likely that the energy-poor will get poorer."

[Whether the lapse is Dr. Lapp's or the printer's, the syntax of the preceding paragraph is obviously faulty. A parenthetical passage has a beginning but no end, and whatever it is that is clear about "any realistic projection of world energy supply" never emerges.]

Mr. Lovins has made a fundamental mistake in targeting an industrial society like U.S.A. as the candidate for hard-to-soft energy conversion. He could make a far better and more constructive case for a nonindustrial country.

Lovins: Much might be said about energy use in developing countries: in India, for example, about half of total non-solar energy use comes from noncommercial sources that do not show in the statistics; the national average for total non-solar energy use is about 3.8 times as great as Dr. Lapp says; and this average conceals a very skewed distribution (the poorest 60% of the people probably use only about a fifth as much energy per capita as the richest 20%). The distribution of new electricity, as it goes by demand rather than by need, is even more skewed: typically 80% to urban industry, 10% to a modest number of urban households, and 10% to a very limited number of households in rural villages (nearly all to the local elite).

I am glad that Dr. Lapp and I agree that a soft path is appropriate for developing countries; and I intend in my forthcoming anthology *Energy In Context* to include an essay or two about India. But I also think the same logic, with due changes in technical details, applies also to overdeveloped countries such as the U.S.

Lapp: "Yet the genie is not wholly out of the bottle yet—thousands of reactors are planned for a few decades hence, tens of thousands thereafter—and the cork sits unnoticed in our hands."

Comment. Mr. Lovins completely misstates the issue in linking nuclear proliferation with commercial power reactors. He ignores the existing stockpiles of nuclear weapons and the alternative modes of acquisition of nuclear weapons that exist external to commercial nuclear power cycles.

Lovins: I do not ignore existing weapons stockpiles, but seek ways to prevent additions to them, and, in due course, to reduce them with the aim of eventually eliminating them. The linkages between nuclear power and proliferation of nuclear weapons are complex and varied, as are the reasons for believing that the existence of other ways of making bombs does not defeat, but rather reinforces, my argument. I have explained these matters systematically, in detail, and with ample documentation in Chapter 11 of *Soft Energy Paths,* and shall not try to summarize them here.

Lapp: "The United States will phase out its nuclear power program and its support of others' nuclear power programs."

Comment. Once Mr. Lovins commits the error of assuming that nuclear proliferation is singularly tied to commercial nuclear power, he is set on the wrong path. Even if one assumes that he is correct, then unilateral action by the United States would not solve the problem since other nations, less adept at safeguarding nuclear material, would pursue the nuclear path.

Lovins: Evidently Dr. Lapp does not share my political judgment that the three-part policy I suggest in section VIII of the *FA* article would remove the domestic political support necessary for other countries to persist. That is his privilege. I do wish, however, that he would not cast our differences of political judgment in terms of my being mistaken on a point of fact, as our difference here is not one of fact. [The three-part policy Lovins recommends consists of (1) phasing out the U.S. nuclear power program and U.S. support of other countries' nuclear power programs; (2) U.S. commitment to a soft energy path, and commitment freely to help other countries follow a soft path; (3) U.S. psychological linkage of these efforts with attempts at mutual strategic arms reduction.
lated parts of the same problem.]

When I have tried these ideas on a sophisticated foreign-policy audience, *e.g.*, at the Council on Foreign Relations, half the people have said it's absurd, while the other half, on reflection, have agreed it feels right and probably would work. (I suspect that now, a year and President Carter's initiatives later, the first group would probably be rather smaller than the second.) Both sides agreed—as I fear Dr. Lapp does not—that at least I have identified an intriguing and constructive new approach that deserves far more study than it has had so far. Up to now the political levers affecting proliferation and nuclear power have been virtually ignored in favor of analyses of blunt instruments like fuel supplies.

Lapp: "The United States will redirect those resources into tasks of a soft energy path and will freely help any other interested countries to do the same . . ."

Comment. Here the U.S. gamble, pursuing a soft energy path of unknown tortuosities, could be harmful to other nations, if its economy faltered as a result of inadequate energy supplies.

Lovins: This reflects Dr. Lapp's oversimplified view of my thesis and his subjective judgment that a soft energy path is unproven, risky, and likely to fail, with severe economic consequences. I disagree, and do not think he has borne the burden of proving his case. Would he care to argue, on the contrary, that the U.S. nuclear policy and energy policy of the early 1970s was constructive for other countries? Or that U.S. actions have no political effects abroad?

Lapp: "The United States will start to treat non-proliferation, control of civilian fission reactors, and strategic arms reduction as interrelated parts of the same problem with intertwined solutions."

Comment. Progress on two of these issues has been excruciatingly slow and a recipe for stalemating further progress would be to follow Mr. Lovins' advice.

Lovins: I believe that the historic stalemate is a direct result of trying to draw a subtle distinction between the two halves of an umbrella—the peaceful and military uses of atomic energy. It is illogical (1) to promote nonproliferation while retaining and brandishing the domineering capacity that nuclear weapons give to countries like the U.S.; (2) to expect perceived (if not real) sacrifices of others that we are unwilling to ask of ourselves; and (3) to expect to promote a world-wide nuclear (eventually plutonium) economy without promoting proliferation. These inconsistencies are often pointed out by representatives of developing countries (point 1), Europe and Japan (point 2), and the arms control community (point 3). *Not* treating civilian denuclearization, promotion of soft energy paths, and strategic arms reduction as interrelated and interlocking aspects of the same problem guarantees that these elements of policy will be antagonistic rather than synergistic. See *Soft Energy Paths,* Chapter 11.

Lapp: "Policy tools need not harm life-styles or liberties if chosen with reasonable sensitivity."

Comment. Since Mr. Lovins has not really set forth scenarios for the energy supply sectors in his soft technology, this statement means very little. Whenever a major shift is made in an economy there are bound to be lifestyle changes.

Lovins: I try here to distinguish the technical changes desired (roof insulation, cogeneration) from the policy instruments used to encourage them (tax policy, financing institutions, exhortation, revision of utility practices), and to argue that it is the latter, not the former, that are politically sensitive if mishandled. I suggest repeatedly that whatever social problems would arise in a soft path would, in a hard path, be bound to be worse, become harder, and last longer.

Lapp: "In contrast to the soft path's dependence on pluralistic consumer choice in deploying a myriad of small devices and refinements, the hard path depends on difficult, large-scale projects requiring a major social commitment under centralized management."

Comment. All energy systems, large or small, hard or soft, require management. True, hard energy in the form of electricity delivered by a utility involves centralized management, but great freedom in energy choice at the consumption point (wall switch) and zero personal involvement in managing the energy source. On the other hand, resort to soft technology is certain to burden the home-owner with myriad management and maintenance responsibilities.

Lovins: The management required is an initial set of policies to get the ball rolling, not continuing detailed direction. As for electricity, the convenience to which Dr. Lapp refers is paid for—in money, freedom, and various kinds of risk. Nor is flexibility free from this double-edged ambiguity: an electric house, for example, can only be run by one or another kind of electric grid, whereas one heated by a circulating fluid can be heated by any source of heat whatever.

I address in the Oak Ridge paper and in *Soft Energy Paths,* Chapters 5 and 8, an erroneous but widespread notion that soft technologies are onerous to manage and maintain: my cost estimates assume a standard of reliability and freedom from household maintenance at least equivalent to that of conventional energy systems.

Lapp: "In an electrical world, your lifeline comes not from an understandable neighborhood technology run by people you know who are at your social level, but rather from an alien, remote, and perhaps humiliatingly uncontrollable technology run by a faraway, bureaucratized, technical elite who have probably never heard of you."

Comment. To be convincing, Mr. Lovins should present an analysis of the management and interpersonal problems of a U.S. suburban community of, say, twenty households sharing the same friendly energy source. Does Mr. Lovins believe in technology to the extent that he thinks such a community energy supply needs no round-the-clock management? How does he compensate for the high management overhead in such a small system? How is the energy-commune regulated to deal with peak loads? Do families switch on air conditioners in a programmed manner? Do housewives flatten out energy demand for their clothes dryers and ranges by staggering dinner hours? If they do not then they will have to be burdened with peak load generation capacity.

Lovins: Soft technologies, especially on a house or neighborhood scale, do not in general require management, because they are so simple and generally are not electrical. (In contrast, Dr. Lapp seems automatically to equate "energy" with electricity.) Neighborhood energy systems—generally meaning heating systems—are of course designed to cope with reasonable peaking.

Contrary to Dr. Lapp's tacit assumption, I have nowhere proposed doing away with existing electric grids (which have peaking problems of their own and are in many cases trying to manipulate end-use patterns in just the way he describes); nor have I proposed "energy communes" or running household electrical needs on neighborhood-scale supply systems. (See *SEP*, Chapter 8.2.) In the asymptotic soft energy economy, air conditioners would generally be unnecessary because buildings would be designed to keep themselves comfortable in all climates, and clothes drying and cooking would in general not be done with electricity, but rather with a combination of solar heat and biomass fuels.

Lapp: "The scale and complexity of centralized grids not only make them politically inaccessible to the poor and the weak, but also increase the likelihood and size of malfunctions, mistakes and deliberate disruptions."

Comment. I would think that the very reach of the electric grid brings distribution to the poor and the weak. A survey of television set distribution would certainly prove my point. As for industry performance in assuring constant supply of electricity to the nation, it is a record of unsurpassed constancy. It's true that hurricanes occasionally interrupt electric service, but how vulnerable is Mr. Lovins' soft technology to the hard impacts of natural violence? If a community finds its solar collector blown away, it may experience real energy pain for a protracted period. Here again, Mr. Lovins has given us no insights into the vulnerabilities and drawbacks of his technology. He argues his case like a zealous lawyer using any argument to support his case and any argument to damage his opponent's.

Lovins: Supplying electricity to most people does not necessarily mean being influenceable by the same people—hence the fuss over high utility bills, disconnections, and utilities unresponsive to consumerism. It is true that utility electric supply is remarkably reliable. This is extremely costly to arrange, and whether you need it or not, you must pay for it anyway because the centralized grid cannot distinguish between different people's differing needs for reliability (electric water heaters vs. hospital operating theaters). Moreover, the pervasive nature of the grid guarantees that any substantial failure will be catastrophic—which is exactly why such extreme standards of reliability are required. If your solar collector breaks down (which is very unlikely, as there is practically nothing to go wrong with it), you can always go next door. If the grid fails, there is no next door, and not even your oil furnace will work.

It is interesting that the parts of the U.S. least hard-hit by the gas shortages last winter were those that did not rely on a pipeline grid, because they did not run out of gas all at once and were not subject to systemic effects (*e.g.*, gas pressure dropping below threshold, pilot lights going out, nonexistent armies of service personnel having to go into millions of buildings to turn the gas off and then restore it over a period of weeks afterwards, etc.). Both for these structural reasons and because of technical simplicity and robustness, soft technologies would be unusually resilient and resistant to failure. And if your solar collector did fail, you would hardly know the difference, because you would, on my assumptions, have an extremely well-insulated house and some weeks of stored heat to fall back on; whereas in a hard-technology world you would indeed freeze in the dark.

Lapp: "Already in individual nuclear plants, the cost of a shutdown—often many dollars a second—weighs heavily, perhaps too heavily, in operating and safety decisions."

Comment. This is a very serious accusation. It is also an area where I have some experience and I believe Mr. Lovins to be in error. In any event, Mr. Lovins should make known to the U.S. Nuclear Regulatory Commission even a single instance which he believes to represent a safety decision override.

Lovins: Perhaps Dr. Lapp is not familiar with the many published instances in which the AEC and NRC have found various utilities guilty of regulatory infringements of this sort. A salient example occurred in the Palisades reactor when a failed radioactive gas treatment system began releasing excessive radioiodine. The operators guessed—correctly—that it would be cheaper to keep running, and pay the fine if caught, than to shut down and correct an awkward and chronic problem. (The fine proposed by the AEC was $19,000: see USAEC letter to Consumers Power Co., 13 August 1974, docket 50–255, NSIC reference 094490.)

There is little doubt that the cost of a shutdown weighs heavily in the judgment of a utility which, like the Vermont Yankee operators, finds an unsatisfactory feature of its reactor (in that case, potential torus jump, studied in 1976), but, after an initial inspection shutdown, decides to keep running anyway.

I did not mean to imply, however, that this sort of rationalization is peculiar to utilities. It is characteristic also of the NRC to license despite many serious unresolved generic issues and uncertainties, on the unsupported assumption that these problems can and will be fixed later. If the NRC stuck strictly to its statutory duties, not a single power reactor, in my opinion, would be operating in the U.S. today.

Lapp: Mr. Lovins itemizes a number of charges relating to nuclear risks. This recital includes highly technical issues such as waste disposal. Refutation of the contentions cannot be made here within the confines of allowable space. However, each contention made by Mr. Lovins has been addressed in the past and he introduces nothing new to the nuclear controversy.

Lovins: These are not "charges" but a partial list of issues, and "addressed" does not mean "resolved." I describe these issues more fully elsewhere (*e.g.* in *Non-Nuclear Futures,* FOE/Ballinger, 1975, and its many citations). I do not think the main point of controversy is technical in any significant nuclear issue; the big questions are trans-scientific.

It is not up to me to introduce new issues, but up to the nuclear establishment to resolve the same old issues that have been unresolved for several decades.

Lapp: "There is no scientific basis for calculating the likelihood or the maximum long-term effects of nuclear mishaps . . ."

Comment. In effect, Mr. Lovins is saying that the WASH-1400 report, the Reactor Safety Study, is without a scientific basis. To be credible, Mr. Lovins would have to display credentials which establish his competence in the area of nuclear safety on a par with those of Dr. Norman Rasmussen who headed the safety study. The specific criticisms which Mr. Lovins submitted to the Nuclear Regulatory Commission were quite readily answered by experts within NRC.

Lovins: Dr. Lapp correctly construes my views of the Rasmussen report. Going further than Dr. Weinberg (" . . . what is the probability of a major accident . . . is by its very nature unanswerable in a strict scientific sense"—*Bull. Atom. Scient.*, p. 56, April 1977), I think the Reactor Safety Study is essentially fraudulent and has severely damaged the credibility of the NRC.

Dr. Lapp's statement that my criticisms of RSS are not "credible" unless I can "display credentials which establish . . . competence in nuclear safety on a par with those of Professor Norman Rasmussen" seems to me an attitude that has no place in a democracy, evades responding to my arguments on their merits, and, within Dr. Lapp's terms, is somewhat misplaced since prior to directing the Reactor Safety Study, Professor Rasmussen appears to have done no professional research and published no technical papers in the area of reactor safety.

My criticisms of RSS do in fact rest in large part on published analyses by very highly qualified people whom I cite; but what concerns me, and what should concern Dr. Lapp, is whether what they say is valid, not who they are.

Dr. Lapp's statement that my criticisms of RSS submitted to NRC "were quite readily answered" is also incorrect. In RSS Appendix XI, the only published NRC response, the staff responded to some minor points, but sidestepped or ignored the more penetrating criticisms—as will be clear to anyone who reads the relevant documents. [In September 1978 the NRC published the report (NUREG/CR 0400) of a review committee it had appointed, chaired by Professor Harold Lewis, to study the published criticisms of the Reactor Safety Study. The Lewis report largely vindicates Professor Rasmussen's critics.]

Lapp: ". . . since GNP in turn hardly measures social welfare, why must energy and welfare march forever in lockstep?"

Comment. GNP may not be a perfect indicator of social welfare, but it is a known index of a value for goods and services and, as such, has real bearing on social welfare. The fact is that GNP and GEC (Gross Energy Consumption) do march in lockstep with an association of about $17 billion (1972$) per q of energy. The record of the past few years, following the oil crisis, shows a dip in GEC and a corresponding decrease in GNP and an increase in unemployment. Mr. Lovins, at this stage of his argument, asks "Why?"—yet he prescribes flattening out of energy growth with no explanation of how he will keep the economy from flattening out as well.

Lovins: GNP measures not only goods and services (except some important ones like being a housewife) but also bads and nuisances (except some important ones like all the externalized risks of energy systems). GNP has no clear relation whatever to social welfare: that is, if there is a relationship, not only its magnitude and functional character but its sign [plus or minus] is obscure. In the U.S. it is not clear whether an increase in GNP increases or decreases net welfare. There is a large professional literature on this subject.

Correlation in a historical time series does not entail causality. If it did, one could readily conclude, I suspect, that a decrease in the consumption of Coca-Cola would lead to national depression and massive unemployment.

Virtually all U.S. energy economists of any professional stature have abandoned GNP-energy correlations as a method of analysis or forecasting. Dr. Lapp will find the reason explained in the 1977 report of the distinguished CONAES economics panel, which concluded in 1976 that within very wide limits, energy and GNP need have nothing to do with each other. [See the CONAES Demand and Conservation Panel report summarized in *Science*, 14 April 1978.] Nor do oil embargoes and efficiency improvements have the same economic effects.

I do not happen to think that further economic growth along historic lines is a good idea in the U.S. My views on this point are between the lines of *FA* section X and of *SEP* Chapter 10, and explicit in the Oak Ridge paper. But these views have nothing to do with my argument that primary energy use and economic activity can readily be decoupled. This has since happened. [The ratio of annual primary energy growth to real GNP growth in the U.S. was about 1.0 in 1974; 0.65 in 1975-7; and about 0.49 in 1978.]

Lapp: Personal preference in energy choices.

Comment. To date the American people have shown little inclination to practice conservation on a voluntary basis. This is regrettable but one cannot argue with the American preference for energy. Mr. Lovins would like to see more conservation and this is understandable—he is not alone in this desire—but voluntary measures are not working and mandated action will probably only follow the onset of crippling energy experiences.

Lovins: The "American preference for energy" has been assiduously promoted, not least by Dr. Lapp's clients in the electro-nuclear industry. It is hard for people to believe in conservation, too, so long as their own government tells them that its Energy Independence program, fast breeders, etc. will solve the problem. I trust that with the new [Carter] Administration we have got past that point. I think people will rapidly improve their end-use efficiency as soon as they are given the tools to do so and as soon as they see more clearly how much it is in their own interests to do so. I think most people already consider it necessary and desirable.

Lapp: The issue here is not a stabilization of or departure from affluence as represented by energy use, but satisfaction of the as yet unfulfilled energy aspirations of millions of Americans. It is fine for Mr. Lovins to believe that America should not continue growing, but it is hardly humane as related to many millions of its citizens who are far from affluence. It would appear that Mr. Lovins is expressing an elitist view.

Lovins: My Figure 3 is consistent with having *everyone* in the U.S. live much better than he or she does now—remember that per-capita delivered function increases more than 4-fold in 50 years. If people choose distributional mechanisms that result in a more egalitarian society and the elimination of poverty, so much the better, and I think this desirable result is more likely to be achieved in a soft than in a hard path. But doing this would not take vastly more energy: Mr. Carl Hocevar has calculated that even with *no* conservation at all, raising the poorest 60% of Americans to the level of per-capita primary energy use of the upper middle class and assuming a population of 265 million in 2000 would imply primary energy use in that year of only 117 q. Thus is appears that equity and population growth are largely a smokescreen in the argument about the necessity of energy growth: the numbers just don't add up.

Lapp: Mr. Lovins stresses a one-path decision.

Comment. The only way one can reach a go-one-path-only solution is to distort the energy options other than soft options so that soft technology becomes the only permissible solution. And this is what Mr. Lovins does to coal and nuclear power. His argument that we are running out of fluid hydrocarbons is not new and should compel him to examine the non-soft energy options more carefully. After all, in his own words Mr. Lovins is urging a "radical" solution to energy ills. Radical solutions tend to be high-risk ones. Yet Mr. Lovins glides by the risks inherent in his advocacy of soft energy.

Lovins: My preference for the soft path rests on a thorough comparison of its implications with those of its alternatives. I do not think coal (presumably meaning large coal-electric plants) and nuclear power are either necessary or feasible on a nationally important scale. It is precisely the short lifetime of the fluid fuels that leads to my sense of urgency about replacing and husbanding them.

In a fundamental sense, the hard path is far more radical in its technical leaps of faith, economic handwaving, and social threats, than is the soft path: it is the soft path, not the hard, that is consistent with the resource base and the political traditions of this country. Neither a soft energy path nor any other is perfect or free of risk; but my published analysis suggests that it is likely to be more nearly perfect, and more free of risk, than alternative policies so far proposed.

Lapp: *Observations.* I would sum up my reaction to Mr. Lovins' article in one word—irresponsible.

Lovins: I agree: Dr. Lapp's reaction is indeed irresponsible. Responsible criticism requires a level of care, detail, explicitness, and documentation consistent with the quality of the material being criticized. These features are notably absent from Dr. Lapp's critique.

Lapp: Mr. Lovins certainly attempts to make a technological assessment of every single energy option, but he fails to bring sound judgment to his evaluation. Too often, he simply selects from a wide range of judgments about a given technology the end-of-spectrum view which suits his case. Granted that it is an exceedingly difficult assignment to bring judgment to so many specialized fields of expertise, Mr. Lovins consistently opts for high-risk energy source possibilities, risky either in terms of practicality, social acceptability, economics or scale. For example, his suggestions that small user-groups employ small diesel-electric or fluidized coal furnaces bear little chance of being adopted for reasons cited in my criticism.

Lovins: I appreciate Dr. Lapp's acknowledgment of that which he previously denied—that I have tried to assess "every single energy option"—and I accept that he and I have differences of judgment. That he chooses to characterize my views as "end-of-spectrum" is a judgment one might equally well form about his own views. I have tried consistently—unlike Dr. Lapp—to state and document from the professional literature the grounds for my conclusions, so that interested readers can dig deeper and judge whether what I say is valid. In this way I have supported the view that the energy technologies I suggest have far lower technical, social, economic, and logistical risks than those which Dr. Lapp suggests.

I have earlier corrected Dr. Lapp's misconceptions about diesels and domestic fluidized beds.

Lapp: Even after careful study of his paper, I find myself unable to define his "soft technology" except to call it some variant of solar energy. He does skip from one type of "softness" to another and becomes quite inconsistent in his advocacy, as for example, when he decries reliance on oil and gas, but then goes on to champion diesel-electric power. He deplores the climatic effects of coal, but goes on to propose coal-fired home furnaces.

Lovins: Soft technology is defined and discussed at length in section V, with appropriate references both directly and via the Oak Ridge backup paper. I did not "champion" or even advocate diesel-electric power (though diesels are indeed useful in some transitional technologies—for cogeneration and other total-energy systems). On coal [also a transitional technology] note that the soft path addresses the serious CO_2 problem by minimizing the integrated fossil-fuel burn compared with that of any other coherent policy.

Lapp: Of all present power sources, nuclear power is most capable of fulfilling set goals by the year 2000; yet this energy source is cast aside without substantial basis for this rejection. One finds only an array of implications. Primarily, it appears, nuclear power must be rejected because of its potential for nuclear proliferation; yet Mr. Lovins fails to deal with the real world of existing nuclear arms.

Lovins: I reject nuclear power because of its "array of implications": it is in many respects undesirable (see *Non-Nuclear Futures*), infeasibly capital-intensive, unnecessary, and an encumbrance. Any one of these four general objections would by itself be conclusive. [Nuclear power is also too slow to meet its goals. To meet a quarter of the lowest ERDA projection (124 q) of U.S. energy demand in 2000 with nuclear power, we would need to order a 1-GWe plant every 4.7 days (assuming that 1 kW-h of electricity displaces 2 kW-h of fossil fuel throughout the economy). Source: Lovins's testimony before the Ryan Committee in hearings on nuclear power costs (U.S. House of Representatives), 21 September 1977.]

Lapp: The fact is that the United States is in transition from its present fuel dependence, three-fourths reliance on fluid fossil fuels, and it must resort to solid fuels plus whatever geophysical energy can be tapped as its future fuel supply. This transition necessitates greater dependence on central station power and on more intensive electrification. Coal and uranium thus become the mainstays of the U.S. energy supply. Mr. Lovins makes no mention of the immense energy potential available from uranium—energy reserves far greater than those available from coal.

Lovins: I agree we are in transition; I disagree about what we are in transition *to.* It would be interesting to see Dr. Lapp's response to the demolition by von Hippel and Williams (*Bull. Atom. Scient.*, December 1976, and GESMO testimony, March 1977) of his and ERDA's case for central electrification, and his detailed analysis of how his electric system is to be financed.

The energy potential of uranium, like that of heat in the Atlantic Ocean, is vast, but the attractions are insufficient to make that potential worth realizing.

Lapp: It's rather silly to talk of hard or soft paths to our energy future. What we need most is a prudent path—one that connects the present with the future in a realistic, as opposed to a radical, way. One that recognizes that reality of existing transmission and distribution systems and one that can satisfy America's energy demands without inflicting mortal harm to its economy. Instead of prudence, Mr. Lovins proffers a triple gamble—decreased growth, rejection of nuclear power, and dependence on untested energy sources.

Lovins: The soft/hard metaphor is, I believe, useful in helping to illuminate the range and implications of possible choices and in assessing their realism—including their social realism, a sphere often neglected by technicians. In my testimony and in *SEP* I have emphasized the important role of existing energy systems, including electric grids, and the likelihood that a soft path would help but a hard path harm the economy.

It is my view of the energy future that is conservative, low-risk, and empirically verified; it is Dr. Lapp's that is truly radical, high-risk, and hypothetical. Until he points out some mistake or illogicality in my argument, I do not see any reason to change my view. Meanwhile, the burden is on him to show that his sketchy, obscure, and wholly undocumented view of the energy future is preferable. He has not done so.

[The Rasmussen Report, referred to above, no longer has the complete backing of the U.S. Nuclear Regulatory Commission. The NRC repudiated the Report's Executive Summary in January 1979 and announced it no longer considers the RSS risk elements reliable.]

7

FORBES vs. LOVINS

Critic Asserts Lovins Is "Myopic and Reckless"

Ian A. Forbes, formerly Chairman of the Nuclear Engineering Department of the University of Lowell, is Technical Director of ERG, the Energy Research Group. Initially disinclined to allow us to reprint his critique of Lovins's article, Dr. Forbes relented on condition that we publish it in its entirety (which we take to mean not only complete, but uninterrupted). Accordingly, the following chapter is in parallel-column rather than "dialogue" format. A further condition was that we call readers' attention to ERG's publication of "our more detailed analysis for ORNL," *Exclusive Paths and Difficult Choices: An Analysis of Hard, Soft and Moderate Energy Paths*, and "other papers such as *The Economics of Amory Lovins' Soft Path,* or *The Soft Path and New Energy Sources,* etc., etc." (At Mr. Lovins's urging, we have included extracts from the first of these papers in an appendix to this volume.) Dr. Forbes's critique follows.

Forbes: What follows is an attempt to state in one reasonably coherent piece my thoughts on energy strategies, and in particular on Amory Lovins' "soft path" as advocated in his paper "Energy Strategy: The Road Not Taken?" If my statement seems lacking in specificity, data, references and general good hard facts, it is because I have accepted and followed Lovins' approach of attempting to deal more with the philosophical and conceptual aspects of future energy strategies than with the technical and numerical details. The flaw in this is that unfortunately much of energy policy does in fact hinge on the hard details of implementation. I have tried to counter this weakness where possible, but the approach is imperfect. It is planned to follow this statement in several months with a paper that will be addressed in more detail to future energy needs, energy supplies and the

Lovins: Dr. Forbes implies that his paper is meant as a dispassionate contribution to scientific and policy discourse. I would find this more persuasive if he had followed the process—customary in the scientific community—of sending me a prior draft for review. This is not only a professional courtesy but a wise precaution against hastily publicizing one's own mistakes or misunderstandings. If Dr. Forbes had adhered to this prudent custom, I believe he could have saved himself embarrassment.

Dr. Forbes's second paragraph is an articulate statement of the criteria I have tried to apply to the energy problem. It is for that reason that I question whether hard energy paths are "proven or viable."

Forbes

requirement for implementation of a viable energy strategy.

What I hope the reader will discern is my conviction that energy strategies must flow from the past to the future along orderly lines of transition, based on minimum possible societal risk (economic, technical, public and environmental) with a just and equitable distribution, while developing through research and understanding those new options that could change society for the better. This is not a proposal for a laissez-faire strategy, for that is what we have had for over a decade and that is what has brought us to our current mess. Action in energy policy, and decision-makers who are capable of making decisions, were never more badly needed. But neither can we experiment with the lives of the world's people by launching into strategies that are not proven or viable. We must be concerned not only with visions of the future but also with the consequences of being wrong. There is a better path.

Lovins

O

The most disturbing aspect of Amory Lovins' "Energy Strategy: The Road Not Taken?" and other papers that recommend a "soft," renewable source energy strategy is that, in the midst of an energy problem that grows steadily worse because of inaction, they offer false hope of simple, almost utopian solutions. To the public at large Lovins offers a vision of energy that is clean and safe, cheap and plentiful; to the policy-makers who have fiddled time away for over a decade he provides an excuse for further inaction by suggesting that a complete rethinking is required; he offers more and better employment if the "hard" path is forsaken; and to those who insist in the face of all evidence that nuclear power is dangerous and solar cheap, he gives a rationale for what they had decided anyway.

O

Dr. Forbes says that I claim a soft path is "simple" and "cheap." I make no such claim. Rather, I claim a soft path is simpler and cheaper than its alternative. I call not for "further inaction," but for vigorous action on a broad front; inaction would imply support for the pattern of actions we had been taking (and failing to take) for the past few decades. Decisions are being made *now* that will have profound long-term consequences, and I call for them to be made differently.

I do not undertake in my article to discuss nuclear risks, but I think the "public risks" which Dr. Forbes rightly identifies as important rest mainly on trans-scientific issues that cannot be resolved on technical merits ("evidence"). Evidence is of course relevant, and I carefully consider it in other publications, but it tends to be Delphic.

Forbes o

Although the distinction between the current "hard" path and his own "soft" path is very fuzzy, Lovins' message is clearly that the soft path is desirable and practical, that it is clean, cheap, and adequately abundant (in fact it will give more for less), that it is more "human" and that the implementation of the soft path cannot be delayed. That it is cheap is untrue, and contrary to the actual current cost data for most of the "soft" energy sources that he discusses. That it is plentiful relies on the assumption that the soft technologies are available at acceptable cost and that current inefficiencies are so great that for very little effort or cost we can obtain much more work energy output from our raw resource inputs. While there is much energy wastage in our system and while conservation of fluid fuels in particular must be our highest priority, the potential for savings is neither as large or as easy as Lovins' numbers game would lead his readers to believe. That it is clean is doubtful since it presumes the truly clean sources (principally solar) can provide the bulk of energy supply. To the extent that he promotes the use of diesel electric generators, domestic use of coal, methanol, wastes, fluidized-bed boilers, etc., his soft technologies are still polluting technologies, only dispersed and less amenable to control.

Lovins o

Dr. Forbes calls "very fuzzy" the soft/hard distinction which I spend 32 pages meticulously drawing. Perhaps he means—correctly—that the distinction is ultimately social, not quantitative or technological: whether a path is hard depends on whether its polity is dominated by the structural problems (autarchy, centrism, vulnerability, technocracy, etc.) described in section IX of my article. This may indeed be a difficult concept for technical people to grasp, but I do not see any purely technological substitute for it.

Dr. Forbes then repeats, twice, the misrepresentation that the soft path is claimed to be cheap; my data make clear that it is only cheaper than not doing it. The same is true of increased efficiency (though that is generally cheaper than increasing supply). Dr. Forbes's assertion that I overstate the scope for increased efficiency is unsupported, and it is hard to understand his characterization of the detailed assessments I cite (*e.g.* that of Ross and Williams) as a "numbers game."

The transitional fossil-fuel technologies I propose (which do not include diesel electric generators) are much *less* polluting than those now in use. I agree that any energy conversion process, even a solar collector, has some environmental impact, but suggest technologies which are generically more benign than current ones.

Forbes o

One of the difficulties in following Lovins' arguments is the use of rather magical mathematics, with some numbers far off the mark, other critical numbers missing, and frequent contradictions. At one point, cogeneration of electricity and steam or hot water is an important step to improving energy efficiency, at another point electricity is dismissed as a technology that is constrained to waste two-thirds of the energy content of fuel.

Lovins o

Dr. Forbes does not justify his first sentence. The second sentence fails to note that my brief mention of Carnot losses on the last line of p. 68 of the FA article refers to the style of generation that characterizes the hard path, namely big central power stations. The fuller discussion ten pages later refers specifically to "a power station." I nowhere generalize that the high-entropy output of a heat-to-electric converter cannot be used at all: indeed,

Forbes

Lovins

with cogeneration and combined-heat-and-power district heating I suggest the opposite.

o

Energy flows are juggled between input and output values to the continual benefit of the soft path, and energy savings and dollar costs are manipulated relentlessly. He rightly professes concern for the global buildup of CO_2 yet proposes the use of coal, oil and methanol. He raises the very real problem of fluid fuel shortages but advocates the use of diesel generators and the abandonment of central electric power generation. He is concerned about the risks of the high technology path but dismisses those of the soft path. All along, discussion of the soft path remains so vague as to its detailed makeup, its costs and how it can be implemented, that it cannot really be called a strategy but qualifies more as some thoughts on possible future goals.

o

There is no contradiction in my CO_2 argument: as I point out on *FA*, p. 88, this problem is a fine reason for the soft path, which minimizes the integrated fossil-fuel burn and uses carbon that is already in the global carbon cycle in biomass conversion. I am not sure what Dr. Forbes refers to in his strictures about juggling and relentless manipulation, but I do try to quantify energy savings, dollar costs, and primary/end-use distinctions (the kinds of details he earlier accused me of omitting). I do not advocate diesels, and advocate not that we abandon central power stations, but that we not build more of them—partly because they are so wasteful of fuels.

I do not "dismiss" soft-path risks, but believe they are much smaller and more tractable than hard-path risks. In the Oak Ridge backup paper, and even more in *Soft Energy Paths*, I give perhaps the most comprehensive soft-path cost survey published anywhere, and as much discussion of implementation as can be justified in an exploratory policy treatise for a sophisticated audience. I nowhere pretend to provide a comprehensive blueprint of laws and other policy instruments, but rather an outline of their content. Putting flesh on the bones is the next step, not the first.

o

The underlying presumption seems to be that the soft path leads naturally and inherently to a future that is not only cleaner and more benign, but also freer and more human. This is not axiomatic. It rests on the assumption that the soft technologies are publicly acceptable and that massive conservation will be painless, for otherwise the path must be coercive. Even if this assumption were correct, it is not appar-

o

It is a false dichotomy to suggest that if massive conservation is not "painless," the soft path must be coercive. If, as I believe, it is far less painful than the alternative, I suspect people are smart enough to choose it without being coerced—which is why those of us who do not want to lose all our teeth get an occasional checkup at the dentist's even though it may mean a momentary inconvenience. I think soft tech-

Forbes

ent that a change to decentralized energy systems will, per se, make for a freer, more "human" society. Some low-energy societies are by no means warm, friendly and free of coercion, while others are co-operative and human in the midst of adversity. (For those who might think this to be a new concept, a review of the anti-industrial movement in 19th-century England is in order.) Certainly we cannot view energy policy as the sole mechanism for "de-humanizing" society. Rather, we need to think out our social goals carefully and in detail, deciding on those attributes we can and wish to acquire, those we wish to be rid of, and the actions and subtle compromises required. In this regard, a careful examination of our own history and the trends of development in different societies cannot be casually disregarded. If we forsake proven "hard" options for an unproven "soft" path, we run a high risk of having to disrupt or coerce society to maintain the soft path. In this sense, the soft path society could be a great deal colder and less free than today's.

Lovins

nologies are publicly acceptable and that this attitude is often publicly manifested. I do not claim that a shift to less centralized energy systems (not quite the same as "decentralized" ones) will *by itself* automatically make a more humane society, but I think the whole package of policies (not merely technologies) that must constitute a soft path would have this effect if it were not gratuitously promoted by the same coercive methods traditionally used to promote a hard path. Coercive enforcement of the technical ingredients of a soft energy path is possible—but is also unnecessary and silly, and the result would not meet the political criteria that define a soft path.

I do not think that unpleasant or coercive low-energy societies in the past or present (in which "low energy" often means no more than "low end-use efficiency") are necessarily any kind of social or technical guide to high-efficiency societies in the future, and I deplore the persistent attempt to identify energy (presumably primary energy) use with social welfare. Nor do I think that what Dr. Forbes misleadingly calls the anti-industrial movement of 19th-century England is directly relevant to our problems today. (Any identification of the themes of that movement with my own analysis is facile and incorrect.) Of course energy policy is not "the sole mechanism for 'dehumanizing' society"; but I argue it can be an important one. I also agree that while social comparisons with history and with other cultures must not be overdrawn, they do have some relevance—which is why I consider them. In representing hard "options" as "proven," at double today's U.S. energy levels, and a soft path, at roughly present per-capita levels, as "unproven," Dr. Forbes is assuming that which was to be proved. If anything, I think he has it backwards: both the technologies and the social conditions of a soft path are, broadly speaking, empirically verified, while

Forbes

those of a hard path have never existed anywhere and are wholly hypothetical.

Lovins

O

Perhaps the most disturbing aspect of Lovins' development of the road-not-taken is that he apparently feels little obligation to discuss the specific implementation of his possible future, the detailed policy mechanisms that he feels would lead us onto the soft path. Yet this is what it largely hinges on. If the soft path really consists of options that are technically viable, available at reasonable cost, and environmentally sound, it would seem that our policy would flow quite easily in that direction with only gentle pushing. Conspiracy theories of energy supply notwithstanding, energy policy (over the longer run at least) consists in essence of developing those options and strategies that are technically and economically viable and publicly acceptable; ideally, they should also be well thought out, socially just, environmentally sound and durable. This involves significant compromise since our goals are frequently incompatible. But to suggest that we commit now to a path that does not satisfy even the first of these criteria, and whose practicality and desirability as social goals are uncertain, is reckless at best. The more prudent strategy would be to apply these criteria in the implementation of policies that stress conservation and the application of current technologies to near-term needs, while investing properly in the development of new technologies and efficiencies that can supplant existing methods.

O

Dr. Forbes begins with a difficulty in deciding what is "the most disturbing aspect" of my work and continues with a repetition of the unsupportable statement that I do not discuss implementation. As much of this material as would fit is in my article, and more is in *Soft Energy Paths*, which was not under the same constraints of space. Dr. Forbes then proceeds to ignore the real institutional factors he accuses me of ignoring—social inertia, vested interests, baroque pricing policies, etc.—and imply that the early phases of the hard path have flowed quite easily "with only gentle pushing," a travesty of recent history. I suggest that on the contrary the central intervention we have seen so prominently displayed in the political and economic process is an attempt to keep fundamental social and market forces from rejecting a hard path in favor of a softer one. Dr. Forbes's remaining arguments in this paragraph are judgmental and unsupported.

O

I urge adopting a soft path soon, not instantly, and pursuing it with due deliberate speed, not recklessly. My article was intended to stimulate the kind of public discussion that must precede such a choice. Dr. Forbes's opening lines do not consider the possibility that some options may exclude others

O

Lovins sketches a hard path with continued high growth and a soft path with massive savings. It is very unlikely that either is realistic. His hard path is a "straw man" that shows future growth well in excess of what most thoughtful projections see as being likely or necessary with feasible conservation efforts;

Forbes

his soft path is simply infeasible in terms of the assumptions it makes for the rate at which efficiency improvements and soft technology can be introduced. And Lovins claims that his two paths are mutually exclusive. This flies in the face of history, logic, and practical necessity. Hard and soft have co-existed since the dawn of the industrial age; in future transitions they must coexist with each other and with conservation. No amount of argument for philosophical oneness can allow us to take an unproven path. Soft technology must first be tested and then brought into the system to the extent that it proves viable and preferable.

This is not to suggest that our future path should be but a heedless extrapolation of the past. To the contrary, conservation is inarguably necessary and the development of renewable sources is desirable. Our demands need questioning and our social goals need care [sic] re-examination. Most of all we need responsible action from public leaders whose negligence has grossly abused the public's trust. We need better recognition of the risk of inaction as well as the risks of taking the wrong action. Prudent policies must recognize the exigencies of the present while approaching the future in an open-minded yet realistic way. Rather than followers of imagined paths we should be builders of roads that are capable of sustaining our (enlightened) requirements for their use.

Lovins

through resource competition or more subtle social mechanisms. He seems to advocate incremental adhocracy, in defiance of the long-term goal-seeking he earlier advocated.

I make it clear that my two graphs [one illustrating a hard-path future and the other a soft-path future] are not forecasts nor precise recommendations, but qualitative sketches embracing infinite variations. Nonetheless, the hard path is quantitatively similar to many official and industrial forecasts which have recently prevailed and which still have some prominent advocates (*e.g.* Chauncey Starr, Americans for Energy Independence, and the high-growth advocates in the CONAES study). The soft path is quantitatively similar to, and perhaps higher than, President Carter's energy program—if we note that the program, despite rhetorical support for LWRs, did nothing to bail them out and everything to compete with them.

But such comparisons of how much energy is used do not go to the heart of the soft/hard distinction, which is structural. Further, Dr. Forbes's emphatic dismissal of exclusivity seems to rest on a purely technological interpretation that ignores cultural and institutional exclusivity. U.S. energy history since World War II illustrates these factors: a certain pattern of investment and attention and political commitment has left us with a restricted range of technical options and a rigid set of institutions, even though both reactors and solar collectors can and do exist today in the same country.

I have already referred to Dr. Forbes's curious habit of regarding as unproven the soft technologies for which my analysis rests on empirical operating data.

Forbes

o

Past energy strategies may have developed through an "incremental adhocracy," but yet the trends in supply

Lovins

o

There are no past energy strategies, only improvisations and tactics. Trends have not been steady since about 1970,

Forbes

and demand have been remarkably steady and the transitions in resource uses and technologies quite consistent. There is a tendency in the "invention" of the "energy crisis" to believe that we have discovered much by way of the new alternatives in the last several years and that energy systems can be molded in short order in the hands of a few capable strategists. Some review of the past suggests that neither point is correct.

Lovins

and some important trends (*e.g.* official expectations for the U.S. nuclear capacity to be installed by 2000) have dramatically *reversed*. I nowhere claim that energy systems can be "molded in short order"; I suggest a transition takes about 50 years.

o

One of the most incisive books ever written on energy is Palmer Putnam's *Energy in the Future*. It serves as an exceptional reference not only for its great depth, but also because the date of its writing (1952) allows us to see how our present was viewed a quarter-century ago—the same time span as is used in most current projections (to 2000). As the author of *Power from the Wind* and *Solar Energy*, and as the engineer of the famous Smith-Putnam 1250 kilowatt wind machine on Grandpa's Knob in Vermont, Putnam was a believer not in adhocracy but in careful thinking. He foresaw much that we should have heeded—growth and conservation, the crunch in fluid fuel supply and imports, the renewable sources and their limits, the need for nuclear power and other hard technologies, the impacts of cost increases, and much more.

He predicted a 1975 U.S. energy demand of 63 quads (rather below the actual 71) and a 2000 demand of 150 quads (rather higher than most current estimates), accounting all the while for improving efficiencies and higher costs. On fossil fuels he states that:

o

The quotations from Palmer Putnam are historically interesting, but the costs, demands, technologies (in many cases), and social conditions he assumed would be considered quaint by modern analysts. It is hardly persuasive to quote his 1952 assessment of the role of renewable sources in 2050, state that today "there is no reason to believe that Putnam was unduly pessimistic"—and let it go at that.

Putnam's 1952 view of 2000 is like that of *some* observers in 1977, but certainly not all, and can hardly be considered authoritative.

Forbes

Lovins

> Domestic peak production of each of the fossil fuels may occur sometime in the next 50 years, perhaps in the order: gas—oil—good strip-mined coal in the East—all coal—oil shale. Ayres (1952) suggests that gas and strip-mined coal will have begun to decline by A.D. 1970, and that the bulk of our oil will then be imported or synthesized. It is plausible to conclude that some of the associated increases in real costs will become significant before A.D. 1975. Imports of oil would, of course, postpone for a few years the need for synthetics, while creating a temporary vulnerability.

Such words strike remarkably familiar chords today, and Putnam's only significant misjudgment was on the extent of Western coal resources. On renewable energy sources (which he calls income sources) he writes:

> Fuel wood, farm wastes, water power, wind power, solar heat collectors, heat pumps, solar power collectors, temperature differences in tropical waters, tides and natural steam are in use or have been tried. The total plausible contribution from these sources over 100 years, and at costs no higher than two times present costs is . . . 7 to 15 percent of the cumulative hypothetical (world energy) demand by A.D. 2050.

Twenty-five years later there is no reason to believe that Putnam was unduly pessimistic. And, writing before the first commercial nuclear power plants were built, he projects a role that is not substantially below what many observers foresee today:

> Thus it appears that nuclear fuels might be suitable for carrying about 15 percent of the total maximum plausible United States energy system in A.D. 2000.

In short, Putnam's predictions for 1975 were quite accurate, and his view of 2000 is not very different from that of today and certainly not devoid of most

Forbes

of the factors that many think we have just discovered.

o

Possibly the only really "new" discovery of today is an examination of growth and a questioning of consumption, manifested in discussions of conservation and the need for energy. While there is a tendency to preoccupation with conservation to the extent that some feel no need to worry about energy production, it is evident that the details of demand are not as well understood as the details of supply. This is evident in debates over energy and jobs, lifestyle and conservation versus growth. To the extent that these debates can move away from posturing for favored strategies toward a careful researching of the basic roots of energy demand, this trend would be a healthy one in the understanding and the options it would provide to match supply and demand in a shrinking world. But we must accept that present relationships of economy and growth do exist and that they have a momentum of public acceptance and advocacy, even while we seek those viable strategies that will permit us to begin unlinking them.

Lovins

o

Thermodynamic perspectives (end-use structure, Second Law efficiency), sociopolitical and geopolitical perpectives, and appreciation of scale issues are other new conceptual developments that deserve mention. I do not understand Dr. Forbes's reference to "present relationships of economy and growth" and am not sure he does either.

Forbes

o

Lovins' view of two distinct and exclusive paths (his Figures 1 and 3) is puzzling. To view the future as a "hard" path with high growth or a "soft" path with low growth, for which we must make a clear choice today, is myopic if not irresponsible. Our energy future is not two paths but a plain, and while our goal may be some distant mountains, our course must be chosen with care, avoiding the potholes as best we can and keeping an open mind about what lies beyond the next bend. In more prosaic terms, the hard path cannot be forsaken till the soft path is tested and found viable, and even then the transition would be a gradual one.

Lovins

o

The first paragraph repeats the misconceptions that the hard/soft distinction rests on how much energy is used and that my two graphs were meant to be definitive. I agree we must design for resilience. I believe soft technologies are, broadly speaking, "tested and found viable," like the conservation and transitional-technology elements of a soft path. It is logically impossible, however, to "test and find viable" a soft *path* [as opposed to soft technologies] while at the same time persisting in a hard path, since it is a category mistake to suppose that the sociopolitical structures and conditions which define both paths can prevail simultaneously in the

Forbes

Why do we so often assume that modest policy decisions at one point in time can produce major changes in an energy system that is so complex and interdependent? Again, some "backcasting" shows trends in the system that are remarkably steady through major social, economic and political transitions. Cesare Marchetti has analyzed historical trends and transitions in technologies and energy sources. As shown in Figure 1 (Figures 1 through 5 are from his paper, "On Energy Strategies and Fate"), the fractional contributions of the principal energy sources show amazingly steady trends over long periods of time even through events like the Depression and the Suez Crisis. [The figures referred to have been omitted.] In fact, Marchetti shows that one can predict current energy input fractions from just the 1930-1940 trend lines. Using the same approach he projects future input fractions, with year 2000 values close to current estimates on oil and nuclear, but much higher on natural gas and much lower on coal. To turn these latter two trends around will be no minor feat in policy and strategy, as current gas shortages and problems in increasing coal production might suggest.

Lovins

same society.

The final reference to "modest" policy decisions is fair if it refers to the inertia of energy systems and of nations, but inadequate if it refers to such initiatives as President Carter's energy policy. The proposed conservation policy, though imperfect, would probably suffice to keep primary energy demand in 2000 to about 90 q or less (though ERDA, apparently not having done the sums, still projects 120); the difference between 90-odd and 120 represents the superfluity of most of ERDA's supply programs.

o

Forbes

A central problem is that rate at which policy can effect change and at which new strategies and technologies can penetrate (and concomitantly, the costs of forcing a discontinuity of trend). Again Cesare Marchetti shows that, beyond the point where new technologies are available and competitive, there is a striking consistency to the rate at which they build into the system even in dramatically different economies. Visions of energy "moon missions" notwithstanding, there is a limit to the rate at which new sources can build into a system that requires affordability. In the meantime, the needs of a whole generation must be met with current technology.

Lovins

Dr. Forbes calls attention to the important problem of relative deployment rates. I consider this problem in *Soft Energy Paths*, especially in Chapter 5.4. As I note there, Marchetti diagrams and classical theories of market penetration tell us nothing useful or valid about soft technologies, which, being numerous, diverse, small, simple, and of short lead times, have properties completely different from other modern energy systems. I agree that "in the meantime, the needs of a whole generation must be met with current technology"—which I take to include the proven and rapidly emerging transitional fossil-fuel technologies that I advocate.

Forbes

In this light, Lovins' assertion that soft technologies could grow to about 5 quads by 1985 and to about 35 quads in 2000 (35% of his energy demand estimate at that time) would take miracles or revolution. More careful studies, such as those by Arnold Cohen and others, suggest that solar heating for example might provide 1% to 3% of year 2000 energy needs *if* major government support is provided to develop the market before economic viability is achieved.

One of the major problems in energy strategy lies in the squeeze on fluid fuels that shows every sign of worsening as domestic production drops and demand increases. Lovins' strategy does little to solve this problem and even makes suggestions that would exacerbate it. Solutions will have to come basically from a combination of conservation, increased production of electricity from non-fluid fuels and from production of synthetics.

Lovins

The last paragraph should also assume "miracles or revolution" in the assumptions of analysts who project more rapid solar growth than I do (*e.g.* Wolfe, Morrow, and CONAES—see *Soft Energy Paths*, Chapter 5.4, for references and discussion). In fact, a central research problem in soft-path analysis is figuring out how to estimate deployment rates for devices and refinements whose rate constraints are chiefly institutional rather than technical. In *SEP* I suggest some approaches to this problem. Analyses such as Arnold Cohen's depend entirely on the assumptions made. If one assumes long development times, high costs, asymmetric cost comparisons, cheap subsidized fuels, rigid institutions, little or no retrofitting, and small ultimate technical potential, then it is indeed possible to derive discouraging solar estimates. [In 1978, the President's Council on Environmental Quality proposed and President Carter endorsed a feasible goal of 25% solar penetration in 2000, Secretary Schlesinger stated that 15-20% was plausible, and the Domestic Policy Review of solar energy considered solar contributions in 2000 as high as about 30 quads—nearly the same as Lovins's 35-quad estimate. The DPR also estimated that renewable sources in 1978 were already contributing about 4.8 quads to U.S. primary energy—nearly the 5 quads that Dr. Forbes says would take "miracles or revolution" to achieve in 1985.]

Forbes implies that Cohen assumes solar heating to be uneconomic. This in turn implies (I have not yet been able to obtain his rather obscure paper) that, like other analysts in this field, he is making solar heating compete with historically cheap fuels rather than with other marginal alternatives capable of doing the same heating job. This assumption is asymmetrical and is bad economics. Nuclear power could not possibly pass the same test—which is why it is receiving "major government support" right now and why the same is proposed for synthetic fuels. It is better economics to price fuels properly, or make economic comparisons symmetrically, or both, than to subsidize some (or even all) energy sources.

Dr. Forbes rightly stresses oil depletion, but wrongly says a soft path would do little to help and might even make matters worse. Conservation, better use of all fossil fuels, and soft technologies (especially biomass alcohols) would contribute more, faster, to replacing oil than a hard path could.

o

In this regard, Lovins' arguments on the impossibility of "massive electrification" are confusing. Electrification has been a stable trend since the first introduction of electric power. Electricity has grown from 5% of energy supply in 1935 to about 28% today (effective input). That is, in the past 40 years, electricity has increased as a *fraction* of energy supply at an average of about 4.5% per year. In fact, the most massive electrification occurred between 1945 and 1955 when the fractional input increased at nearly 7% per year. Clearly, future "electrification" will be much lower, if only through saturation. If the *fractional* electricity supply growth for the next 25 years were just *half* that of the past 25 years (4.4%), electricity would supply nearly 50% of U.S. energy needs by 2000.

o

Dr. Forbes neither shows that my electrification arguments are "confusing" nor addresses any of them. He correctly states that penetration of electrification must slow down—otherwise *all* our fuels would be going into power stations by about the turn of the century—but he does not consider my macroeconomic, logistical, or social arguments that further central electrification is bound to encounter formidable and, before long, insurmountable obstacles, even with lower-than-historical growth rates.

o

In looking back at the past it is worth considering the projections made a few decades ago of energy demand today. In the mid-1940s, both Westinghouse and Edison Electric Institute were projecting an electric demand growth for the ensuing 30 years of 5.5% per year—leading them to forecast an installed U.S. electrical capacity of 250 gigawatts in 1975. The actual capacity required turned out to be *twice* that amount.

Forecasts over the past several dec-

o

Dr. Forbes's account of electrical growth fails to mention that the real price of U.S. residential electricity fell fivefold from 1940 to 1970—owing in part to remarkable technical advances—and that rate structures and other incentives were manipulated throughout this period to promote rapid demand growth. [For example, half the profit on a new nuclear plant is its negative income tax arising from federal tax subsidies.] Now, with marginal cost much higher than historic cost and ris-

ades have frequently underestimated demand growth. While declining population growth rates, rising energy prices and energy conservation will serve to decrease demand, there is a distinct possibility that the economic recession of the past several years has colored our vision and led to forecasts of low future demand growth without provision for major deterrents to a return to high growth as the nation moves out of the recession. Forecasts of low to moderate energy growth factored into national supply policy decisions, coupled with a demand environment that does not preclude high growth, runs the risk of inadequate supply contingencies.

ing steeply, the world has changed. Dr. Forbes might also have mentioned that the trend towards ever larger stations has greatly increased requirements for reserve margin, especially in recent years, so that nowadays nearly a third of capacity is being built simply to provide reverse margin for itself. Such pathology says nothing about real needs.

The pattern as I see it, not only in the U.S. but around the world, is that utility forecasters, reluctant to admit a fundamental change of circumstances, are having to postulate ever more extravagant "catching-up" growth r es for the 1980s in order to avoid adjusting their long-term projections drastically downwards. (I happen to think Professor Forrester's arguments about the possibility of a Kondratiev cycle also have merit, but need not invoke them to suggest that long-term electrical growth rates will be low, zero, or negative.) Of course, if one builds lots of power stations "just in case," then one must certainly find ways to subsidize and sell all that electricity if one is not to go bankrupt from carrying charges. We have been doing this for a long time, and I suggest we cannot continue.

Dr. Forbes might also consider that if economic growth quickens, capital stocks will turn over (and be augmented) faster, so that old, energy-inefficient stocks of cars, buildings, machines, etc. will be diluted or replaced faster, and therefore energy conservation will be faster. People who try to assume rapid economic (or population) growth *and* correspondingly rapid energy growth are trying to have it both ways.

o

For too long we have avoided decisions on problems that were apparent a quarter of a century ago, and we cannot use visions of a different path as an excuse for further delay. It is time to confront reality and use the tools we have available to solve supply and demand

o

The first sentence sounds a bit like "Forward, lemmings!" Russell Train reminds us, wisely, that we should be suspicious of people who tell us there are no alternatives—and, anyway, no time to consider them—because such assertions generally mask serious weak-

Forbes

problems now. New supplies and strategies must be brought into play, but none can provide easy or quick solutions.

o

Much of Amory Lovins' strategy is based on the beguiling propositions that "the United States could probably improve energy efficiency by a factor of three or four." In this way he reasons that through efficiency improvements alone we could get twice as much energy out of the system for the same input by 2000, and more thereafter. This "more for less" seems like an offer that can't be refused, especially when combined with the statement that energy is lost "in elaborate conversions in an increasingly inefficient fuel chain." Lovins claims, for example, that in Britain energy input has doubled since 1900 while output has increased by only one-half. This propositions is contrary to the facts.

In *Energy in the Future*, Palmer Putnam analyzed the efficiencies of world energy systems from 1800 to 1950 in great detail. On the United Kingdom he concludes that although energy input from 1900 to 1950 grew at only about the same rate as the population, the overall conversion efficiency more than doubled in the same period so that the energy output per capita also more than doubled. As to the United States, Putnam calculates that the overall efficiency increased [from] about 11% in 1900 to nearly 30% in 1950. Thus he concludes that while U.S. per capita input increased by about 75% in this period, energy output per capita increased by about 350%. It is evident, therefore, that the historical trend has been toward more efficient energy use, not less efficient. Examples vary from the replacement of the fireplace by the stove and the basement furnace to more complex efficiency improvements in industrial processes. The efficiency of electric generating plants increased

Lovins

nesses. I do, however, agree with Dr. Forbes's last two sentences.

o

Dr. Forbes contests my statement that fractional losses in conversion of primary energy to end-use energy have increased and are increasing. (The 23 March 1977 energy policy statement of the West German government projects that over half the primary energy growth to 1985 will go to conversion losses. See also the remarkable projection of conversion losses given by the Flowers Commission's famous report of September 1976.) Unfortunately, Dr. Forbes confuses this type of inefficiency with the separate question of end-use efficiency in cars, factories, stoves, locomotives, etc., and treats the two concepts more or less interchangeably. Of course power stations and some major end-use devices have become notably more efficient since 1900, but it is not true that the gap between primary and end-use energy is narrowing.

My British example can be readily confirmed by looking at the official statistics. I did not say that the useful heat derived from a pound of coal, say, has not increased as furnaces replaced open grates; I said that though primary energy in the U.K. has doubled since 1900, delivered end-use energy ("secondary" energy) has only increased by half (or by a third per capita), with the other half going to conversion losses. It is true that the delivered fuel (or electricity)—which in 1900 was mainly coal—was often used inefficiently by modern standards; but my point was that the energy input needed to mine, treat, and deliver the coal to the point of end-use was much smaller than the equivalent energy requirement to deliver its successors—oil, gas, and secondary electricity. For this reason, and because electricity has been dramatically replacing direct fuels (yield-

Forbes

nearly ten-fold between 1900 and 1970, and the efficiency of rail transportation increased dramatically (in energy terms) after 1950 with the phasing out of steam locomotives. It is evident, too, that the room for further improvement, while substantial, is hardly as large as Lovins suggests nor will it be as easily won as past gains.

There are, however, some important areas for efficiency improvements and in this regard Lovins' points are well taken. The principal emphasis needs to be an increased efficiency and conservation of fluid fuel use, since oil and natural gas account for three-quarters of U.S. energy consumption, imports account for nearly half of oil consumption, and demand is still rising.

The internal combustion engine is a prime example, particularly since transportation accounts for about one-quarter of U.S. energy use and over one-half of oil consumption. Since 1920 the efficiency of automobile engines has increased only from 22% to 25%, but increasing size and number of "options" has decreased today's mileage to below that of 1920; there is obviously room for improvement (though in quoting a 27% improvement from 1974 to 1976, Lovins neglects to mention that 1974, with its pollution control devices, was a low ebb in auto efficiency). With appropriate efforts, steady gains can be made, but no overnight doubling can be expected. Of course, savings can be effected by switching to smaller cars, or away from cars and planes to buses and trains. This is no longer in the realm of "something for nothing" but requires changes in habits and lifestyles. (In this regard, recall that the oil embargo resulted in demand for small cars that manufacturers moved to match, only to be faced with surpluses as the recession eased and the public moved back to bigger cars.)

Electricity is another area in which efficiency improvements are possible. The steady increase in generating effi-

Lovins

ing the dominant term in the increasing conversion losses despite better power station efficiencies), conversion losses have taken half the growth since 1900. Putnam is correct that U.K. per-capita primary energy was about the same in 1950 as in 1900, but it has increased about a third since 1950 owing to electrification and the influx of cheap oil and gas. Putnam's "efficiency" is not measuring the quantity I was describing.

Dr. Forbes suggests that I have exaggerated potential improvements in energy efficiency (especially end-use efficiency), but provides no evidence for this view. He does not dispute, or even mention, the many quantitative case-studies I cite. His further statement that future improvements will be harder to win than past ones is true for minor improvements in, say, power-station thermal efficiency in the traditional Carnot sense, but is certainly not true for thermal insulation, shifts to total-energy systems, or use of heat-pumps.

Dr. Forbes suggests my example of 1974-6 car efficiency improvements is unfair because "1974 with its pollution control devices, was a low ebb in auto efficiency." The pollution control devices are even more numerous and complex in 1976 than in 1974, and mileage in 1974 was not only at a low ebb but threatening to decline further were it not for higher oil prices. There is thus nothing unfair in the comparison—though there is in Dr. Forbes's implication that I have claimed an "overnight doubling" is possible; what I think realistic is rather a threefold improvement over about 50 years. (Extensive technical studies, and experience with some current German and Japanese cars, suggest we can do the same at the margin in nearer 5 years than 50.) [In 1978, a fivefold improvement had already been obtained with an experimental diesel-electric hybrid car weighing some 3400 pounds.

Forbes

ciency can be expected to continue, but co-generation and the use of heat pumps have the potential to provide for marked improvement. Heat pumps are considerably more efficient than resistance heating and are already making inroads in the domestic market, although further development is needed for reliable operation in the northern U.S. and Canada. Co-generation is, as Lovins suggests, largely a problem of institutional barriers. (This is true of many things in energy.) It is certainly feasible and desirable to use the "waste" heat from power plants for industrial steam or for district heating, but siting policies, rate structures and many other factors will need to be modified to provide the incentives for this. These are not trivial problems.

Lovins

Making it lighter (but still big) could have yielded a further factor of two or three—all within the present art. And in 1978, the Volkswagen Diesel Rabbit was already 2.4 times as efficient as the U.S. fleet average of 17 mpg; the about-to-be-introduced turbocharged version was 3.8 times as efficient.]

I did not assume the modal shifts that Dr. Forbes decries, but rather noted, for example, that a diesel Mercedes, which is hardly a small tinny car (and is not as efficiently designed as it might be), normally gets about 27-30 miles per U.S. gallon right now.

Dr. Forbes refers to the "steady increase" in Carnot efficiency of power stations. It is true that with great effort and expense, new fossil-fuelled power stations are wringing one or two tenths of a percent greater efficiency out of fuel than they did a few years ago. But in general the design efficiency of the U.S. population of power stations has been going *down* (and the empirical efficiency has often dropped even faster) since about 1970, owing partly to poor performance and partly to the introduction of nuclear stations, which have inferior steam conditions. The "marked improvement" to which Dr. Forbes allludes can only come from combined cycles and various sorts of total-energy systems—all of which are much more cost-effective than slight improvements in the Carnot efficiency of classical power stations.

I agree that institutional problems are not trivial. The new [Carter] Administration is starting to address them though, and I expect large payoffs.

o

Forbes

There are many other areas where improvements can be made (home insulation, industrial equipment, etc.), but the rate of efficiency increase will not be as fast as before. Many of the areas require technical developments, and beyond that must wait for old equipment to be retired in order for it to be

Lovins

Dr. Forbes repeats the absurd statement that "the rate of efficiency increase will not be as fast as before" in home insulation, industrial equipment, etc. We never really tried before; fuels were cheap and our attention was elsewhere. We have not yet begun to tap the biggest areas for savings.

Forbes

replaced with the new and more efficient; this factor is a significant limiter yet it is frequently ignored. And of course as systems close in on ultimate efficiencies the gains become more difficult to make.

Lovins

I did not ignore the need for new technical developments, but rather limited my calculations to what is now available, and did not forget the rate constraints of retrofitting. It is true that as we approach ultimate Second Law efficiency limits, improvements will become harder, but we are so far from that point—typically a factor of five or ten away from *practical* limits—that we shall not need to worry about ultimate limits for decades.

o

There is much room yet for efficiency improvement, but it can come only at a steady pace and not at the rate or magnitude that Lovins' vision offers. Nor should it be assumed that this is some magic finding that is not already included in good energy demand projections—though unfortunately many projections fail to factor past efficiency gains into future demand estimates. To slow supply growth beyond what can be done through efficiency requires voluntary and mandated demand reduction. Would that there were better recognition that conservation means more than efficiency improvements by industry or savings by "the other guy." The policy decisions for mandating conservation in a way that is publicly acceptable and equitable will not be at all easy.

o

Dr. Forbes says yet again that I have exaggerated the rate and magnitude of conservation gains. He has provided no support whatever for this view. Nor is he necessarily correct in stating that "good" energy demand projections already take proper account of the scope for improved end-use and energy-system efficiency. That is right only if "good" models are taken to exclude virtually all those now used in government planning. A conflict between a really good model and ERDA's models is nicely illustrated by von Hippel and Williams (*Bulletin of the Atomic Scientists*, December 1976, pp. 20–32), and it would be interesting to learn exactly where Dr. Forbes disagrees with its data and conclusions.

o

Much of Amory Lovins' soft strategy is based on the proposition that the economic cost of continuing the hard path is unacceptably burdensome, and that the soft path is well within an affordable range. Yet current and projected costs lead to the opposite conclusion. If Lovins were indeed correct, it would seem that this path would be a very natural one that would require relatively straightforward policy decisions to give it impetus. But his path would be more expensive, due both to the direct cost of the soft technology pro-

o

Dr. Forbes assumes that my calculations require the mandatory early retirement of unamortized energy facilities. This is not correct. I do not assume early retirement for any significant facilities (though I leave open the option of phasing out existing reactors, for other reasons, and am prepared to argue that doing so would result in a net economic benefit). In general I assume that existing big energy systems would be replaced not prematurely but through normal attrition. Some oil- and gas-fired power stations might be

Forbes

posed and to the cost of the early retirement of existing energy conversion technology his strategy would impose.

With regard to the hard path, it is a matter of record that fuel costs and capital costs have risen dramatically over the past several years. These cost increases are putting a substantial strain on our energy system and our society, including the dramatic impact of imports (up from $3 billion in 1970 to about $35 billion in 1976), and will continue to do so for some time. Increases in the cost of fluid fuels can be expected to continue well into the future as these resources become harder to tap (although not at nearly the same rate as in the post-embargo period). Fluid fuels present probably the most difficult aspect of energy strategy, and the current shortage of capital exacerbates this since fluid fuel use generally involves lower capital outlay and higher fuel costs than do solid fuels. But costs have begun to stabilize and the prospects are that costs will increase in the future, not at the pace of the post-embargo recession period, but nearer the slower pace of the pre-embargo era—though costs themselves will never return to that level. In short, future energy investments will require a higher proportion of capital outlay than in the past, but not dramatically so, and certainly not the debilitating levels that Amory Lovins suggests.

Lovins

converted to coal—whereas in a hard path they would have to be prematurely shut down if nuclear power were to displace imported oil. Dr. Forbes's incorrect statement that soft technologies cost more than hard ones is considered below.

It is hard to separate the discussion of fuel costs and capital costs, but apparently Dr. Forbes does not agree with either my detailed capital-cost calculations for particular technologies or my straightforward use of the Bechtel model output. He states no grounds for this disagreement. Such a firm dismissal deserves quantitative discussion, not mere assertion.

O

Forbes

Lovins' juggling of capital costs for energy supply is confusing at best. He claims that "the capital cost per delivered kilowatt of electrical energy emerges as roughly 100 times that of the traditional direct-fuel technologies." Since capital costs are factored into consumer prices, one would expect that electricity would emerge as having a market price on the order of 100 times that of direct fuel use. While electric resistance heating is more expensive than oil or natural gas, as anyone using

Lovins

Dr. Forbes's conclusion (that if marginal capital-intensity is 100 times greater for delivered electricity than it was for historic direct fuels, then the former should cost 100 times as much as the latter) ignores several well-known economic principles: for example, the difference between marginal and historic cost, the interfering effect of subsidies, the difference between unit technical cost of producing fuels and retail price of fuels (price = cost + rent), and price differences in inhomogeneous

Forbes

it knows, it is certainly not anywhere close to being that much more expensive, nor is the capital component. First, Lovins' costs are not consistent 1976 dollar costs as he claims; for example, for a nuclear power plant the capital cost investment is about $80,000 to $90,000 per barrel of oil per day delivered in 1976 dollars (equivalent end use as resistance value, at 65% plant capacity factor and including transmission losses), not the $200,000 to $300,000 that Lovins gives. Second, direct fuel use investment must include not only the investment in (for oil, say) exploration, drilling, pipelines, tankers, refineries, etc., but also the investment in final delivery to dispersed locations and in end use. Third, Lovins plays an interminable game with energy efficiencies: electricity's end-use energy content is not always the resistance heating value, as for example in refrigerators, air conditioners and heat pumps, or with co-generation; the direct use of fuel is never at the full energy potential of the fuel, even at the end-point, for which the prime example is the automobile engine. Fourth, it is surprising that a believer in life-cycle costing would miss the point of replacement rate for investment. No source of energy or energy conversion equipment lasts forever, and must eventually be replaced; the useful life of a mine, well, power plant or solar panel is therefore an important factor in the rate at which capital investment is needed (power plants, for example, have a useful life of about 40 years—well beyond that of many parts of the energy system). This would be particularly significant in a strategy that would require the early retirement of existing equipment in order to replace it quickly with more efficient equipment of soft technology.

o

Due to the necessity of replacing fluid fuels where possible and to con-

Lovins

markets (electricity in fact commands a premium price). These factors account for the residual difference between the 1960s fuel price and 1980s electricity price.

Contrary to Dr. Forbes's assertion, my capital cost calculations are indeed in constant 1976 dollars, as can be ascertained from the calculations set out, as cited, in the Oak Ridge paper. Like Professor Bethe, Dr. Forbes neglects many important elements of a complete nuclear *system* and counts only the power station, thus underestimating by a factor of order 2.2-3.7. Moreover, and again contrary to his implication, I do count whole-system costs for all other energy systems, including end-use-device costs where appropriate (*e.g.* solar heating)—precisely what he does not do for nuclear power. Nor do I neglect, as he suggests, to consider the First Law efficiency of end-use; this is a central theme of the comparative section of the Oak Ridge paper, and is taken fully into account in the *FA* conclusions, such as the solar-heating-*vs.*-other-long-run-heat-sources comparison in note 22. Nor, finally did I fail to consider replacement investments where relevant. The soft and transitional tchnologies I assume generally should not have a shorter lifetime than competing hard technologies (to the extent that one knows either), and in the rare cases where they should, I have taken care that the effect is not so large as to disturb the robustness of my economic conclusions. [See also note 5 in Lovins's *Science* letter of 22 September 1978, which is reprinted as an Appendix.]

The assumption in the last four lines [that existing equipment would be retired before its useful life was over] is Dr. Forbes's, not mine.

o

The paragraph begins with an unsupported assertion [that demand for elec-

Forbes

sumer demand, electricity demand will continue to grow. Contrary to Lovins' assertions, future electrification will be less than in the past and the capital costs will not be unduly burdensome. In reviewing the "spiral of impossibility" of utility financing, he misses the very basic point that many utility problems stem directly from the regulatory process, rates and allowed return, particularly in an economy that has been both recessionary and inflationary. In this regard, Lovins alludes to "reduced (even negative) demand growth" resulting from electricity and energy cost increases. Demand elasticities calculated from the events of the last several years must not be used without caution (as is too often the case). First, it is difficult to separate true elasticities from the general economic climate of recent years; second, current trends of economic recovery suggest that the elasticities of demand are themselves elastic (people react to a sudden price hike, then readjust as they come to accept it). Elasticity does exist, but the post-embargo period should not be used without considerable qualification to suggest major future demand growth reductions.

Lovins

tricity will continue to grow] (contradicted by *e.g.* the excellent Ross and Williams analysis cited in FA note 12), continues with an uncontested truism [that future electrification will be less than in the past], and follows that with a restatement that "capital costs will not be unduly burdensome"—welcome news for utility executives such as Bob Taylor, Chairman of Ontario Hydro, who has been saying for the past year that capital shortage—not demand—is *the* governing factor in his long-term planning.

Dr. Forbes is correct that regulatory delays have caused many past utility finance problems—considered by many short-term analysts, including Bankers Trust, to be largely a thing of the past. But the point of the Kahn study (FA note 10) is the *future* dynamics of utility finance: owing to high capital intensity, long lead times, and extreme sensitivity to assumed price elasticity of demand, by the time the utility discovers the elasticity is higher than expected, it is already too late, overexpansion has already occurred, and the downward spiral has begun. Subsidies, by suppressing an early-warning signal in this unstable behavior, increase the danger of collapse. My views of the behavior of the electric sector over the next few decades rest not on debatable elasticity estimates but on elementary macroeconomics—or on a control-theoretic view of the inherent instability of cash-flow in an enterprise where it takes far longer to build a plant than it takes people to respond to the higher prices one must charge to finance the plant's construction.

o

Forbes

But all this might mean little if the "soft" path was inherently cheaper and less capital intensive. With relatively straightforward incentives and policy directions, businesses and consumers could be expected to leap at the opportunity to save investment and operating

o

Lovins

This paragraph correctly implies that with institutional barriers removed and with proper economic signals, a soft path will tend to implement itself if it is cheaper than a hard path. I agree, and believe it is cheaper. Dr. Forbes disagrees, but does not say why. Perhaps

Forbes

costs. Unfortunately, much of the soft path is more costly than the hard, for the foreseeable future at least. Because a number of key parts of the vaguely defined soft path are not well developed, cost estimates are uncertain (this alone should warrant concern). But enough is known to make some observations.

Lovins

his conclusion is based on his mistaken nuclear cost estimates.

o

Lovins dismisses economies of scale, for which very reason power plants, refineries and other supply and manufacturing processes have been getting built in larger units over the years. And it is for this reason, combined in some cases with technical complexity, that the "hard" approach has been suggested by some for *renewable* energy sources. (It might be well to note, too, that the definition of "hard" and "soft" is user-dependent; what is soft for industrial use might be hard for the individual consumer.)

I do not "dismiss" economies of scale, but suggest they are often outweighed by diseconomies of scale that do not enter classical scale calculations. This argument, sketched in the FA and Oak Ridge papers, is set out more fully in Chapter 5 of *Soft Energy Paths.*

o

Lovins suggests "backyard" diesel generators. True, they can be inexpensive (though not nearly as low as Lovins' $40 per kilowatt if they are to withstand more than a year or two of continuous operation), but ask anyone living in a remote area supplied by diesel generators what they pay for electricity and one quickly finds out that it is hardly a cheap route (even when the diesel fuel comes from a "hard" refinery). Suggesting converting breweries to alcohol production to satisfy fluid fuel demand leads from the sublime to the ridiculous (Pabst Blue Premium? Chablis Premier Crude?), not least of all at current stove alcohol shelf prices of $4 to $7 per gallon. Besides, alcohol still produces CO and CO_2 that we are concerned about.

I did not "suggest" diesel generators, but think they can be economically attractive. (Mass-produced car engines currently cost about $2/kW. Prices up to 20 times higher would seem to leave a good deal of room for greater durability.) Nor did I suggest converting breweries to fuel alcohol production. Most authoritative estimates of production cost for biomass fuel alcohols fall around the production cost of gasoline, and generally below its taxed price: it is important not to confuse fuel alcohols (which can, and indeed should, contain some higher alcohols) with purified ethanol or methanol, which are inferior as fuels but cost more to make. Most higher cost estimates assume high collection and transportation costs for unrealistically centralized conversion, plus brute-force techniques such as distillation for separating ethanol from water (which can be done cheaply and with almost no energy requirement by using branes).] And buying alcohol in small retail cans says nothing about its marketing costs on a gasoline-like scale. hydrophobic plastics or synthetic membranes).

Finally, in stating that alcohol "still produces the CO and CO_2 that we are concerned about" Dr. Forbes forgets that producing fuel alcohol from products of photosynthesis, then burning that alcohol, merely produces the oxidized carbon which we would otherwise have gotten rather promptly by leaving the feedstock to rot, or by letting some animal eat it (a slower form of burning it). No manipulation of carbon that is "on current account" in the photosynthetic carbon cycle adds to atmospheric CO_2 inventories; only mobilizing carbon that was locked up in fossil fuel can do that. Thus carbon dioxide released from any use of bioconverted fuels has no climatic significance. (Carbon monoxide and other emissions also tend to be lower from fuel alcohols than from petroleum hydrocarbons, so reducing pollution.)

o

Lovins claims that solar heating is available now and solar cooling imminently. This might well be questioned in terms of the durability of collector systems in the northern U.S. or Canada, and the reliability of adsorption cooling systems. This is a major factor in life-cycle costing, but let us assume that the target lifetimes are achievable. Lovins claims that "they are cheaper than present electric heating virtually anywhere, cheaper than oil in many parts, and cheaper than gas and coal in some." A recent study for ERDA finds, in essence, that solar heating may be "cheaper" than electric *resistance* heating in *most* of the U.S.—for new houses with 12 inch attic insulation and brick veneer siding—a somewhat different conclusion.

o

This paragraph ignores the extensive operating experience with solar collectors in all U.S. regions (and, often even more extensively, abroad). This experience is far longer than that of nuclear power reactors, and offers a sound basis for concluding that decently engineered solar systems last for many decades.

The ERDA/MITRE study to which Dr. Forbes refers concludes that solar space and water heating installed at an equivalent price of $20 per square foot of collector "is competitive today against electric resistance systems throughout most of the U.S. If the system cost is reduced to $15 per square foot solar systems become competitive against oil hot water heating and/or oil and electric heat pump space heating in many cities. Finally, if the cost should be reduced to $10/ft^2 by 1980 through a combination of technical innovations and incentives, solar hot water and

Forbes

Lovins

heat would be economically competitive against all fuel types."

Several points about this study should be emphasized:

- it assumes house insulation far inferior to what is economically justifiable, so increasing both the size and the "peakiness" of heat requirements and making solar systems look worse [it may increase the required collector area as much as tenfold and the storage volume by even more];
- it assumes a solar system price that is twice ERDA's 1980 target (of $10/ft²) and, worse, is twice the price at which field-erected solar systems of good quality are being commonly installed today (and about four times the price at which the cheapest good systems can be installed today);
- it assumes single-family houses even though these are the least favorable for solar economics and are not the dominant type of housing being built;
- it assumes real price escalation of only 4%/y for fuels and electricity;
- it assumes that each solar system requires a conventional backup system of *undiminished* capacity—so that the capital cost of the solar system must compete with the saved fuel cost alone of the fossil-fuelled system even though, with a cost-effectively efficient building, it is cheaper to have a 100% solar system with no backup than a partly solar system;
- it assumes that the cost comparison of interest is with cheap fuels, not with all other ways of heating the house at the margin;
- it assumes a solar lifetime of only 20 years and a 2%/y maintenance cost;
- it assumes that a system is economic if "positive savings occur in 5 years or less *or* payback occurs in 15 years or less."

Forbes

Put in terms of Lovins' own cost parameter, if a solar heating system at a cost of $5000 could supply 45% of a U.S. average annual home heating requirement of about 20 barrels of oil (an optimistic assumption), the capital cost translates to about $200,000 per barrel of oil per day delivered (current cost). Since solar heating systems can cost twice this and more for high quality systems and/or professional installation, they are certainly not cheap; solar cooling promises to be even more expensive. Solar heating is a desirable trend, particularly in solar-designed housing, but it is not a low-cost path. And to say that "If we did this to all new houses in the next 12 years, we would save about as much energy as we expect to recover from the Alaskan North Slope," requires a calculator with a sliding decimal place. The 25 to 30 million new houses and mobile homes projected to be built in the next 12 years would save 0.1 to 0.15 billion barrels of oil in that time if they were all equipped with solar heating; the Alaskan oil fields in the same 12 years are expected to produce 8 or 9 billion barrels. Or, put another way, the 25 to 30 million homes would take close to 70 years to produce savings equal to the total recovery expected from the Alaskan North Slope. (This is not an argument against, just a correction of perspective.)

Lovins

Taken together, these assumptions are unrealistically adverse to solar technologies. A parametric analysis also shows the results are highly sensitive to assumed maintnance costs, discount rate, fuel escalation rate, etc., and the method of optimizing the solar system is grossly inadequate.

O

Dr. Forbes's $5000 and 45% assumptions are far out of line with the best current practice—on the ERDA/MITRE data, a factor of two worse than the average design parameters for a $10/ft^2 system competing with oil. [By 1978, a 100%-active-solar-heated house had been built in Saskatchewan with a solar system so small that, on Lovins's assumptions, it would cost only about $1-2000. The heat-conservation measures that allowed this high solar performance are all cost-effective against present fuels in new buildings and against long-run marginal alternatives for retrofitting old buildings.]

Dr. Forbes's solar system cost is about 2.8-4 times too high (*Soft Energy Paths*, Chapter 7.4). Solar systems *can* cost more than twice his figure, but if prudently bought, *will* normally cost a quarter to a half as much and can cost less (see, *e.g.* U.S. Congress, Office of Technology Assessment, "Application of Solar Technology to Today's Energy Needs," 1977), even with professional installation and full warranty. Solar heating is not cheap, and I did not claim it was; it is only cheaper than not having it. The discrepancy between Dr. Forbes's and Dr. Bliss's calculations of energy savings in houses arises mainly from the period of savings (and oil production) assumed—12 years for Dr. Forbes, 30 for Dr. Bliss—and from a large but indeterminate difference in the level of energy savings assumed. If Dr. Forbes consults Dr. Bliss's article (*FA* note 23), he will find that Dr. Bliss is talking about rational design of energy-efficient houses, not active solar

Forbes

Lovins

heating. [This controversy is analyzed further in a forthcoming (1979) Working Paper from the MIT Energy Laboratory by Mark McKinistry et al. Its August 1978 preliminary draft showed fairly good agreement with Lovins's 100%-solar system size computations (a price difference of 34% using the same price assumptions). The difference may be narrowed or resolved by analytic improvements suggested by Lovins. In contrast, Forbes had obtained solar system costs more than six times Lovins's.]

o

And again, on windmills, costs for ERDA/NASA's recently installed 100 kilowatt Sandusky wind turbine and those currently on order, are around $5,000 per rated kilowatt. At a 30% capacity factor, this translates to a capital cost of about $1.2 million per barrel of oil per day, without storage and transmission losses. Significant cost reductions can be expected in production machines, but it is still an expensive route. The more practical application would seem to be the approach of Canada's National Research Council in developing wind turbines for use in combination with diesel generators in remote or isolated areas where fuel costs are high.

o

Dr. Forbes extrapolates from the $4700-10,000/kW peak capital cost of the NASA prototype wind turbines. The "significant cost reductions . . . in production machines" have already been achieved: my calculations are based on a currently commercial Canadian 200-kW Darrieus machine with a run-on production price (small lots) of about $650/kW peak installed. In *Soft Energy Paths,* Chapter 7.2, I cite machines costing up to 80% less, or 38 times less than Dr. Forbes's figure. [Lovins's 1978 *Annual Review of Energy* article cites currently commercial Danish and American wind machines with installed prices of $760/kWp and $205/kWp respectively—6.6 and 24 times lower than Dr. Forbes's figure. A recent design study of a free-wing turbine predicts an installed price of about $65/kWp.]

o

The cost of solar-generated electricity will likely be still higher. The capital cost of ERDA's 10 megawatt solar thermal electric power plant, planned for completion in the Mojave Desert (an ideal location) in 1981, is estimated at about $100 million, or about $2 million per barrel of oil per day delivered. Solar photovoltaic cells are currently two to three times more expensive than this.

The definition of the soft path is too

o

I nowhere assume solar-electric systems, least of all solar-thermal-electric ones. I do not consider this approach necessary nor particularly attractive, though its technology might be useful in high-temperature process heat or in some special electrical applications (e.g. long-term alumina smelters in the tropics).

Once more, Dr. Forbes uses a first-of-a-kind prototype cost as indicative. Moreover, he overstates the price of

Forbes

unclear as to its details to permit a thorough cost accounting. But it is clear that it is very likely to be more capital intensive than the hard path. Dispensing the capital in smaller lumps at a time does not make it any less so.

Lovins

silicon photovoltaic arrays, and wholly ignores the existence today of technology to make CdS photovoltaics for a commercial price of about $300/kW peak (see *Soft Energy Paths*, p. 143, note 28). (Again, however, I assume no solar-electric technology.)

The components of a soft path are clearly outlined in the *FA* paper and its citations, and the cost accounting I have published is fairly comprehensive. It is regrettable that Dr. Forbes has offered no specific criticisms of my data or sources.

I do not assume that capital costs can be reduced by breaking them up—although the shorter lead times of soft technologies do yield a significant saving in interest, escalation, etc.

o

Forbes

A view that a soft technology path is environmentally preferable to a hard path must be predicated on the availability of clean, "renewable" energy sources. Yet the projections of Energy Research Group are that renewable energy sources are unlikely to provide more than 4% to 6% of energy supply in 2000, and we have no reason to believe that other detailed methodical estimates come to significantly different conclusions. Energy supply for several decades at least will thus have to come principally from fossil fuels (hard and soft, natural and synthetic) and from nuclear fuels. It is our opinion that nuclear power is unavoidably necessary, but no attempt is made in this paper to debate the issue in detail.

Lovins

The Energy Research Group projection is unsatisfactory in several respects: it is stated as a percentage of a homogeneous and unspecified number, it rests on very restrictive assumptions, and it is of poor scope and low technical quality. One "reason to believe that other detailed methodical estimates come to significantly different conclusions" is that they do: *e.g.* the preliminary CONAES estimates, which, despite being based on asymmetrical cost comparisons (comparing renewables with cheap fuels rather than with marginal nonrenewables), project over 25 q of fossil fuel displaced by renewables in 2000 and about 70 q in 2010. (The soft-technology contributions are 20q in 2000 and 56-60 q in 2010.) See also *Soft Energy Paths*, p. 98. [An even better reason is that the Domestic Policy Review found renewable sources already supplying 6% of total U.S. primary energy in *1978*!]

I agree that fossil fuels will dominate the energy mix for the next few decades, but not that renewables cannot be important over that period or that nuclear power is necessary anywhere.

Forbes

Embarking on a soft path with no immediate prospects of significant input from clean, renewable sources means in essence the dispersed use of fossil fuels—oil in diesel generators, coal in basement furnaces and a myriad of other polluting applications, including present ones like the automobile. With dispersal, pollution control is more difficult, less amenable to regulation, and hardly more environmentally sound than the "hard" path. Nor would Lovins' path be totally "soft," since mining, refining and delivery would still be essentially "hard." Without assurance of major solar energy input, the soft path would not be preferable in terms of our environment. Rather, it would be a return to negligent ways of the past in a much more heavily populated world.

Lovins

I do not agree that there is "no immediate prospect of significant input from clean, renewable sources" over the next few decades (including a large contribution from fuel alcohols in the 1980s). The transitional fossil-fuel technologies I assume (which, again, do not include diesels) must indeed play an important role, but are *less* polluting than conventional fuel-burners, as my references make clear.

Dispersal might make regulation more difficult—though centralization implies the political power to resist regulation—but my transitional technologies are sufficiently more efficient and technologically superior to outweigh this potential disadvantage. (For example, the 25-MW(t) district-heating fluidized bed being commissioned in October [1977] by Enköpings Värmeverk in Sweden is warranted to meet the rigorous Swedish air-quality standards on all fuels, including high-sulfur coal, without a scrubber.

It is idle to argue that my soft path would not be completely soft during the transitional period; the transitional technologies would have many soft attributes (other then renewability), and the whole energy system would be steadily evolving in a soft direction. I nowhere assume "a return to negligent ways of the past," but the use of combustion technologies and energy systems designs far superior to those we now rely upon.

o

Forbes

Lovins' suggestion that coal can provide a ready bridge to a completely soft energy mix relies on his claims that energy consumption in 2000 can be held to about 25% below current projections (declining thereafter) *and* that soft technologies can provide 35% of the 2000 demand. With these two assumptions, coal production would require only a modest increase of about 65% by 2000 (and then be halted by

Lovins

Dr. Forbes omits oil and gas from the transitional fuels, and acknowledges my attention to transitional fuel needs (contrary to earlier implications that I ignored the need for a transition). He fails to state *whose* current projections energy consumption must be held below (apparently he means ERDA's). I have cited projections by Dr. Weinberg's group and by CONAES and the March 1977 GESMO-testimony pro-

Forbes

2025). But if just *one* of these two assumptions failed, there would have to be about a four-fold expansion of coal production by 2000, which would be difficult to accomplish and not very "soft." Stating that "We are developing supercritical gas extraction, flash hydrogenation, flash pyrolysis, panel-bed filters and similar ways to burn coal cleanly" is correct if the emphasis is on the word "developing." To suggest that these will be readily available for small-scale, "soft" use in no time at all is incorrect. The same can be said for Lovins' enthusiastic endorsement of fluidized-bed boilers. Their reliability with caking coal leaves much to be desired and their ability to remove sulfur pollutants needs improvement. Pending further development work it is difficult to project the future scale of fluidized-bed system use in larger commercial applications, let alone in individual home units.

Lovins

jections by Drs. von Hippel and Williams of Princeton University (89 q primary energy in 2000 with modest conservation—in some respects less conservation than the President's 20 April 1977 program—and 112 q with a no-new-initiatives "status quo"). I think these studies strongly suggest that ERDA's 120-125 q expectations for 2000 are about a third too high in the light of the President's current program of conservation.

I assume 1 q of coal use in 2025 (and 1 q of oil and gas)—nominal levels which could be covered by renewables if need be.

Dr. Forbes's supposition about a quadrupling of coal production by 2000 rests on the assumption that my conservation or soft-technology program would not only fail, but fail *completely*. This tacit assumption does not seem realistic. My use of "developing" with respect to flexibly scaled coal coversion technologies and panel-bed filters was deliberate; these technologies are now at pilot scale (including the new Stoke Orchard continuous-process demonstration of supercritical gas extraction). I did not claim that any of these technologies would be available "in no time at all," but believe they will be ready when needed if vigorously pursued now.

The state of the fluidized-bed art is much more advanced than Dr. Forbes implies; all kinds of coal have been burned successfully, and sulfur removal has ranged from over 80% in atmospheric beds to over 98% in pressurized beds. Fluidized beds are already being commercially applied in substantial sizes (*e.g.* the Enköping plant is being built turnkey, with strict performance warranties and a complete two-year guarantee, after stiff competition by five vendors), and domestic fluidized beds have done well in field trials. The references in *FA* and the Oak Ridge paper cover all these points.

Forbes

o

In an age of energy hype and polarization, it is difficult (in the U.S. at least) for the proponents of hard *and* soft, of nuclear and solar, of synthetics and conservation, to find an audience that will believe them to be sincere on both points. This is sad commentary on our times and a disruptive barrier to the development of sound policies. There is no reason why a belief in the necessity of maintaining current "hard" technologies and developing new ones cannot coexist with a belief in the importance of conservation and an excitement for the potential of certain softer technologies—solar heating and photovoltaic cells, wind turbines, waste conversion, energy storage and much more. Stating that they cannot conquer the energy world tomorrow or even in twenty years is not a denial of their value, only a recognition of reality.

Such a view might look to a future akin to that sketched in Figure 6 [omitted], with energy demand growth reduced substantially from historic trends by conservation, with coal and nuclear displacing oil and natural gas in the nearer term and with renewable energy sources building in substantially beyond 2000. (Conversion technology transitions are assumed within individual primary sources.) It is emphasized that the figure is illustrative because substantial policy actions and programs would be required and because there is some latitude to the path. It is uncertain beyond the next ten years or so and speculative beyond 2000, but it is feasible.

Lovins

o

I agree that it is hard to know whom to believe. It is for this reason that I have been at pains to document my analysis so thoroughly.

I have not stated that soft technologies can "conquer the energy world" (which I take to mean becoming the dominant energy source) "tomorrow or even in twenty years"; but neither do I agree with Dr. Forbes that renewables cannot exceed about 30q in 2025, about 5-8 q in 2000. That is not "reality" so much as self-fulfilling prophecy.

Forbes

o

There can be no doubt that there is a need for action and a need for that action to be based on sound goals and social policies. But energy strategy cannot be used as the sole instrument to manipulate a path for society. Rather, energy policies should be developed in consonance with social objectives that

Lovins

o

I agree with the first four sentences. Beyond that, Dr. Forbes appears to make the same assumptions about the hard path that he accuses me of making about the soft path, *e.g.* that the former is "clean, available and affordable." (If hard technologies are available, why does ERDA spend so much money de-

Forbes

have been openly agreed upon. Lovins' claim that a soft path society would be a "freer" one would seem to make this a desirable goal. Again, the assumption is that the soft technologies are clean, available and affordable; otherwise the strategy is of dubious value and probably coercive. But even if the assumption is accepted for the sake of argument, is it then a desirable goal? Possibly, but not axiomatically, for it flies in the face of centuries, if not millenia, of human transitions and the desire for less individual effort (or for much of the world, less hardship). It recalls familiar cries for "the good old days" by people heading willfully in the opposite direction. A reverence for the traditional and a desire for change must be reconciled in a need to keep the best from the past while seeking better futures. An *ideal* soft path might be a desired goal. Perhaps.

Lovins

veloping them?) Dr. Forbes's statement that a soft path implies greater "individual effort (or for much of the world, . . . hardship)" is his idea, not mine, if he is referring to labor—though a soft path will often require individual initiative to be exercised more freely, and I consider this a psychological advantage.

o

Forbes

But it gives pause for concern when we hear that "Soft technologies are ideally suited for rural villagers and urban poor alike" and the soft technologies "do not carry with them inappropriate cultural patterns or values" for poor countries. For this seems hardly less than the "noble savage" concept resurrected in modern dress. We can point out to others the mistakes we feel we have made, but we cannot dictate the path they should take, for the freedom of choice must be theirs.

Lovins

The last paragraph fails to note that soft technology is compatible with modern development concepts (ecodevelopment, New Economic Order, etc.) while hard technology is not. The "noble savage" concept, if by it Dr. Forbes means a desire to fob off second-rate technologies on developing countries lest they acquire the comfortable life he and I enjoy, is the opposite of my intention—though I concede an interest in cultural diversity, in urgently meeting basic human needs as a prerequisite for any further development (or even for survival), and in using scarce resources to advantage. See the recent publications of Dr. A. K. N. Reddy (Indian Institute of Science, 560012 Bangalore, India) for perhaps the best introduction to the arguments for capital-saving, inequity-reducing development technologies.

I agree developing countries must have the freedom to choose and that we should not force them to repeat our mistakes. Many are waiting for our

Forbes

The concept that centralization and industrialization dehumanize society is by no means a new one, but dates back to the protracted debate over getting on the "hard" industrial path in the first place. The writings of William Cobbett, William Morris and others attacking industrialization in 19th century England bear remarkable similarities to the philosophical renderings of Amory Lovins, Barry Commoner and other current writers in the same vein. More careful thought remains before we can conclude that we are fundamentally on the wrong track.

o

There are glimpses throughout Lovins' paper that his desire to halt nuclear power is not only intense but undoubtedly one of the seminal foci from which his statement grew. A strategy based on preoccupation with finding a rationale to prove nuclear power unnecessary, without thorough quantification of the criteria by which it is judged to be so obviously unacceptable, is likely to be flawed in its coherence and its adhesion to the practical. Energy strategies must work a different way, based on what is necessary, available and acceptable rather than on *a priori* assumptions about what is inadmissible. It would be fruitless to attempt detailed comment on such fundamental errors as "a nuclear temperature of millions," and diversionary to re-run the existing data on the safety and environmental impact of nuclear power relative to other energy sources. The issue of future energy paths is more fundamental than the nuclear debate, although it has lain quietly behind that debate for many years.

Lovins

soft-path example, or, too impatient to wait for it, are setting a better example themselves.

o

Dr. Forbes's allusion to William Morris and William Cobbett is misdirected: I raise questions about current patterns of industrial development, but my analysis in no way depends on these questions or their answers, and can readily be construed as a "pure technical fix" by those who consider present values and institutions to be wholly satisfactory.

o

My analysis is not an outgrowth of a desire to find "a rationale to prove nuclear power unnecessary"—though I think nuclear power bears grave environmental and social risks, and that it would do so even if it involved no toxic or explosive materials. I am rather motivated by a desire to find an energy policy that makes sense. Being non-nuclear is only one element of being sensible. Arguments about nuclear risks can get rather involved and sterile, so I consider it advantageous that if one can show nuclear power to be hopelessly capital-intensive, or unnecessary, or an encumbrance, then one need not argue about whether it is safe or not. Evidently, for different reasons, such arguments do not trouble Dr. Forbes, since his Figure 6 [not shown here] corresponds roughly to 500 GW(e) of nuclear capacity in 2000 and to 1000 GW(e) in 2025: a GW(e) commissioned every 19 days from now on.

The "nuclear temperature of millions" is not a "fundamental error" but a misreading on Dr. Forbes's part: see my response to Dr. Lapp. I have surveyed elsewhere the problems of nu-

Forbes

Lovins

clear power (A. B. Lovins and J. B. Price, *Non-Nuclear Futures: The Case for an Ethical Energy Strategy*, FOE/Ballinger, 1975, and various technical papers) and the even harder problems of assessing the costs, risks, and benefits of various energy sources ("Cost-Risk-Benefit Assessments in Energy Policy," 45, *George Washington Law Review*, 911–943, August 1977).

O

But some points are in order. Lovins trots out a weary list of risks in operation, human fallibility and malice, risk through greed for profit, violence and coercion, abrogation of civil liberties, guarding of long-lived wastes, establishment of an elite priesthood, long-term effects of nuclear mishaps and, in general and without evidence, a shopping list for a technology where everything is to be as bad as possible, in contrast to soft technology where everything is to be as good as possible.

O

I am sorry Dr. Forbes wearies of my list of nuclear problems—a very condensed list—but the items are important to many people. I did not feel that the *FA* article was the right place to canvass the arguments in detail, any more than Dr. Forbes cares to do so in his commentary. I therefore devoted a mere half-page to nuclear problems because they are important enough to mention, not because they are central to my thesis. What we should do instead is more interesting.

I do not assume "everything is to be as bad as possible" in nuclear technology, but rather that the same manifestations of human nature that have already marred many enterprises, including nuclear power, will persist. In short, I am being realistic, not utopian.

O

More than 40 countries have seen what Amory Lovins fails to see—that nuclear power is an essential part of their future development. To suggest that "in almost all countries the domestic political base to support nuclear power is not solid but shaky" is contrary to the fact of a growing world commitment to nuclear power that is outpacing that of the United States. It is ridiculous to quote Norway, which will not need nuclear power for some time, or Australia or New Zealand which have not needed it till now, as examples of countries that have "rejected" nuclear power. That it "may soon be stopped in

O

It is imprecise to talk of "40 countries" having seen "that nuclear power is an essential part of their future development" without mentioning that this sentiment is shared decreasingly by certain elites (and often not shared by substantial numbers of their colleagues and constituents) after an intensive worldwide promotional campaign. It is, I think, significant that no power reactor seems to have been sold to a developing country without very generous financing by the vendor country—sometimes so generous as to amount to paying the customer to haul the reactor away. Such subsidized export programs

Forbes

Sweden" is unlikely, given Mr. Fälldin's rapid about-face after winning the last election. International commitments to nuclear power will continue to grow out of necessity.

Lovins

are merely a convenient way to bail out domestic nuclear vendors.

My view of the shakiness of the domestic political base for nuclear power in many countries is amply supported by the trade press and several careful surveys; further, the "growing world commitment" is faltering, particularly in those countries where political dissent is not wholly suppressed or where some semblance of market mechanisms can still operate.

I am familiar at first-hand with the energy and fuel resources of Norway, Australia, and New Zealand, and think even Dr. Forbes would be hard put to make a persuasive case for ever having nuclear power in any of them. He conveniently ignores my references to Japan, the Netherlands, West Germany, France, Switzerland, Italy, Austria, Canada, Britain, and other countries less favorably endowed. The Swedish position, as I can confirm from a visit in late May 1977, remains fluid. [In 1978, the Swedish government fell for a second time owing to the nuclear controversy—and Austrians voted to prohibit the operation of a recently built nuclear station, their first, which cost more than half a billion dollars.] I believe international commitments to nuclear power may continue to grow for a short time—as long as vendor countries continue to foot the bill—but that political and fiscal realities are rapidly starting to assert themselves.

o

Forbes

International nuclear proliferation is a matter that requires serious efforts for solution. But it is naive to suggest that a unilateral cessation of domestic *commercial* nuclear power activities (neglecting research and military programs) will bring a halt to proliferation. If the U.S. were to halt its nuclear energy programs, others might falter temporarily but they would not halt. This is hardly to suggest that the export of nuclear technology should go uncon-

Lovins

This paragraph misstates my position and misses the point of a rather complex and subtle analysis. Please see *Soft Energy Paths,* Chapter 11, for full details.

Forbes

trolled, but it is well to note that no nation has yet produced a nuclear weapon through its commercial nuclear programs. As is often the case, the simple solutions sound attractive but they are unlikely to work. Control of international nuclear proliferation will mean hard work to develop comprehensive safeguards and international agreements, with the possibility of sanctions.

Lovins

o

It is not infrequently suggested that certain energy conservation measures could in the future save an amount of energy equal to that produced by nuclear power. That is indeed possible, but it is hardly a justification for halting nuclear power. Conservation will and should have its greatest effect in the reduction of consumption of oil and natural gas—as does nuclear power. At a time when we are importing about half of all the oil we consume, it is hard to see that we do not need both.

o

Nuclear power tends not to save oil and gas at the margin but to compete with coal, and cannot displace much oil without deeply penetrating the heat and transport markets—both requiring astronomical investments for purposes with no engineering or economic justification. [In the *Bulletin of the Atomic Scientists*, November 1978, Lovins cites calculations by Vince Taylor that if every oil-fired power station in the OECD (non-communist industrial) countries were instantly replaced by nuclear power, OECD oil consumption would decline by only 12%, and the fraction of that consumption that is imported would decline only from 65% to 60%, simply because most oil is used for heat and transport, not electric generation. The oil saving in countries whose electricity is made mainly from coal (West Germany, Britain) or hydroelectricity (France) would be especially small.]

o

Logic and reason are supposed to be a basis for decision-making, even if they do not always prevail in practice. Unfortunately, illogic and faulty reasoning are becoming institutionalized, particularly in the energy field. Thus a writer in a recent edition of the *Bulletin of the Atomic Scientists* says "the safer the system, the greater the risk" and a member of a State energy commission states at a technical meeting that he prefers his

o

The Harden quotation [the safer the system . . .] makes sense in context—at least as much sense as Edward Teller's remark that reactors are dangerous, which is why they are safe. Dr. Forbes's remarks on intuitive judgment suggest to me a faulty conception of the role of scientific expertise in the democratic political process.

Forbes

"intuitive judgment to hyped quantification." Other examples are legion. To be sure, the "public" do tend toward intuitive judgment by necessity, but surely elected or appointed decision-makers can manage better. And it is also true that hyped quantification abounds in private and government reports alike (e.g. Project Independence), but the unhyped kind is still available.

O

In addition to this, it would appear that decision-making is a vanishing art, for verification of which one has only to look at the status of U.S. energy policy, with oil imports up 15% in the past year, natural gas in short supply, and no significant advances on any domestic front—conservation, coal, nuclear, solar, etc. This seems to stem in large part from the emotionalism and high degree of polarization in the energy debate. It is almost axiomatic that avoiding controversy means avoiding decisions of any import. In their eagerness to avoid arousing displeasure, government officials increasingly trot out thick reports that deal in little beyond inarguable generalities or dubiously-derived numbers. One recent draft report (unreferenced only because it seems unfair to single one out of so many) states that: "The collection, treatment and intermediate storage of radioactive wastes shall be performed in a manner that provides reasonable assurance that the public health and safety will be protected." No one is arguing otherwise; nor does it lead anywhere. Decision-making will require defining what is meant by "reasonable" and *how* the task should be performed, but that is also what will arouse controversy (from pro or con).

It is this question of "how" that is so fundamental to decision-making in general, and energy in particular, yet it is absent from many current deliberations in all but the most simplistic

Lovins

O

I have already addressed the issue of implementation and of impacts, costs, and resilience: These matters are indeed the point of the *FA* paper and of *Soft Energy Paths.*

A preoccupation with the "whats," the possible futures however improbable, has led to faulty visions and even to arrogant visions; faulty because many of our decision-makers are so anxious to please a constituency or avoid controversy that they delay and deal in generalities; faulty because facts are adjusted, exaggerated or glossed over to push a personal vision or pander to someone else's; faulty because some imagine that changes can be made rapidly, flying in the face of any careful examination of the relatively slow process by which real, stable change occurs; arrogant because some would treat this as a crossroads in time, as if other peoples or other generations will not have different attitudes or choose their own innovations (not that we don't badly need decisions, but this is not The Chosen Year). And the avoidance of the detailed "hows" and the dearth of thoughtful mechanistic analyses of viable policy alternatives has left the public feeling that the choices are broad and the solutions not particularly difficult—if they believe there is a problem at all. Unfortunately, the public has not been made aware of the risks of inaction or the risks of taking the wrong action.

This paragraph describes with pecu-

Forbes

sense. It is common in current studies (Project Independence, National Plans, Amory Lovins' "Energy Strategy: The Road Not Taken?" etc., etc.) to state a goal or a *possible* future without a clear delineation of how to get there—perhaps in a strict technical sense, but not in terms of the policy requirements for implementation. That is, they give a possible future scenario, but do not analyze its political and social viability, let alone its impacts, costs and durability. In fact, the concept of scenarios has been abused and distorted beyond recognition from Herman Kahn's concept of a "surprise-free future" starting-point. A draft of a current major study contains a scenario showing less energy use in 2000 than today, which might be possible; but (regrettably in vogue) it in essence absolves itself of the need to examine the decisions required for implementation—rendering it almost worthless for practical application. If we could focus more in our strategizing on the "hows," the "whats" of our energy future would be apparent as being much more tightly circumscribed than many would have us believe.

Lovins

liar fitness, I think, the hard-path, nuclear-oriented hard path that I criticize.

o

Forbes

We need energy strategies that are viable—politically viable at least, and preferably publicly, economically and socially viable. It is also desirable that strategies have the potential for durability. Since energy involves long lead times, comparable to the life of a full political generation, it is costly if not dangerous to develop policies that are likely to be reversed by a relatively small shift in political or public attitudes. Examples of what would appear to be durable trends are those for cleaner air and water (even if specific policies require modification) and the scrutiny of energy expenditures. Examples of the reversals are the switch in electric power plants from coal to oil in the sixties for cleaner air and back to

Lovins

I agree, in general. I do not, however, agree that there are such things as value-free "facts" (in any sense relevant to policy).

coal in the seventies for resource reasons (as opposed to working toward cleaner coal burning in the first place), or the shift to compact cars following the oil crisis and the subsequent switch back to bigger cars in the absence of any policy efforts to hold the trend.

Insistence that we go beyond the "whats" to the hard "hows" and to the development of viable and durable policies could produce several desirable results: it could well lead to a better analysis and understanding of demand (and thereby of energy/economy links, conservation and demand growth); it could lead to better understanding of the choices and the compromises involved, whatever future we head for; it could even lead to responsible action.

There are contradictions in public attitudes on energy, not only between different demographic groups but within individuals, in terms of desires for energy to be clean, plentiful and cheap, in terms of a belief in conservation despite the small impact of voluntary efforts to date, or in terms of a belief in solar energy and an attraction to low capital costs. Such contradictory desires are part of human nature and will not disappear. But more honest, detailed information can permit choices to be made.

We need action, and decisions will have to be made in the midst of controversy. That is *not* the same thing as saying counter to the public's wishes; on the contrary, decisions must reflect the public will or be modified. But the facts behind the decisions should not be manipulated to reflect the preconceptions of any interest group or even the public at large.

Most of all we must recognize that there are no overnight answers, no easy ways out. Rather, we must tackle the problems of today while looking to the future with realism and an open mind.

Forbes

o

Until I was 16, I lived in Northern Ireland in a fairly low energy society. The house I lived in was heated by coal fires, had no refrigerator, no clothes washer or dryer. We did own a small car, although I travelled mostly by bus. In short, it was an energy-frugal life. Even though I believe I am still very careful about my energy expenditures—a re-insulated house, no air conditioning, no clothes dryer and a compact car—I use much more energy than in my childhood. I do not regret this. I believe in conservation, but I have no desire to return to hauling coal in buckets or rising in a 40-degree bedroom. In passing, I would suggest that Northern Ireland is living proof that a soft-energy society is not necessarily a "freer" one.

Lovins

o

The society described uses primary energy inefficiently (*e.g.* in coal fires), has curtailed functions, and has nothing whatever to do with soft technology. Dr. Forbes's identification of modern Ulster with "a soft-energy society" strongly suggests that he does not understand the thesis he is criticizing.

Forbes

o

When I was 16, I moved to Newfoundland where I went to college. Living in the capital city of St. John's my existence was very comfortable. But in the fishing villages of that Province I had another opportunity to see a low-energy society first-hand. For many of these villages have everything that a soft future might need: one of the world's richest fishing grounds just beyond the doorstep, forests stretching hundreds of miles into the interior, water being tapped in small (soft?) hydro plants as well as large ones, and plenty of wind. While Newfoundlanders are amongst the friendliest people I have ever met, their life in these "outports" has not been an easy one. And as they grow and build their future, it is toward centralization and greater convenience—with the familiar trappings of cars, central heating, greater electric generation, etc. The old life is disappearing, and it is regrettable I suppose for tourists and others that the quaint and traditional is giving way to the familiar and modern. But I do not think the residents regret most of the

Lovins

o

The description of "a low-energy society" in Newfoundland misses the mark: how would life be in Newfoundland if the renewable energy flows there were tapped in an intelligent way and used efficiently? Such an experiment is now being tried in nearby Prince Edward Island, and I think the results so far are very encouraging. They have nothing to do with condemning people "to live with past hardships," but rather with living better than ever with the resources they have. In this spirit I agree with the last sentence.

Forbes

changes, and surely no one who believes in equity and freedom of choice would condemn them to live with past hardships for the sake of maintaining a "simpler life."

I do not believe in the squandering of resources, but neither can I accept that we cannot build toward a society that is both freer from hardship and more civilized, in the true broad sense of that word.

Lovins

8

WOLFE vs. LOVINS

GE Executive Dubs Soft Path a "Shangri-la"

In its July 1977 issue, *Foreign Affairs* published an exchange of letters between Bertram Wolfe, Vice President and General Manager of General Electric's Nuclear Energy Programs Division, and Amory Lovins. Their exchange is reprinted here.

To the Editor:

Amory Lovins describes his energy Shangri-la in the October 1976 *Foreign Affairs*. But, as with the original Shangri-la, the author never gives the directions necessary to reach his better world. The difficulty in solving our energy problem is not in developing the broad idea that we would all like to live better with less. The difficulty is in finding solutions which can be implemented so as to achieve the objective. Here, Mr. Lovins leaves us in the dark. He does not propose a course of action; he proposes a course of lack of action. He argues that we should stop doing what we know how to do in order to do something better that we don't yet know how to do. He gives no specific data for implementation of any of the soft technologies proposed, only the assertion that they will be better. Admiral Rickover once described the difference between real nuclear reactors and paper reactors. Paper reactors are simple, cheap, and use off-the-shelf technology. Real reactors, on the other hand, are complex, over budget, behind schedule, and are slogging through unforeseen engineering obstacles. Similarly, soft technologies are attractive, cheap and clean; while hard technologies are hard.

Lovins: I share Dr. Wolfe's concern with problems that will, as my article stated, "pose a supreme challenge to the adaptability of democratic institutions and to the vitality of our spiritual life." I therefore outlined an approach—Dr. Wolfe calls it a "philosophy"—that can make those troublesome choices easier, less disruptive, and more equitable. I am sorry Dr. Wolfe found the outline unspecific. Within the constraints of space, audience, and intention, the text and citations were densely packed with practical proposals. These were naturally framed at a level appropriate to policy discussion rather than to the detailed engineering design that must come later. But I proposed vigorous action on a broad front, not "a course of lack of action" as Dr. Wolfe claims. I merely do not share his apparent belief that efforts "which might alleviate the energy problem" consist only of ways of "developing" (using up) scarce resources faster.

Wolfe: It is difficult to analyze Mr. Lovins' article in detail because of its own lack of detail, but let me discuss a few of his points. Mr. Lovins points out that the capital costs of power generating plants are high. He states that the 1976 to 1985 energy program proposed by President Ford will cost over a trillion dollars, of which 70 to 80 percent would be for new replacement power plants. He then makes a reference to a study by the American Institute of Architects which concludes that by 1990 improved design of new buildings and modification of old ones could save a third of our present total energy use. But he forgets to mention that the architects estimate that the added costs of these better buildings would amount to as much as 1.4 trillion dollars. It may be worth it. The architects do an economic analysis and argue that overall it will save money as well as fuel. But, without further detailed information, field tests and actual construction, should we not be suspicious that (as with the Alaskan pipeline, nuclear plants and the metro subway system in Washington) the cost estimates may be optimistic? It might also be noted that the architects' analyses indicate that even with the 12 million barrels of oil per day equivalent which their trillion dollar investment will save, the nation will by 1990 still require 50 percent more energy than we are using today and that oil importation will rise.

Lovins: The American Institute of Architects does compute for a "nation of energy-efficient buildings" a marginal capital cost of $0.73–1.46 trillions, which Dr. Wolfe compares with my estimate from Bechtel data of over $1 trillion for increasing U.S. energy supply. But the former range of figures is for 1973–87 and is in current dollars (assuming inflation at 5.6 percent per year for residential and 7.8 percent for other buildings), while the latter figure is for 1976–85 (five years fewer) and is in constant 1976 dollars. These differences inflate the former figure roughly threefold relative to the latter. In fact, the AIA program saves fuel, its cost (several trillion discounted constant dollars through 1990 at constant 1977 fuel prices), and capital. In cash flow terms it does even better. Conservation repays investment two to four times as fast as supply, even at the pre-embargo prices assumed by the AIA, so the capital invested in conservation can be used at least twice over in a self-supporting revolving fund. The AIA computes that such a fund would need only $86–568 billion (current dollars) in working capital—vastly less than for equivalent supply. Nor are the AIA cost data speculative as Dr. Wolfe implies. Unlike analogous projections for the truly speculative hard technologies, they rest on much empirical knowledge and understanding. Finally, the AIA projection of increasing U.S. energy needs is borrowed from others as a basis for discussion—e.g., for showing that the AIA program could provide more energy more cheaply than nuclear power could—and is not an independent AIA conclusion.

Wolfe: Consider the $40-a-kilowatt diesel proposed by Mr. Lovins. His cost is derived from the cost of an automobile. But automobile engines, when well maintained, operate for a hundred or a couple of hundred thousand miles. At 50 miles an hour, this is less than a year of continuous operation. Central station power plants are designed for lifetimes of 30 or more years and some have operated for much longer than this. How do the overall costs compare when one takes into account, not the initial capital costs that Mr. Lovins used, but the costs including costs of replacement and repair? Or if one builds the diesel to last ten years, what would its cost be then? Is the 35 percent efficiency projected by Mr. Lovins in fact realistic? The efficiency of internal combustion engines changes greatly with the load. What would its actual efficiency be when one averages its minimum load at night with its few hours maximum load during the day? And, what is the cost of the pedestal to support the machine, the anti-smog devices and stacks to accommodate the exhaust—and what about the noise?

Of course the hooker in the whole discussion is the fuel. If we took Mr. Lovins' suggestion seriously and converted all of our central station electrical plants to home diesels and generated all of our present electrical usage from these diesels with 35 percent efficiency, we would have to import over 9 million barrels/day to fuel them. (Unfortunately, during the *last* reform movement we converted our electrical generating plants from coal to oil and now import 2 million barrels of oil/day to fuel them.) So, plainly this can't be a serious proposal by Mr. Lovins and is meant only to make the point that small is good and there are ways to make small things economic.

Lovins: Moving now to diesel generators: my response to Hans Bethe in the April issue of *Foreign Affairs* makes clear—as Dr. Wolfe eventually realizes—that I mentioned diesels as an analogy, not a proposal. Nevertheless, modern diesel total-energy systems have for the past few years proved cheaper than central generation. According to vendors such as the Cummins Engine Company (Columbus, Indiana), a three- to six-engine (say, 350 to 3,000 kilowatt electric) installation, including at least one engine in reserve plus all buildings, antivibration mounts, heat recovery devices, and silencers, currently costs about $250 to $350 per installed kilowatt electric. Lifetime is typically 12 to 20 years, with 20–25,000 hours between major overhauls. Forced periods of unavailability can reportedly be measured in minutes per year or less. The thermal efficiency of the diesel generator alone surpasses that of most central power stations now operating (indeed, some experimental high-technology diesels in Europe have exceeded 45 percent efficiency), and the efficiency of diesel total-energy systems is about 75 percent. Gross costs are typically about 2.8 cents/kW-h(e) for fuel, 0.7 cents for operation and maintenance, 0.5 cents for capital amortization, and minus 1.4–3 cents as credit for recovered waste heat, depending on whether it replaces a cheap fuel (residual oil) or an expensive one (liquid gas). Compared with utility generation, modern diesel total-energy systems can save 50 percent of the fuel and yield 35 percent or more annual after-tax return on investment. Thus if I *were* proposing such devices, it would not be an idle proposal.

Wolfe: But is small always good? We have after all a perfect example of Mr. Lovins' approach. We have our mass-produced individual transportation systems. They're known as automobiles. But every environmentalist worth his salt knows that these are evil. What we really should have is that big, impersonal, inconvenient but efficient mass transit system. Big is now soft and small is hard; or is it that soft is hard? I get a little confused. However, I have found a way to keep things straight: there is one common definition that we can hang our hat on in Mr. Lovins' article and that is: if it's in use now and we know how to do it it's hard, and if we don't yet know how to do it it's soft.

Lovins: Perhaps Dr. Wolfe correctly describes the state of knowledge at General Electric when he says that hard technologies are what we know how to do, soft technologies what we don't yet know how to do. But this is not true of the many cited technologists who have been at pains to devise and demonstrate conservation methods and soft and transitional technologies. Dr. Wolfe is unfortunately not familiar with what they have done (much of which I cite). Were it not for this unfamiliarity one might suspect that he has got his own version of the soft/hard distinction exactly backwards: after all, the $64-million hard-technology write-off at Morris, Illinois, for which he is responsible, has amply demonstrated that there are some hard technologies that *nobody* knows how to do. [Having built a nuclear fuel reprocessing plant at Morris, General Electric was unable to make the plant work.] [The remainder of Lovins's response to this point occurs below.]

Wolfe: Biomass conversion is soft and good, and perhaps this can fuel our diesel. Melvin Calvin has estimated that we can obtain gross yields of alcohol of 140 million Btu/acre-year using sugarcane (sugar plus stalks). Under intensive cultivation our corn fields produce 32 million Btu/acre of food, but 16 million Btu/acre of energy is used to get this yield. Sugarcane doesn't grow very well in much of the United States, and sugar beets, whose productivity is a third that of sugarcane, may have to be grown in colder climates. What will be the average net yield in practice? Could it be as much as 80 million Btu/acre-year? If so, each barrel of oil per day equivalent requires 26 acres or an investment of $13,000 in land alone. According to Mr. Lovins this is in the range of investments for North Sea and Arctic oil (including their transportation systems). To produce the amount of liquid fuel now consumed by our transportation systems alone (9 million barrels oil/day) would require 240 million acres of cropland. We currently use about 400 million acres for food. Investment would be over $1 trillion. Where do we get the additional agricultural land? What are the environmental impacts of this additional farming? A major problem of expansion of our coal supplies is transportation. Wood has about half the density of coal and half the heat per unit weight. Other biomass fuels would be less dense. It takes a 100-car freight train each day to fuel a 1,000-MW coal-fired plant and a 40-car freight train each day to take away the ashes. Will it take four 100-car freight trains each day to bring the biomass fuel to a fluidized bed gasifier?

Lovins: My biomass estimates assume the use of agricultural and forestry residues that we already harvest, not intensive monocultural plantations that compete for land, water, and fertilizer with existing food crops. Displacing crops and using the methods of chemical agribusiness just to grow feedstocks for bioconversion would be unnecessary and silly. Trying to imitate today's elephantine power stations by building centralized bioconversion plants with input equivalent to 3000 thermal megawatts, as Dr. Wolfe suggests, would be even sillier, since a local plant using local crop residues or woodlots and giving local jobs would work just as well. It would be the height of absurdity to burn the hard-won bioconverted liquid fuels in stationary electrical generators, as Dr. Wolfe further implies, rather than in the vehicles that uniquely require them. I assume, too, that U.S. transport in 2025 can run not at today's U.S. efficiencies but at the best current European efficiencies, which are about three times as good.

Wolfe: And, what about those fluidized bed combustion chambers in Lovins' scenario? Congress should investigate why ERDA (Energy Research and Development Administration) is spending so much development money in this area when the technology is already at hand.

Lovins: Spurred by favorable studies at Oak Ridge, the House of Representatives Science and Technology Committee, and elsewhere, ERDA is starting to take fluidized beds more seriously. Two major ERDA design studies were commissioned in early 1977. But other technologists are not delaying their own commercialization. The Tennessee Valley Authority has announced plans to build a 200-megawatt fluidized bed with or without ERDA's help. After a joint study, Stal-Laval, Babcock & Wilcox

(U.K.), and American Electric Power Company are to decide in mid-1977 whether to build a 64-megawatt (electric) fluidized-bed coal-fired gas turbine similar to that described in my article. [They decided to proceed to the next phase—detailed engineering design—and the program is said to be going well.] The 25-megawatt fluidized-bed boiler of Enköpings Värmeverk in Sweden is to be commissioned in late 1977. Several companies and the President's Council on Environmental Quality have domestic-scale fluidized-bed projects. Part of ERDA's delay is simply explained: in September 1976 their international coal experts had never heard of Fluidfire or Stal-Laval. I fear that ERDA will continue to reinvent the wheel if it does not speed and simplify its procedures to European levels. So far it has been left behind—and hasn't even known it.

Wolfe: Windmills are soft, but the problems are the same. We can get energy from windmills but it is questionable whether we can get enough energy to significantly impact on our energy problems at acceptable costs, reliability and environmental effects. Mr. Lovins gives us no details, but Goldsmith and his colleagues at the Energy Research Group, using data from Heronemus, project that a single thousand-megawatt power plant, using windmills, would occupy an area of 1,800 square miles with a windmill on each square mile sector. Each windmill would consist of a structure 75 stories high supporting 20 wind turbines each with 50-foot diameter rotor blades. Many people complain about the visual pollution of a standard electrical transmission line, but I presume that with some re-education we can all learn to admire the esthetics of the wind towers when they are located in our neighborhoods.

Lovins: Dr. Wolfe's doubts about wind machines are not, so far as I know, shared by anyone familiar with the technology. Characteristically, ERDA/NASA's high-technology approach is incurring capital costs seven to 70 times those demonstrated in similar or more advanced projects, mainly in Canada and Europe. My analysis assumes, here as elsewhere, the best technologies already working in the field, not the best used by ERDA. There is often a vast difference.

Mr. Goldsmith's bizarre numbers have nothing to do with soft technology, but rather with irrationally using wind machines to imitate reactors. I would not propose a 75-story structure to support 20 small wind machines when a tower less than a tenth that height can support a single machine with the same 0.55-megawatt total average capacity. (Even that capacity per square mile is about seven times the average U.S. electrical load density and hints at overcentralization.) Yet, interestingly for those concerned about Goldsmith's 1,800 wind machines, a new thousand-megawatt power station in the United States requires approximately 1,800 towers to hold up its 140 miles of high-voltage transmission lines, and, with its distribution and fuel systems, would sterilize much more land, more lastingly, than would any equivalent population of wind machines in the most useful sizes (tens of kilowatts to a few megawatts).

Wolfe: There is no doubt, as Mr. Lovins argues, that increasing the efficiency of energy usage could have large payoffs. The National Academy of Engineering estimated in 1974 that by diligent efforts we could improve the efficiency of energy utilization by as much as 20 percent by 1985. Mr. Lovins' interpretation of the American Physical Society study that we could improve efficiency of energy utilization by "a factor of at least 3 or 4" is a caricature of the results of the APS study and has been disclaimed personally to me by one member who wrote the report. However, Mr. Lovins' statement serves as an example of the problem with his thesis. The APS study was, in fact, an examination of the areas where research in physics might pay dividends in reducing our energy needs. A number of areas were identified and theoretical analyses were presented which indicated where large savings could be made. To take this report and argue that it "suggests" that we could improve energy efficiency by a factor of 3 or 4 is like suggesting that we will in the future travel between New York and Los Angeles in less than a second because the laws of physics say that we can approach the velocity of light as a theoretical limit. Similarly, to argue that we can *painlessly* reduce our energy requirements to the per capita levels of Sweden, West Germany, or Switzerland is deceptive and in my view incorrect. It is not plain, for example, that the market for secret bank accounts or watches would permit us to follow the Swiss pattern with our much larger labor force.

Mr. Lovins and I do have some areas of agreement. Both of us see the need for a major conservation effort and both of us expect an increase in energy need and use during the rest of this century, even with drastic conservation. However, one gains the impression from Mr. Lovins that the conservation program is going to be socially enriching; whereas I fear that an effective conservation program will not only impact on our life styles and standard of living, but also on our basic democratic institutions. Again, the trouble arises not from the broad generality that it would be nice to cut back on energy, or that it is necessary, but from the specific implementation of plans which will make people live differently than they have in the past and affect their livelihoods and aspirations in different ways.

Lovins: Dr. Wolfe claims that I "expect" increased U.S. energy needs "during the rest of this century, even with drastic conservation." I do not. Figure 3 in my article is a hypothetical illustration, not a forecast. My forecast would be that the United States will be hard pressed to produce as much primary energy in 2000 as it uses now (75 quads), and is unlikely *ever* to use—or need—more than about 85–90 quads. In retrospect, the top curve in Figure 3 now seems unrealistically high. It corresponds to end-use energy in 2000 far above Oak Ridge physicist Alvin Weinberg's preferred estimates, which assume much *less* conservation than I do. It shows primary energy in 2000 equal to the 89-quad "moderate conservation" scenario of Professors von Hippel and Williams (given in their March 1977 GESMO testimony to the Nuclear Regulatory Commission); even their no-new-initiatives "status quo" projection is only 112 quads. And the primary energy that Figure 3 shows for 2010—86 quads—is far above the eminently reasonable 70-quad [later changed to 63- and 77- quad] scenario of the National Research Council's 1977 CONAES (Committee on Nuclear and Alternative Energy Systems) study, let alone the 40-quad [later changed to 53-quad] scenario advanced by the CONAES "Lifestyles Panel." The trend of recent authoritative projections for the United States, as for other countries, is steeply downwards as we realize more clearly the extent of our waste. Does Dr. Wolfe disagree? Can he show in quantitative detail why he disagrees?

The American Physical Society study cited in note 11 of my article shows that the U.S. economy takes on average about 20 times as much energy to do its tasks as the minimum energy theoretically required to do those tasks infinitely slowly under ideal conditions. That is, the average Second Law efficiency of the U.S. economy is about 1/20 = 0.05, whereas Professors Ross and Williams—the former a director of the APS study—have stated that "study of a variety of devices and processes suggests that a goal of (about 0.25 to 0.5) . . . is reasonable for ultimate practical systems." This view, readily supportable by the sorts of case-by-case analyses that I cite, is consistent with my interpretation that U.S. efficiency can be improved by at least a factor of three to four over the next 50 years or so. I doubt that any serious student of the subject will disagree: the closer people look at energy use, the more ways of increasing efficiency they discover. These ways are not painless, but they are far less painful than not adopting them. Those that I assume do not involve the curtailment that Dr. Wolfe fears: insulating one's roof does not mean freezing in the dark.

Wolfe: Mr. Lovins and I agree that large solar electrical generating plants have problems, Mr. Lovins because of philosophical reasons and I because of technical considerations. We also agree that solar heating holds promise. I have in the past discussed solar heating with General Electric people working on it and with Mr. Sheldon Butt, the head of the Solar Energy Industries Association, the association of organizations trying to make solar heating a commercial enterprise. My GE colleagues indicate that economical solar heating is probably 5–10 years off, while Mr. Butt believes that it is very close, if not already in hand if the tax treatment of solar heating were properly handled. Butt indicates that solar heating is probably impractical as a retrofit on old buildings but is practical for new buildings specifically designed for solar heating. He projects that near the end of the century we may have a few percent to as much as eight or nine percent of our energy supplied by solar heating and is disturbed by people who won't recognize that this is a major contribution. But, of course, it will by itself not solve or change the character of our energy problem, and Butt vigorously supports the development of all available energy sources.

And, let me point out that contrary to what is implied by Mr. Lovins, the capital cost is again very large. Solar power isn't free and the reason there is a Solar Energy Industries Association is that members of the free enterprise system would like to participate in a very large potential market. The widespread use of solar heating to supply eight percent of our needs at the end of the century will involve investments of hundreds of billions of dollars. Unfortunately, the sun doesn't shine all the time and even with large heat storage systems which will increase the capital cost, one has to account for the rare occurrences where there are a week or two weeks of cloudy days, or when extended snows cover the solar collectors just when the weather is at its worst. Thus, in addition to the capital cost of the solar units, which will indeed save fuel, we require the capital cost of the backup systems for those dismal periods last winter which we Californians learned about from the TV news.

Lovins: As my cited Oak Ridge paper makes clear, and as Dr. Bethe now accepts after detailed correspondence following our April exchange, the solar space-heating systems that I assume include full seasonal heat storage and require no backup. If one combines the ERDA/MITRE November 1976 analysis (M76-79) of solar economics with the solar collector prices analyzed by the Office of Technology Assessment, it appears that solar heat now competes handily with electricity, oil, or gas practically anywhere in the United States. But that is a misleading type of comparison: it is a far more stringent test than, say, nuclear power could meet. We are now obliged to decide whether to invest in various long-term alternatives to oil and gas, including nuclear power, coal-gas plants, insulation, and solar collectors. To avoid misallocation of resources, we should be comparing all these alternatives *with each other*, not comparing some of them with the cheap but dwindling fuels that they are all meant to replace. On this basis, the completely solar systems I describe, retrofitted into most existing buildings, appear to have lower capital costs than their hard-technology competitors to heat the same buildings. (New buildings should use passive solar systems at zero or negative marginal cost.) As I made clear, retrofitted solar heating is indeed expensive; but it is cheaper, both initially and over its lifetime, than not having it. Likewise, solar process heat—some of which can use the techniques now being misdirected into large-scale solar-thermal electric plants—is often expensive, but is cheaper than the hard-technology alternatives to it.

Wolfe: The soft scenario game is easy to play: my own soft technology is the electric car. Every physicist knows that energy of motion can be converted back into stored energy. A large amount of energy is lost in braking a car and bringing it to rest. But an electric car can run its motor backwards as a generator and save the energy of braking by recharging the battery.

Consider the electric car with a breeder reactor power plant supplying energy. Breeder reactors are already operating in France, Great Britain, and Russia, and the French are ready to export them at any time. The breeder reactor can attain thermal efficiencies of over 40 percent and the electric auto can attain 90 percent efficiency. Therefore, a breeder reactor supplying electricity to charge auto batteries used to drive electric cars should have an overall efficiency of 35 percent or more (the same target Lovins used for his diesel engine). Furthermore, since we recover the energy of slowing down a car, the overall efficiency for driving is much better than that of the diesel auto. Indeed, if a good engineer could work on the problem of friction, the laws of physics indicate that we could approach the point where transportation could take place with negligible net energy requirements. Since the battery could be charged at night this would help to even out electrical loads, reducing the overall capital cost of energy capacity; and, since breeder reactors don't pollute the air, we will at the same time help clean up the environment. It seems that we may be at a crossroad and the Congress should consider legislation outlawing the internal combustion engine. Plainly, continued effort in this direction takes away from the early deployment of the electric car.

I hope that no one will be so coarse as to ask me what kind of a battery I intend to use. There is considerable research on this matter and great progress is being made at General Electric, the National Labs and elsewhere. Besides, I consider this a soft technology. If it is soft, can battery development be hard?

Lovins: Except perhaps for special uses and sites, I do not find electric-battery cars attractive, and feel they are receiving too much emphasis. Suitable advanced batteries, which do not exist and may never exist, will be costly and heavy, producing structural problems in a car crash; they use nasty materials such as molten lithium or sodium; and the electrical systems they entail are prohibitively capital-intensive. Thus it is not very important that their whole-system efficiency is similar to that of an ordinary car and its fuel cycle. They are of no interest save in a nuclear-electrified world.

Obviously I do not agree that breeder reactors—the archetypical hard technology—will "help clean up the environment," any more than I agree that cyanide is more healthful to eat than candy because it doesn't promote tooth decay.

Dr. Wolfe's claim that breeder reactors "are already operating" abroad is likewise written with an economy of truth: those small prototypes have doubling times of the order of 30–60 years; i.e., it takes this long for them to make available added fuel equal to their original fuel input. For all practical purposes they don't breed. The best quasi-breeder, the French Phénix, had cumulatively produced only 51 percent of its full-time, full-power rating through 1975, before its forced shutdown of many months, and the British and Russian machines have done even worse. All are astronomically costly. I see no prospect whatever of a breeder economy in the United States or anywhere else.

Happily, with efficient use of biomass fuel, I also see no need for significant numbers of battery-electric cars or of the baroque hard technologies that General Electric wishes to sell in order to run them. (This does not exclude, however, the possibility of using biomass fuels in portable fuel cells—at even higher efficiency than Dr. Wolfe's breeders—to run electric cars, which could store their braking energy in practicably small batteries.)

More broadly, Dr. Wolfe is wrong to portray ordinary cars as a soft technology. He has not read my definition of "soft technology" as carefully as I wrote it, and so confuses small scale or decentralization (which is what he thinks I said) with *scale matched to end-use needs* (which is what I actually said). Transport needs, like energy uses, have a spectrum of scale. We have dense traffic in and between cities, sparse traffic in the countryside. Trains, trams, subways, and buses—that is, mainly mass transit—are the right tool for the former job; mainly private transport (cars, minibuses, motorcycles, bicycles, walking) for the latter; a more even mixture in the suburbs in between. It does not make sense to have private cars swarming over the inner cities nor subways over the countryside, but rather to match the mode to the need. Likewise with energy systems, we should seek to run smelters with existing large hydroelectric dams and to heat houses with small solar systems, not the other way around.

Now, it happens that both our transport and our energy systems are currently mismatched to end uses: we need to move toward more mass transit *and* toward smaller energy systems. But these needed shifts of average scale are the *consequence* of a sensible principle of matching, not the reverse. Thus, it is internally consistent for conservationists, believing in appropriate tools, to seek simultaneously both smaller energy systems and more public transport, just as their opponents, believing in profitable waste and inappropriateness, seek simultaneously both larger energy systems and the universal primacy of the private car.

Wolfe: The reader should understand that the electric car concept (when approached more seriously) actually holds great promise as do many of the concepts advocated by Mr. Lovins. But, there is a long road to travel between the always attractive conceptual idea and its practical implementation. We are already traveling down this road, trying to develop the concepts that Mr. Lovins proposes, but according to Mr. Lovins this is not satisfactory. He asserts we cannot afford to wait to find out if the development will be successful, for unless we abandon the old ways first we won't diligently pursue the new ways. Boiled down it seems to me that this is the key message of Mr. Lovins' article. The defect of the approach is clear.

Although he labels it soft technology, it is not a technology, but a philosophy with significant ramifications which Mr. Lovins espouses. I have in the past been disturbed and baffled by those who seem to oppose *every* effort which might alleviate the energy problem—be it the development of Alaskan oil, western coal, hydroelectric dams, offshore oil, geothermal power, nuclear power, shale oil, or synthetic fuels. Mr. Lovins has performed a major service by explaining an underlying rationale for this oppostition in his articulate and readable article. I now feel better educated although I'm not sure I feel better.

There is an oft repeated complaint that in its early days nuclear power was oversold by predictions of electricity so cheap that meters would not be needed. I have heard this statement so often that I accept it as true although I have been unable to find it in the technical literature. However, let's accept the argument that nuclear power, or at least its ease of development, was oversold—and ask whether there is a lesson to be learned from it.

Lovins: Finally, Dr. Wolfe rightly says that I think we should give up many present projects, including those that now mainly engage his firm, in order to be able to devote our full resources to implementing conservation methods and soft and transitional technologies whose feasibility is already proved. But while he is at a loss as to how to do these unfamiliarly simple things, others are not. As nuclear advocate Linn Draper remarked in our Senate debate on December 9, 1976, what is needed for soft technologies "is not additional research, but incentives"—I would prefer to say lack of disincentives—"to deploy techniques." If the big hard-technology programs are superfluous, and if, as Dr. Draper agreed, the soft technologies are not only known to work but are cheaper and better for jobs than the hard technologies, then I do not see what we are waiting for. If we are to wait for Dr. Wolfe to find out that there are many able soft technologists who know their business better than General Electric does, we shall wait for far too long.

General Electric and its kin can and will be recycled. Any company that keeps and adapts its skills can remain profitable in a changing world. But it is genuinely difficult for companies that have sought their challenges in complexity to see the challenges in simplicity. For those that can do so, soft energy paths offer not a threat but a remarkable opportunity.

* * *

[Dr. Wolfe initially declined to permit us to reprint his *Foreign Affairs* letter, but later relented provided we also print the following letter after the Wolfe-Lovins exchange.]

Dear Mr. Brower:

As I noted in my letter to you of October 3, 1978, I have a concern with respect to publication of my critique of Lovins' "Soft Energy Path" thesis and his response. The format of your proposed publication presents my points and then interposes Lovins' rejoinders. The reader may gain the impression that Lovins' rejoinders substantively respond to my points. I do not believe that they do.

The fact that Lovins restates his thesis that solar power is now economic, or that fluidized bed combusters are "state of the art," or that biomass conversion could provide significant large scale economic energy, or that diesels make sense as a power source, does not in any way prove his points or close the argument. Indeed, your readers can make their own evaluation of one part of Lovins' thesis by calling their local solar contractor and getting an estimate for solar heating of their own homes.

I am somewhat wounded by Lovins' charge that I use "an economy of truth" in describing the breeder reactor. Evidently Amory and I not only disagree on technical and philosophical points, but our humors also fail to mesh. If the reader will reread my breeder comments he will find that they are placed within the context of "the soft energy game" and will find that my description of the breeder parallels closely Lovins' original *Foreign Affairs* description of the off-the-shelf Stal-Laval fluidized bed gas turbine power plant. I believe that truth is economized in both descriptions, although I at least categorized my scenario as a "game."

Mr. Lovins' response to my critique mentions TVA's plan to build a fluidized bed power plant and American Electric Power's plan for a fluidized bed gas turbine plant. I contacted both of these organizations. Harold Falconberry who heads the TVA project reported that the project is being considered for the same reasons that the Clinch River Breeder project was undertaken; namely, to gain experience on a technology that is not yet mature. He said "I would be irresponsible if I implied that anyone should try to buy even a small fluidized bed power plant on a commercial basis." Mr. John Tillinghast, Vice Chairman of American Electric Power, reported that their proposed Stal-Laval gas turbine fluidized bed venture was also a developmental non-economic endeavor and indicated a number of problems uncovered in early test work. His conjecture was that if the AEP development project went ahead and was successful, pressurized fluid bed power plants might reach a commercial state in the late 80's but he said that he was not able to even make a conjecture at this time about costs.

Thus, as indicated above, although Lovins' responses are articulate, I believe they are misleading and I am concerned that the non-technical reader may not discern that the issues remain open.

Because of these concerns, I indicated in my October 3rd letter to you that I was not able to give permission for the use of my material in your book. However, we both agreed that the issues Lovins raised should be a subject of public discussion and debate.

Thus, let me suggest that a means to accomplish our mutual objectives and yet accommodate my concerns would be to publish the exchange as you propose followed by this letter.

[Signed: Bertram Wolfe]

Does Dr. Wolfe seriously mean that his favorable allusion to breeder reactors was simply good, clean fun in a "soft energy game" context? If so, we suspect that his colleagues among the higher-ups in GE's nuclear hierarchy will be distressed to hear it.

Dr. Wolfe refers to "Lovins' original *Foreign Affairs* description of the off-the-shelf Stal-Laval fluidized bed gas turbine power plant." But Lovins did not describe the power plant as "off-the-shelf." He referred, actually, to "one system currently available from Stal-Laval Turbin AB of Sweden [with] eight off-the-shelf 70-megawatt gas turbines powered by fluidized-bed combustors . . . ." Clearly, "off-the-shelf" refers to the turbines, not to the combustors or to the entire system. If I wrote about "a virile man accompanied by his charming wife," it would surprise me to be accused of calling the wife virile.

We find it hard to believe that AEP and TVA would venture into fluidized-bed technology without a fairly confident (and informed) expectation that commercially important applications will result. (U.S. utilities are traditionally disinclined to risk money on basic research with distant or nonexistent payoffs.) It may be expedient, however, to describe initial fluidized-bed projects as "developmental," just as it has been expedient to classify commercial nuclear power plants as "experimental."

—Hugh Nash

9

COUNCIL ON ENERGY INDEPENDENCE vs. LOVINS

CEI Accuses Lovins of "Gross Inaccuracies"

All members of the U.S. Senate and House of Representatives were sent a letter dated February 16, 1977, which was written by Daniel W. Kane, President of the Council on Energy Independence, and enclosed a five-page paper commenting on Amory Lovins's *Foreign Affairs* article. The CEI critique is of interest for two reasons, primarily: (1) it was very widely distributed, and hence is cited by other critics, and (2) it elicits from Lovins further details of the capital costs and life-cycle costs of some soft technologies.

Congress of the United States
Washington, D.C. 20515

The October 1976 issue of *Foreign Affairs* magazine contained a lengthy article by Mr. Amory B. Lovins entitled, "Energy Strategy: The Road Not Taken?" Such an event would ordinarily not be of practical significance; however, it has been widely reported in the news media that this article is being taken seriously by many responsible elected and appointed public officials. Due to the nature of the article, the matter is of considerable concern to the Council on Energy Independence and hence, the Council has written brief comments on the article pointing out certain fallacies that the article contains. These comments are enclosed.

The Council on Energy Independence (CEI) is a nonprofit organization which is concerned with providing factual information to the public on energy matters; its membership primarily consists of professional power plant engineers. A short description of the Council on Energy Independence is also enclosed for your reference, since you may not be familiar with the organization or its purpose.

The news media have reported that the Energy Research and Development Administration (ERDA) is making a special study of Mr. Lovins' ideas and that Mr. Lovins will be hired as a consultant to ERDA for this study at taxpayer expense. Such an expenditure of public funds would seem most inappropriate in view of Mr. Lovins' apparent lack of technical qualifications and the gross inaccuracies contained in his article. It is certainly to be hoped that ERDA has more beneficial ways to spend money for energy research.

Your comments on this matter would be greatly appreciated.

[Signed: Daniel W. Kane]

Lovins: My qualifications are for ERDA to judge. ERDA approached me, not the other way around; no doubt it can ascertain from my curriculum vitae, published work, and other clients whether my qualifications are satisfactory.

In view of my responses below, I feel CEI could have avoided embarrassment if they had verified their allegations before sending them to all members of the Senate and House. A suitable method would have been to send me a draft for prior comment in accordance with the usual practice in the scientific community.

[The CEI memo enclosed in Kane's letter follows.]

CEI: The October 1976 quarterly issue of *Foreign Affairs* magazine contained a lengthy article by Amory B. Lovins entitled "Energy Strategy: The Road Not Taken?" In the article, Mr. Lovins advocates the replacement of "hard technology" by what he terms as "soft technology." His article is rather specific about what he wishes to do away with but is vague when it comes to what he will use to replace the items eliminated. For example, he wishes to do away with, and feels there is no real need for, central electric power stations regardless of power source (coal and nuclear central power stations at present and solar central power stations in the future). He states that he believes that a modern industrialized society could viably exist without any central station electric power plants at all. Also, he states that large-scale coal gasification and liquefaction plants for producing synthetic natural gas and gasoline should not be built. He would replace the above items with his "soft technology" hardware such as household solar panels, household or local windmills, and household fluidized-bed coal burners.

Lovins: Contrary to paragraph one, I propose dispensing with additional central power stations because we do not need more of them, not shutting down all those which we already have. For an expansion of reasons why in principle we do not need any big power stations at all, see my response to Dr. Lapp; but as my text makes clear, this argument is an illustration, and "in practice we would not necessarily want to go that far, at least not for a long time." [In his reply to Lapp, Lovins says "technical fixes (à la Ross and Williams, for example) can readily reduce the current appropriately electrical end-use needs of the U.S. from about 4 to about 2.5 q." He goes on to say that this amount could be supplied by present hydroelectric production (about 0.8 quads under adverse conditions) and cogeneration (1.7 quads by the mid-1980s).] My text and its citations (including the heavily documented Oak Ridge backup paper) are quite specific about what kinds of soft and transitional technologies I propose. CEI have muddled up soft technologies (which come in gradually during the transitional era) with transitional fossil-fuel technologies (which come in and then go out again) and, as I shall suggest below, appear not to understand either. CEI have also failed to grasp the importance of the end-use structure outlined on pp. 78–9 of my article.

CEI: Much of Mr. Lovins' theory is reminiscent of certain ideas utilized by the People's Republic of China during the years of the "Great Leap Forward." At that time, it was stated by certain Chinese policy makers that new large "central station" iron and steel production plants were not needed to double the Chinese steel production—what was needed were thousands of backyard iron furnaces and steel converters. The Chinese proceeded to build their backyard iron furnaces and steel converters by the tens of thousands. It is rumored that as many as 20 million Chinese may have been involved in the effort. The backyard steel turned out to be unstable due to its poor quality and the "Great Leap Forward" became the "Great Leap Backward." Apparently, Mr. Lovins believes that the United States can succeed where the Chinese failed.

Lovins: The analogy is poor. Backyard steel mills have not been popular since the Middle Ages, and except in the hands of exceptionally skilled medieval Japanese swordsmiths (whose work modern metallurgists sometimes cannot equal) they never worked very well. In contrast, diverse small- and medium-scale energy technologies have a long and successful history and are now being enormously improved by modern designs and materials. CEI have also apparently identified "soft" with "small," which is not what I said.

It is worth noting parenthetically, however, that (1) according to *Fortune*, p. 76, March 1971, the U.S. is getting successful "regional, even local" steel plants; and (2) China today has [five] million biogas plants (mainly at village scale) and over 60,000 small hydroelectric sets totaling about 3 GW(e), which together form the backbone of the energy system outside the main cities: the hydro sets produce most of the electricity for half the production brigades and over 70% of the communes in the country. While I must not be understood to say that I consider the Chinese political system necessary or desirable, the example suggests that a large developing country can benefit from dispersed renewable sources. I believe much the same logic, though with very different technical details, applies also in industrialized countries, including the U.S., without assuming any changes in sociopolitical structures or settlement patterns.

CEI: Mr. Lovins' beliefs are not altogether surprising in view of the fact that he apparently has never worked in the power industry. However, ideas, regardless of source, should be examined; hence, a brief discussion of home heating by solar panels, electricity from home windmills, and home fluidized-bed coal burners is given in the following paragraphs. Hopefully, the discussion will provide a perspective as to the total cost of "soft technology" to the American public (societal and dollars).

Lovins: It is correct that I have never worked in the power industry, though I know a good deal about the business and have some friends in it.

CEI: It is well recognized by experienced engineers that a practical home heating system based on solar panels at most United States locations can at best supply roughly one-half of the necessary household heating during the year. Also, one definitely has to have a conventional (e.g., gas) heating system as well as the solar panel system at most United States locations if one wishes to avoid freezing at certain times during the winter. It is our understanding that such solar heating systems currently sell for $7,000 to $10,000 or more installed. The price is not likely to be reduced radically in the near future since a significant portion of the total cost is due to installation charges, and carpenters and plumbers are not planning to work on solar heating units for free.

Lovins: The generalization that solar heating systems in most of the U.S. can "at best supply roughly one-half of annual heat requirements" is true only under rather special conditions: traditional (*i.e.*, poor) standards of insulation, architecture oblivious to solar heat gain, rather pedestrian and expensive solar collector design, and—most important—an attempt to make the solar heat compete in cost with dwindling supplies of cheap oil and gas. (Even on these assumptions, the MITRE analysis for ERDA finds the solar fraction of heat and hot water loads to range from the 70s to the 90s of percent in virtually every case at solar costs ($10–15/ft^2 collector) which OTA considers realistic. The classical papers, *e.g.* by Tybout and Löf, suggesting about a 50% fraction were all calculated at pre-embargo oil prices and with other restrictive assumptions.) But these criteria are unrealistic, even absurd. In brief, I think I have shown that retrofitted 100% solar heat anywhere in the lower 49 states (and probably in most of Alaska too) is lower in capital cost, using today's technologies in the mid-1980s, than any of the hard-technology alternative systems at the margin for heating the same house. It is this comparison—of solar heat with heat derived from nuclear power and coal synthetics—that is relevant as we try to decide which of these systems to build. It we tried to make all the alternatives to cheap gas competitive with gas, we couldn't install *any* of them except insulation—least of all nuclear power.

The 50–80% solar heating systems that CEI mention typically sell at competitive bid today not for $7000+ but (assuming a durable, high-quality product properly installed) for about $3000 on retrofit for a badly insulated house, or half that for an apartment (with cooling probably thrown in for the latter). There are wide variations below and above this figure depending on design, details of house, climate, and contractor. A retrofitted 100% solar system, if designed skillfully, would cost only slightly more—perhaps of the order of 10% more. A solar system in a new house designed for it can *reduce* capital cost because it saves so much on conventional heating equipment. [In *Soft Energy Notes* 6, March 1979, Lovins gives a striking empirical example of how a properly heat-conserving house makes a 100% solar system very small and simple.]

CEI: If we were to install solar panel heating systems in virtually all U.S. homes and apartments, as Mr. Lovins suggests, then we should certainly have some idea of what such a scheme would cost. Mr. Lovins neglected to provide the cost of this proposal in his article. To determine whether the idea is truly feasible on this large a scale, a representative calculation can readily be performed which will fully illustrate the relative magnitude of costs. For illustrative purposes, there were 40 million homes and 24 million apartments in the United States in 1970. Assuming that each home and every two apartments require a $7,000 solar panel heating system, the resultant cost to the American public would be $364 billion. This cost would be directly borne by the U.S. public since, obviously, the home owners and apartment owners would have to pay for their system and its installation.

Home heating via solar panel systems will undoubtedly see greater use in the future strictly as a natural gas conservation measure and provided that the federal government provides significant subsidies to the home owner for constructing such systems. There is certainly not likely to be any immediate stampede by the public to buy $7,000 solar heating systems without a heavy government subsidy due to the simple economics of the situation. Most people do not have $7,000 to spare and would have to borrow money from a bank under a home-improvement loan. Such loans typically entail around 12% interest these days and require repayment in ten years. Assume that one gets a "real deal" and the bank provides an 8¾% loan and a 30 years repayment schedule. The home owner then has to pay $660 per year to the bank for his solar heating system (for repayment of the loan) non-inclusive of any maintenance costs. Since the home owner will save only roughly half his gas bill, this means that his gas bill will have to exceed $1,300 per year in order to break even. Incidentally, the 8¾% loan over 30 years really is a "good deal." If the home owner gets a more common 12% home improvement loan repayable in ten years, his yearly payment would be $1,205 and his "break-even" gas bill would have to be over $2,400 a year. Even with the 8¾% loan, natural gas prices would need to increase by three times or more before solar panel heating truly becomes barely economical to the home owner. If the U.S. government and/or industry were to invest $364 billion in large-scale coal gasification plants to produce synthetic natural gas for home owner use, the "problem" of natural gas shortages would essentially be solved.

Lovins: My contention is that without a solar system, one otherwise has to pay *more* (via taxes and utility bills) to build hard-technology systems to do the same job. My article proposes at pp. 87–8 a capital-transfer scheme to transfer the investment burden from the householder to the utility—where it would otherwise be anyway—to the benefit of both parties. (Just one of the U.S. gas utilities using such a scheme today, Michigan Consolidated Gas Company, has already insulated over 100,000 houses in this way, at a lower cost to itself and the consumers than it would otherwise have had to pay to find gas for them.) CEI have ignored this discussion and repeatedly stated that "obviously, the home owners and apartment owners would have to pay for their solar systems." I am encouraged that in his 20 April message to Congress, President Carter proposed to require state utility regulators to implement such a capital-transfer scheme, and proposed further measures to remove imperfections in capital markets.

The approximately $3000 retrofit price quoted above for half-solar heating is at today's prices, which are generally expected to drop by about a third in real terms over the next 8 years. This estimate also assumes no reduction in the heat losses of the house. But *after* the house is well insulated (which will be economically advantageous no matter how it is to be heated), its heat demand will be reduced roughly fourfold to an average of at most 1 kW(t), and for a 1-kW(t) heating load the corresponding investment for a 100% solar system with zero backup (1976 $ throughout, mid-1980s installation) will be about $750–1050—say about $1000. (For details, see my cited Oak Ridge paper or *Soft Energy Paths*, Chapter 7.4.) On CEI's other assumptions, this figure would make the total installed price of the solar installations about $32–55 billion, not $364 billion. (The insulation and other alterations to reduce average heating loads to at most 1 kW/house would cost of the order of $300–1200/house, or $20–60 billion for all 52 million units, but this should not be counted as a charge on the solar program, since it should be done anyway. Insulation to 3–4 times better than the 1-kW nominal figure would further reduce the solar costs and would probably reduce total costs below those cited here.) Even lower solar costs would result from neighborhood systems that shared infrastructure, or from simpler, cheaper collectors. Both these innovations are attractive and will undoubtedly occur, but I have assumed neither.

Another way to look at such an investment, as the American Institute of Architects points out, is as a result of operating a revolving fund that requires much less—typically 12–40% as much—working capital because of the rapid payback compared with power plants, coal-gas plants, etc.

CEI propose to heat houses instead with synthetic gas, which is indeed likely to be cheaper than new electricity for this purpose and is several times less capital-intensive (even with an electric heat-pump). Building a coal gasification plant ordered today, with its associated fuel cycle and water supply but without any new pipelines that might be needed to complete the connection to the user, costs about $40,000/(bbl·day) delivered, or about $600 (1976 $) per delivered thermal kilowatt of gas, at 90% capacity factor (see *Soft Energy Paths*, pp. 115–6). At a 12%/y fixed charge rate, 56% coal-to-gas conversion, operating costs less byproduct credits of 10 cents/10^6 Btu, unlimited water, minemouth coal price of 60 cents/10^6 Btu, gas delivery cost (in existing pipelines) of 10 cents/10^6 Btu, and 70% First Law efficiency of end-use in a cost-free furnace, a delivered 10^6 Btu of heat costs $5.23. (These assumptions are generally optimistic, and in technical respects are identical to those used in September 1975 congressional testimony by H. R. Linden, president of the Institute of Gas Technology and a leading advocate of synfuel plants.) In contrast, at the same 12%/y fixed charge rate and with the same operating and maintenance cost and zero backup requirements, delivered solar heat from a seasonal-storage system costs only $3.10–4.30/$10^6$ Btu. Thus a coal-gas system has a capital cost comparable to or perhaps slightly lower than that of an equivalent 100% solar heating system, but a substantially higher life-cycle cost, owing to its need for coal. This comparison, too, ignores several promising ways to decrease the solar costs, and generously assumes that the gas plant is only replacing an existing gas system and so can take advantage of free pipelines and furnaces. For a new gas system, one should add at least $1/$10^6$ Btu for the pipeline and about $0.30 for the furnace and associated plumbing. These substantial conservatisms suggest that the solar system has a robust life-cycle cost advantage under all circumstances and a capital-cost advantage under most circumstances of syngas deployment at a truly large scale (as CEI propose: $364 billion at $600/kW(t) delivered implies 607 GW(t) or about 18 q/y, equivalent to nearly three quarters of present U.S. natural gas consumption). In cash-flow terms the solar cost advantage is even stronger because the solar collector repays investment at least twice as fast as the coal-gas system, so its capital can be used twice over. [In 1978, DOE estimated that delivered synthetic gas would cost about $7/$10^6$ Btu (1976 $).]

The CEI discussion of financing solar installation, as noted above, ignores my proposal for a capital transfer scheme. Such a scheme guarantees that the cost of money for solar projects will not exceed that paid now by utilities on their investments: indeed, it should be less, as the solar projects pay back so quickly and save so much capital compared to electric or synfuel investments that they will improve the utilities' attainable rate of return and thus give them even better access to capital than they now enjoy.

CEI: Mr. Lovins' idea of providing electricity (or some other convenient source of energy) via home windmills is even more "innovative" and the costs are even more immense. The amount of electrical power required by the typical modern home is relatively large. To our knowledge, systems have yet to be marketed in the U.S. which would supply reasonable quantities of electricity (*e.g.*, 600–700 kilowatt-hours per month) at affordable costs. A typical system which is stated to be available will supply only around 375 kilowatt-hours per month and costs $10,000 to $15,000. If for illustrative purposes one assumes that an adequate system can be made available and installed for $10,000 and that one such system will serve either one home or two apartments, then using the assumed number of homes and apartments previously listed, the cost would be $520 billion to the U.S. public. It is perhaps worthwhile to compare the current cost of electricity from a $10,000 windmill system to that of centralized commercial utilities. Assuming again that the bank provides an interest rate of 8¾% on a $10,000 loan over 30 years, the homeowner must pay $944 a year to the bank for his windmill non-inclusive of maintenance. Unless one contemplates an electric bill exceeding $944 a year, most people are not going to rush out and buy home windmill electrical systems. Incidentally, the cost per kilowatt-hour (kWh) from the home electric windmill system, assuming that it generates 700 kWh/month, would be 11.3 cents/kWh. A typical current price that one pays on an electric bill is, for example, 3.5 cents/kWh. Thus, electricity would have to increase in cost-to-the-consumer by a factor of three before it is economically sound for the consumer to invest in home windmills. Government subsidies in the form of comparatively low interest rate loans could, of course, lower required payments and, hence, make windmills more attractive. However, even a loan at the low rate of 6% would require repayment at $719/year for a $10,000 loan repayable over a 30-year payment period.

Lovins: I have not proposed home windmills, and particularly not to make electricity. I proposed wind machines mainly to provide forms of energy of which there is not already a surplus (*e.g.* compressed air and pumped heat), and wind-electric machines today are indeed sufficiently capital-intensive, especially in small sizes, that they would be attractive for single houses mainly in remote areas and in the absence of the cheap photovoltaics that now seem imminent (see *Soft Energy Paths,* p. 143, note 28). More likely, wind machines will be shared among a substantial number of people. It is true that wind-chargers in small sizes and of the types sold today in the U.S. are expensive: CEI quotes an unusually high price of $20–29,000 per average kW(e) sent out, corresponding to imported machines like the Elektro, Dunlite, or Aerowatt. It is not true, however, that nothing better is available. My FA/Oak Ridge figure of about $650/kW(e) peak, or about $3000/kW(e) average delivered, is for an electrical system using a currently commercial Canadian vertical-axis machine of 2000 kW(e). It corresponds in a moderately windy site (average about 8 m/s, or capacity factor about 0.3) to about 3.5–4 cents/kW(e)-h, competitive with normal sources—which is why Hydro-Québec has put such a machine on its grid. A 2-MW(e) machine [17] times cheaper still is virtually completed in Denmark; [the Danish firm Riisager El-Windmølle, at Skaerbek, near Herning, Jutland, is routinely installing 22.5kW(e) machines for about $760/kW(e) peak; the Danish engineer Niels Borre is doing the same with 20-kW(e) machines at under $650/kW(e) peak; and the American engineer Charles Schachle is offering large machines installed turnkey at $205/kW(e) peak and plans to offer ~25-kW(e) versions at a factory cost of about $250/kW(e) peak (both rated at 17.9 m/s]. [Lovins, *Annu. Rev. En.* 3:477–517 at 496, 1978.] (See *Soft Energy Paths,* pp. 121–3, and *New Scientist,* pp. 567:70, 10 June 1976.) Further calculations cited by Ryle (*Nature 267*:111–7, 12 May 1977) confirm that my $650/kW peak estimate is pessimistic—even though it is an order of magnitude below CEI's estimate.

It is unfortunate that the CEI engineers are not up to date on their wind technology. It is doubly unfortunate that they should have chosen for their calculation not only a machine that cost 7–200 times as much as it should, but also the bizarre notion of substituting for the existing electric grid an array of 52 million such wind machines. This fantasy bears no relation whatever to my proposals for the appropriate use of soft technologies, for four reasons: on CEI's data, about three times as much electricity would be generated in wind machines alone as can possibly find a thermodynamically appropriate use in private households, even assuming no improvements in end-use efficiency; CEI have ignored the important role of hydroelectricity and (during the transitional period) of cogeneration in meeting those appropriate electrical needs, instead casting wind electricity in the role of a panacea rather than following the soft-path principle of letting each technology do what it does best; the optimal scale of average wind machines would probably be one or two orders of magnitude above the roughly 4+ kW(e) which CEI have assumed; and wind machines should only be used in good sites, not all over the U.S. CEI's calculation rests not only on out-of-date prices but on a complete misconception of what soft technologies are and how they are to be used to advantage.

CEI: The last of Mr. Lovins' "soft technology" solutions to the energy problem was the fluidized-bed home coal burner. Discussing the economics of this item is somewhat pointless since the item is not available in the United States. However, when and if such an item does become available, it is probably a reasonable guess that it would cost $5,000 or more installed. The arithmetic previously utilized readily demonstrates that the cost of full utilization of such a system in the United States would be in the hundreds of billions of dollars.

Lovins: CEI have mistakenly described a fluidized-bed coal burner as a soft technology. It is, as my text makes clear, a transitional fossil-fuel technology, and cannot be soft because it is not renewable. The domestic fluidized-bed burner I mentioned has been tested in the United States, is the subject of a CEQ project run by Dr. John Davidson with the help of Professor Arthur Squires, typically costs less than a conventional furnace (not "$5000 or more"), and is considerably more efficient and just as convenient. Michael Virr of Stone Platt Fluidfire, its developer and licensor, estimates that in contemporary small-lots production, the installed price of a domestic fluidized-bed system would be about $500 without, or $850 with, bunkers and plumbing for storage and handling of coal and ash. Large-run production would reduce these costs further.

CEI: The sum of the cost to the public of solar panel home heating ($364 billion) and home windmill power systems ($520 billion) as previously computed is $884 billion. Mr. Lovins' main objection to "hard technologies" in his article was that he thought they were too expensive. He estimated that "hard technologies" would require the power industries to make an approximate $1 trillion investment; an amount which he considered to be too expensive. If Mr. Lovins' suggestions were followed, American home owners would have to supply roughly a trillion dollars of investment rather than American industry. If given a choice, the average home owner would undoubtedly prefer that industry should supply the investment capital rather than the homeowner. Also, it is important to note that a trillion dollars spent by the power industry will supply at least several times the additional power that a similar expenditure by American home owners would "save" or produce. For example, to double the present electrical generating capacity in the United States (using a mixture of coal and nuclear power plants), it would cost very roughly around $520 billion, assuming a cost of construction of $1,000 per kilowatt electric, which is certainly a reasonable number for actual power plants being planned at this time. It has been previously estimated that Mr. Lovins' home windmills would provide electricity for $520 billion; however, such electricity would be provided for only the homes themselves—no surplus for other use would be available. Since home use of electricity is approximately 33% of total U.S. electricity usage, Mr. Lovins is in essence suggesting that U.S. homeowners pay roughly $520 billion to increase the U.S. electrical generating capacity by 33%; whereas, industry would spend roughly that same amount to increase current generating capacity by 100%. Mr. Lovins' suggestion would appear to be non-cost effective by a factor of three or more.

Lovins: CEI add a solar cost inflated by roughly a factor of seven to a wind cost inflated by a factor of 7–200 (and misconceived to boot), failing to note that both systems are cheaper (in life-cycle cost and often in capital cost too) than competing hard technologies. They then erroneously suggest that this cost would be borne by "homeowners" whereas the costs of hard-technology investments would be borne by "American industry." Yet American industry is not a charity; it would recoup its costs, with interest and profits, from those same homeowners. They, in turn, could instead turn the same process to their advantage by a capital-transfer scheme or an AIA-style revolving fund. In either case they would not then pay such large profits and interest charges, besides saving heavily on the investments themselves.

CEI's wind-*vs*-central-station investment comparison confuses costs per kW installed, sent out, and delivered. By the last measure—which is what matters—the Canadian wind technology I mentioned is about 1–2 times (not three times) as capital-intensive as equivalent fossil-and-nuclear central *stations alone.* But the wind system is comparable (*vs.* coal-electric stations) or substantially *lower* (*vs.* nuclear stations) in capital cost when compared with the entire electrical *systems* which it replaces. This is because wind machines need no fuel-cycle investments and are small enough to distribute their power cheaply in the locality without requiring the addition of transmission and distribution facilities that can cost as much as, or even more than, the power station. (See *Soft Energy Paths*, Chapter 6.1.) Moreover, this Canadian technology, as mentioned above, is [17–25] times more expensive than a wind system that has been built in Denmark and [24–29] times more expensive than one that has been fieldtested and is now being geared up for mass-production in the U.S. The Canadian data are given in my Oak Ridge paper which CEI apparently did not consult.

CEI: The major items which comprised Mr. Lovins' "soft technology" hardware suggestions (namely, home solar panels, windmills, and fluidized-bed home coal burners) have already been briefly discussed and roughly economically assessed. His article contains a very large number of questionable "facts," statistics, and statements. A detailed list of comments on the article has been compiled and is available to interested parties, if desired. There are a few points, however, which provide some insight as to the general nature of the article and Mr. Lovins' competence in the power field which are worth specifically mentioning and which are therefore briefly summarized below.

Lovins: CEI repeat the erroneous statement that fluidized-bed burners are a soft technology, and add to it the new erroneous statement that they, home solar panels, and home wind machines "comprise" (by which I think CEI must have meant "constitute") my soft-technology hardware suggestions. The whole point of soft technologies is that they are diverse, not limited. In implying that there are only three soft technologies, CEI have ignored all non-domestic solar collectors (including those for process heat), sensibly sized wind machines (including the non-electric ones which dominate the wind mix), and all biomass conversion technologies. And these are only broad categories: a list of the specific types of technologies thus omitted would fill many pages.

In view of the low calibre of the CEI comments here considered, I have not thought it worthwhile to ask CEI for their "detailed list of comments" for further response, though I shall be glad to respond to any particular points in it on request.

CEI: In his article, Mr. Lovins implies that if the volume of U.S. production of beer and wine were scaled up by a factor of 10 to 14 times then one would have enough alcohol to replace one-third of the gasoline annually used in the U.S., assuming, of course, that one distills the alcohol out of the beer and wine. His numbers appear to be off by roughly a factor of 10 or more. To check his values, the following 1973 data from the U.S. Department of Commerce were utilized (undoubtedly, better data could be obtained but the 1973 data should suffice for illustrative purposes): wine production assumed to be 599 million gallons, beer production assumed to be 4,433 million gallons, and gasoline production assumed to be 2,399 million barrels at 42 gallons per barrel which equals 100,800 million gallons of gasoline. One-third of the gasoline equals approximately 33,600 million gallons of gasoline to be replaced by alcohol from beer and wine. For calculation purposes, beer was assumed to be 5.5% alcohol by volume and wine was assumed to be 10%. The higher heating value for gasoline was assumed to be 20,000 Btu per pound and that for ethyl alcohol was assumed to be 12,800 Btu per pound. On this basis, it was calculated that a total of roughly 839,000 million gallons of beer and wine would be required to replace 33,600 million gallons of gasoline. Thus, the ratio of required production to present production of beer and wine is roughly

836,000 million gallons divided by 5,032 million gallons (present production of beer and wine) which equals roughly 166. Mr. Lovins arrived, somehow, at the number 10 to 14 as the ratio. The 166 ratio does not take into account the heat necessary to distill the wine and beer in order to get the alcohol in a usable form, thus the actual ratio is presumably even higher.

Lovins: In this long paragraph, having implied that biomass technologies do not make it onto the list of soft technologies, CEI introduce them, but mar them with a misconception that ignores my carefully phrased caveats: for details, see my response to Dr. Lapp. In summary, CEI agree that the U.S., as I state, produces 5% as many gallons of beer and wine per year as it does gasoline. The discrepancy between CEI's factor of 166 and mine of 10–14 (10 is nearer the truth) is simply a weighted average of the reciprocal of the alcohol fractions of beer and of wine. This is due to CEI's unfounded assumption that I was proposing to make fuel alcohols by the same processes now used to make beer and wine—rather than using the analogy to compare the physical scale (size of vats and plumbing), as measured by fluid output, of two different kinds of plants.

CEI: Mr. Lovins' knowledge of power engineering in general, and nuclear engineering in particular, appears to be on a par with his knowledge of wine and beer production. In his article, Mr. Lovins proceeds to deprecate nuclear power reactors as a source of energy (specifically as possible source of electrical energy for home heating) as follows: "Where we want only to create temperature differences in tens of degrees, we should meet the need with sources whose potential is tens or hundreds of degrees, not with a flame temperature of thousands or a nuclear temperature of millions—like cutting butter with a chainsaw." It would be interesting to ascertain from what source Mr. Lovins deduced that nuclear power reactors operate at temperatures measured in the millions of degrees—one thing is certain, the book was not an engineering textbook. Typically, the highest temperature in a light-water power reactor is the fuel centerline temperature which is roughly 4,000 degrees Fahrenheit. The cladding temperature of the fuel is approximately 600 degrees Fahrenheit, and steam is produced in the 500 to 600 degree Fahrenheit range. One is also forced to wonder just what Mr. Lovins believes he is referring to when he mentioned "with a flame temperature of thousands." Can it be that Mr. Lovins believes that a fossil fuel such as oil and gas burns with flame temperatures of thousands of degrees at electrical power plants but burns at flame temperatures of only tens or hundreds of degrees in a home hot water heater or furnace? His use of terminology throughout the article certainly implies that a power engineering background is lacking.

Lovins: A similar misunderstanding, made also by Dr. Lapp, mars CEI's interpretation of "nuclear temperature." See my response to Dr. Lapp for details. It is likewise axiomatic that fossil fuels have flame temperatures approximately as high in home furnaces and water heaters as they do in power stations—and are just as mismatched to low-grade end-uses in both cases.

CEI: In discussing his "local energy generation" alternatives, Mr. Lovins fallaciously argues against the necessity of an interconnected supply grid. The necessity for such a grid has been adequately demonstrated by the harsh realities of this winter. Many areas of the country would have been virtually without electricity for various periods of time this winter if not for interconnected electric power grids. As most of us are aware, certain regions of the country have been recently subjected to rotating blackouts and voltage reductions. Without such interconnected grids, many areas of the country would have been without electricity during the coldest part of this winter.

Lovins: Contrary to the implication, I have nowhere suggested that existing electric grids should be dismantled. What I have argued against is their pervasive expansion. My references to local energy systems are not confined to electricity, and do not even refer primarily to electricity save insofar as cogeneration and combined-heat-and-power district heating systems would site local plants according to the heat load and leave the byproduct electricity to be distributed elsewhere through the existing grid. The arguments for smaller unit scale in electricity generation (and in other energy systems) are set out briefly in *Foreign Affairs* and more fully in the Oak Ridge paper and *Soft Energy Paths* (especially Chapter 5).

CEI: The last point worth considering and perhaps the most important point of all is the potential impact of Mr. Lovins' theory on the American working man. It has been previously pointed out that Mr. Lovins' energy theories are designed to unload the necessity of capital investment from American industry to the American homeowner—a fact that is not likely to please the homeowner. In his article, Mr. Lovins proceeds to imply that energy, gross national product and jobs are really quite unrelated topics. While this assumption may be quite popular with people in certain academic circles, the idea is, quite frankly, an insult to the American working man. It is suggested that Mr. Lovins should tell the two and one-quarter million workers, as reported by the news media, who were laid off due to energy shortages during the recent cold weather, that there is no connection between jobs and energy supply. Our oil and natural gas supplies are rapidly shrinking and will be virtually exhausted within the next 20 to 30 years regardless of whatever practical conservation measures may be employed. The United States government must rapidly implement the commercialization of coal gasification and liquefaction to provide synthetic gas and gasoline. In addition, it must utilize coal and nuclear fuel to the maximum possible extent for the generation of electricity if an economic disaster is to be avoided.

Lovins: It is my concern at the implications of present policy for the American working man—and woman—that has led me to propose alternative policies. These are designed not to "unload . . . capital investment" onto the homeowner, but to reduce that investment, to augment the reduction by a saving in fuel cost and vulnerability, and to leave the capital burden, as at present, on the shoulders of the utilities and other existing institutions.

It is true that unexpected shortages of fuel, or of any other resource, can lead to disruption and unemployment. I expect that a consequence of pursuing CEI's policies would be to make such shortages endemic rather than sporadic. It is not true that ensuring secure energy supplies from low-risk transitional ways of using existing fossil fuels and from proven renewable resources would be bad for jobs. On the contrary, those technologies are far better for jobs than hard technologies, and, being capital-saving, have a multiplier effect on the economy by saving jobs elsewhere that would otherwise be starved for investment. In this sense, according to Professor Hannon's input-output studies at the University of Illinois (Urbana-Champaign), every big power station we build *loses* the economy about 4000 net jobs—because at about a quarter of a million dollars per direct job, it ties up capital which would otherwise employ more people if invested in almost any other way.

CEI: Mr. Lovins claims to describe two roads. He has not, however, followed these roads long enough to determine where they lead. If he did, he would find that our forefathers demonstrated infinite wisdom in their decision to eat grain and burn coal.

Lovins: As for the last paragraph, I have set out in my article some of the implications of hard and soft energy paths. I notice that CEI have not criticized my characterization of the hard path and its economic and social problems. From that I infer that they think the problems of a soft path are even worse. If they think this only on the basis of the sorts of arguments I have responded to above, their case falls to the ground.

10

YULISH vs. LOVINS

Consultant Describes Lovins's Arguments as Flaccid and Flatulent

Charles Yulish Associates, Inc., is "a management and communications consulting firm specializing in energy and environmental matters." In June 1977, the firm published *Soft vs. Hard Energy Paths: Ten Critical Essays on Amory Lovins' "Energy Strategy: The Road Not Taken?"* Charles B. Yulish was the editor. As a condition of permitting us to reprint his introduction to *Ten Critical Essays*, Mr. Yulish asked for several changes and amplifications. First, he asked that he not be identified in the chapter subheading as a "publicist," saying "This is incorrect." We stand corrected, but point out that the primary definition of "publicist" in *Webster's* unabridged (second edition) is ". . . any writer, as a journalist, on matters of public policy or political interest, including sociological and economic topics."

Second, Mr. Yulish pointed out "some rather obvious errors":

a. "Lovins says I do not give his qualifications in my publication. In point of fact I stated his qualifications nearly exactly as they were given in the *Foreign Affairs* publication." [Lovins is indeed identified correctly, if somewhat laconically, in an easily skipped-over two sentences between the table of contents and the introduction. What he actually said had been omitted were his *professional* qualifications. They were.]

b. "Dr. Bethe was asked to be a contributor to my publication, and he said he did not have the time to complete it within my schedule, although he did say he was corresponding with *Foreign Affairs* on Mr. Lovins' article."

c. "Mr. Lovins and I carried on quite a correspondence regarding his displeasure at my not permitting him to review the critiques before they were published, and he again makes the point 'as is a normal practice in the scientific community.' *Foreign Affairs* is hardly a scientific publication, and to the best of my knowledge, he has not applied his own logic to his own writings, but then again, perhaps his interpretation of what constitutes a 'peer' is quite restrictive." [Whether or not *Foreign Affairs* is a scientific journal is beside the point. What matters is that Lovins and many of his critics are members of the scientific community. Contrary to Mr. Yulish's supposition, we have Mr. Lovins's word for it that he sought (and benefited from) extensive peer review of early drafts of his article.]

Third, Mr. Yulish wants it known that "I have read Mr. Lovins' responses to the selected quotes from my article and find them weak, self-serving, and entirely consistent with his earlier blind spots and tunnel vision." We are grateful that Mr. Yulish has sufficient courage of his convictions to allow readers to judge for themselves.

Speaking of tunnel vision, we find it amusing that Mr. Yulish considers the format of this book unfair because "what you appear to be doing is taking the criticisms of Mr. Lovins' statements and writings and allowing him to rebut them in whatever manner he wishes. . . ." What Mr. Yulish has done with his *Ten Critical Essays* is encourage attacks on Mr. Lovins's professional and scientific competence without permitting Lovins so much as a pre-publication review of the attacks, much less any form of rebuttal. Further, unlike the simultaneous EEI edition, which reprinted the FA article, Mr. Yulish included instead his own summary of it, noting that it "was not reviewed for accuracy by Mr. Lovins." Actually, Mr. Lovins had reviewed it, found it "seriously misleading," and recommended that it be omitted and the original article substituted for it.

Yulish: In the marketplace of ideas and the struggle to influence public thinking, poets and elegant prose are as useful a resource as are statistics. Amory Lovins is quite comfortable and resourceful using them all for greatest impact.

While virtually none of the ideas or information presented by Mr. Lovins in his article is unique or new, the scope, *chutzpah* and *panache* of his article seem to present the reader with a fresh, coherent and bold unification of energy-related considerations as a basis for a complete overhaul of our existing approaches. As with most aggressive social reform documents, his article contains clear cut villains, heroes, and a single exclusive path to glory. It appeals to the discontent of the masses as well as to the uneasiness of their leaders. It offers a creed and togetherness and promises a better life for all, with, in this particular case, no serious sacrifices, other than complete allegiance to the "true way." It is a social and political Manifesto.

A single piece of writing on the subject of energy rarely achieves the kind of impact that Amory Lovins' "Energy Strategy: The Road Not Taken?" has. The article, which first appeared in the October issue of the respected journal, *Foreign Affairs*, has since been widely reprinted. The Friends of the Earth newspaper has editorialized that Lovins should be awarded the Nobel Prize for his work. According to news reports, the article has stimulated considerable interest throughout the world, and Mr. Lovins has embarked upon an active schedule of speaking engagements and meetings with government and other leaders.

Lovins: This 166-page booklet was published on or about 11 June 1977 by Charles Yulish Associates, Inc., though draft copies were made available some weeks earlier to selected journalists (not to me). Apparently the print run was small and Mr. Yulish does not expect to reprint—possibly because the Edison Electric Institute simultaneously published a "slick" (typeset, illustrated, glossy-paper) edition of *Electrical Perspectives*, an EEI periodical, that is entirely devoted to a condensed rearrangement of the Yulish material. Evidently the two editions were prepared in parallel by close cooperation over a period of weeks, sharing even the cover artwork. The EEI production was especially elaborate and time-consuming.

The Yulish edition, which does not mention the EEI edition, carries a disclaimer: "This publication was neither paid for nor sponsored by any entity other than Yulish Associates, Inc. Each of the authors' contributions was provided as a public service." Mr. Yulish has stated that he paid for the printing out of his own pocket because he feels so strongly about the issues. His pocket is lined partly by the U.S. electronuclear industry, for which he is generally regarded as a principal public relations consultant. His booklet describes his firm as "a management and communications consulting firm specializing in energy and environmental matters"; its many services to the electronuclear industry are not mentioned. Under the circumstances, any claim of disinterested objectivity on Mr. Yulish's part seems somewhat disingenuous. EEI, as an overt trade group, needs (and can bear) no disclaimer.

Mr. Yulish's booklet is a remarkable social document. I cannot recall—at least since *The Limits to Growth* was published—a case in which the mere exposition of a thesis has evoked such a battery of implacably hostile attacks from persons anxious to appear objective. The *ad hominem*

Mr. Yulish has summarized at some length the professional qualifications of his authors while omitting my own. (Readers concerned about such matters might find it relevant, for example, that my clients include three organizations—OECD, ERDA, and OTA—with which Mr. Yulish's authors are affiliated.) His biographies also omit much interesting information. For example, Mr. Kane's employer, Sargent & Lundy, is a well-known nuclear architect-engineering firm, noted chiefly for its work on the Commonwealth Edison reactors. Mr. Kane's less well-known organization, the Council on Energy Independence, is distinguished mainly by its pointed attacks on various critics of nuclear power (especially the late David Comey). Ralph Lapp, Ian Forbes, and Margaret Maxey seem to spend much of their time serving in various public fora as staunch and outspoken defenders of nuclear power in general and the plutonium economy in particular. Their active speaking and writing careers in this cause seem a curious omission from Mr. Yulish's biographies.

tone of most of Mr. Yulish's booklet suggests that my *Foreign Affairs* article touched some raw nerves. Some powerful institutions apparently perceive certain patterns of ink on paper as a real threat to their present activities. This perceived vulnerability to the pinprick of ideas, freely exchanged in an intellectual marketplace, has led Mr. Yulish's authors to rely heavily on the devices of polemical rhetoric: caricature, sarcasm, hyperbole, misrepresentation, straw men, vituperation, innuendo, guilt by association, and many more. Such tactics suggest that the authors cannot or will not respond at the substantive level traditional in professional discourse.

That most of Mr. Yulish's authors substitute character assassination and demagoguery for the normal mechanisms of criticism (such as letters to *Foreign Affairs*) bespeaks the regrettable evolution of a siege mentality. It also says much, I think, about the genuine difficulty of defending intellectually the arguments on which the electronuclear industry rests. Two leading proponents of nuclear power have attempted such a defense— unsuccessfully— in the letters column of *Foreign Affairs* (Professor Hans Bethe of Cornell in April 1977 and Dr. Bert Wolfe of General Electric in July 1977). It is suggestive that both these efforts have been discreetly omitted from Mr. Yulish's collection.

This response will address comprehensively the substance of Mr. Yulish's introduction and his ten essays. The booklet raises no issues where my original article erred in fact or logic, and none (save perhaps a few of Dr. Pickering's misreadings) that even merit substantive response. This will, I believe, become clear to any open-minded reader who carefully compares Mr. Yulish's booklet and this response side by side. Such close comparison requires much time and patience, but I fear it is essential to form a sound view.

While I must ask readers, in fairness to all concerned, to bear with me in this unavoidably tedious exercise, I have tried to reduce the tedium by addressing only specific points about things I actually wrote, rather than commenting on every objectionable word, phrase, or nuance. The first page of Mr. Yulish's introduction, for example, characterizes my article as "a social and political Manifesto" that demands "complete allegiance to the 'true way,' " offers "a creed and togetherness," and contains "clear cut villains, heroes, and a single exclusive path to glory." Carefully read, my article has none of these features, but in general I shall not trouble to say so, lest I approach here the fatuity of the material to which I am responding.

Yulish: When I first read "Energy Strategy: The Road Not Taken?" I was impressed with its scope, yet skeptical. My skepticism came from areas of his writing which I found either argumentative or erroneous. This led me to wonder if others might find equivalent reservations in areas of their own expertise. I invited a wide spectrum of authorities to critique Mr. Lovins' article, and was surprised at how many were eager do so as a public service. Had there been more time, others I contacted said they would also have written essays. The critiques we have in this report provide a useful starting-point for an appraisal of the Lovins piece.

Lovins: This paragraph does not specify what is "argumentative or erroneous" in my work. Parts of my article were "argumentative in the sense of stating my opinion and arguing in support of it, and these parts were carefully distinguished from statements of fact wherever the distinction would not be obvious to a dispassionate reader.

Unfortunately I cannot agree that Mr. Yulish's collection provides "a useful starting-point for an appraisal"; it is of grossly inadequate technical quality for that purpose. Until the critics practice the scholarship whose lack they impute, we cannot hope to begin the serious appraisal, criticism, and refinement that my article sought.

Yulish: In his polemical article, Mr. Lovins has recast the Cartesian dictum, "I think, therefore I am," into a more serviceable, "I say it, therefore it is." Such self-esteem hardly brooks criticism. In a recent interview, Mr. Lovins is quoted as saying: I think it's fair to say that so far no one's found mistakes [in the article, although] people have found respects where I should have explained things more clearly."

Lovins: Miscasting my article as "polemical" by a mysterious sort of psychological projection, this paragraph simultaneously miscasts it as dogmatic. If Mr. Yulish were better acquainted with my writings, he would find in them a profound skepticism and antidogmatism, applied as much to my own work as to others'. Accordingly my *FA* article, unlike Mr. Yulish's anthology, is at pains to document controversial points as thoroughly as the editor permitted. The statement of mine that Mr. Yulish decries—that no mistakes have been found in my article so far—does not reflect unreasoning dogmatism; rather, it reflects my considered judgment that any objective reader of all the relevant material will form the same view. (In the one case I know of where a neutral scientist—Dr. Gene Rochlin of the University of California at Berkeley—has been asked by another reviewer to read the article, critiques, and responses, and to form a view, he has indeed concluded that though he cannot personally vouch for the precise value I assign to every number, nobody has shown any of my numbers to be wrong. I consider that a fair and correct statement.)

Yulish: The ten critiques in this report should give Mr. Lovins pause, for they call into question most of the important aspects of his article—from the very fundamental assumptions to the technical details.

Mr. Lovins has said that the social implications of the soft path are "the guts of the issue," and he urges his readers to focus upon those matters, rather than the technical details. I believe the authors of this report have complied with Mr. Lovins' wishes, with rather disastrous results for him to contend with.

Because of Mr. Lovins' unrestrained advocacy of "soft" technologies *über alles*, various experts have felt compelled to point out some of the practical and social limitations of his "soft path" goals.

In this report, Sheldon Butt, President of the Solar Energy Industries Association and the solar researchers, Aden and Marjorie Meinel, express profound shock and dismay at Lovins' cavalier notion of the role of solar power in our energy future. Butt states:

> Mr. Lovins has allowed advocacy of his own concept of the future to lead him to support this position with a lengthy series of distortions and even misrepresentations of physical and scientific fact.

He then goes on to cite those errors, and to say:

> Since the basic knowledge of solar technology which Lovins evidently possesses must tell him that his final conclusion is not only false but very substantially so, the only reasonable conclusion which I can reach is that his statement is not only untrue, but deliberately so.

Lovins: I shall demonstrate that Mr. Butt does not point out any "distortions and even misrepresentations of physical and scientific fact," but rather shows that his own view of solar technologies is narrower than is justified by solar buildings now in operation—a point on which even Professor Bethe has accepted my analysis. Accordingly, Mr. Butt's repeated conclusion that I am wrong, therefore deliberately lying, is not only defamatory and unworthy, but wrong in fact on both counts. Had Mr. Yulish troubled to read my exchange with Professor Bethe—concluded and widely available since mid-April—or simply to ask me, he could have found this out.

Yulish: The Meinels open their critique with the statement that they are "chilled at the actions advocated by this seductive and well-written article and appalled that a person claiming to be a physicist could resort to what appears to us as distortions of technical reality." Strong words from leaders in soft technology research, development and commercialization.

Lovins: Though the Meinels have done much interesting work on solar energy (see e.g. *Soft Energy Paths*, p. 126), little if any of this work bears any relation to soft technology. It is possible that their displeasure with my article stems from their unhappiness that I have drawn this distinction.

Yulish: Mr. Lovins' desire to avoid nitpicking notwithstanding, technical criticisms abound in these critiques, including detailed ones by Dr. Ralph

Lapp and Dr. Ian Forbes, who take on Mr. Lovins' assumptions as to growth and the philosophical as well as numerical focal points. Aside from the harsh criticisms of Lovins from the technical point of view, we must also take into account the criticism of ethicists Pickering and Maxey, who find Lovins' notions of a soft world about as substantive as a rotl of baloney. They see the ethical, societal and sociopolitical arguments advanced by Lovins as having little, if any, accuracy or merit. The "friendly Fascism" which Lovins sees in the wings, waiting to go on stage, is challenged in concept, as is his new act which might be called "Friendly Collectivism."

It is also interesting to note that, in spite of the welter of Lovins' economic statistics and conclusions, many of the authors contest both, and reject, like Dr. Arnold Safer, Mr. Lovins' economic arguments as flaccid and flatulent. Presumably, organized labor would welcome Mr. Lovins' approach with open arms as he purports to offer a full employment economy directed towards rebuilding and recentralizing [sic] our cities and nation. One is taken aback then by the comments of the eminent labor leader and former U.S. Secretary of Labor, Peter Brennan, who says:

> I find that, although Mr. Lovins and his article offer a rather majestic and sweeping view of the subject, nevertheless it's still the old story of "the Emperor has no clothes on." In spite of the hypnotic impression of finery, elegant plumage and excellent workmanship, the truth is that Mr. Lovins is the purveyor of naked nonsense.
>
> After all is said and done, one gets the impression that Mr. Lovins' soft path is quite like the Yellow Brick Road. Both lead to "never-never land."

Lovins: The arguments cited in the first paragraph are of course fully addressed below. My views bear no relation to collectivism, friendly or otherwise. Likewise, Mr. Yulish's characterization (in the second paragraph) of Dr. Safer's article is his own, not Dr. Safer's.

The quotation from Mr. Brennan is judgmental rhetoric unsupported by analysis. It is not I but the anthologized authors whose credibility may suffer from our exchange, for none of them took the obvious precaution of checking a draft with me (as is normal practice in the scientific community) to ensure against publicizing their own embarrassing mistakes.

Yulish: If more is better, then indeed there is more, much more, criticism to be found in the ten essays which follow. If less is more, then let's forego further quotes and leave it to you, the reader, to proceed at your own pace through the report.

In reading these critiques, I conclude that Mr. Lovins has launched a fresh and vital debate, but at the same time he may have lost significant standing as a credible fount of knowledge.

The most obvious and pervasive defect in his philosophy and article is the fundamental belief that Robert Frost's "the road not taken" can be extrapolated from the decision-making of one man at one time, to the experience of the world for all time. Common sense makes it obvious that many roads must be taken to reach various points in the future, and that notions such as "mutual exclusivity" smack of arrogant and arrant nonsense.

Lovins: My article suggests not extrapolation "from the decision-making of one man at one time . . . to the experience of the world for all time," but rather the better use of existing market and social processes to adapt to changing circumstances. This adaptation is a continuous, organic process that nobody can foresee in detail, and certainly neither I nor anyone else can lay out a detailed and unalterable blueprint even for one country. I believe, however, that my ideas may help democratic societies to understand the energy problem better and to devise more fitting solutions through the give and take of normal politics. As for "mutual exclusivity"—which Mr. Yulish regards as "arrogant and arrant nonsense"—just as the hard/soft distinction is not primarily one of hardware, so exclusivity is not primarily a technological incompatibility. Where we are today is a fine example of the sort of exclusivity I mean: decades of lopsided commitment to one sort of technology have left us with a limited range of perceptions and technical options and a rigid set of institutions.

Yulish: The wellspring of Mr. Lovins' cry for a Claes Oldenberg world sculpted with "soft technologies" appears to derive from, and pivot upon, his fear of nuclear weapons proliferation and the conviction that nuclear power is to blame. If all nuclear power plants were to disappear from the earth and no more ever built, nuclear proliferation would still not be halted, for simpler, cheaper and unclassified reactor technologies are in common use world wide to make nuclear weapons materials. In fact, none of the existing nuclear powers used a power reactor to produce their bomb materials. But one suspects Mr. Lovins already knows this and chooses to beat the straw man unmercifully as the modern energy scapegoat upon which to heap all our fears. Like the witch hunters and burners of old, the bonfires did not solve their problems, but certainly made them feel a lot better. The destruction of nuclear power may provide psychological relief for a few, but it will not solve their basic problems, and will only make for a more uncomfortable and risky world for most of the rest of us.

As more and more interest is generated in discussions and debates regarding soft vs. hard energy paths, perspective will become more and more important. I trust that these ten critical essays will make their contribution to that badly-needed perspective. It is my opinion that we will find ourselves in need of, and exploring, *many* energy paths, and that in doing so we will find they can be harmonious and productive paths for the generations that follow as they find new roads and choices to explore.

I expect that we have only heard the beginning of Mr. Lovins and his ideological converts. I hope that those who find his messages so intriguing will also want to measure what they read and hear against other points of view. The search for an energy Manifesto all or most people will follow surely does not end here.

Lovins: Mr. Yulish's view of the nuclear proliferation problem is fashionable but, I think, naive. In Chapter 11 of *Soft Energy Paths*, I argue that nuclear power is a peculiarly convenient proliferative route and that phasing it out, in closely linked combination with initiatives to promote soft energy paths and strategic arms reduction, could indeed go very far to put the nuclear genie back in the bottle. As it happens, fear of nuclear proliferation—a fear I consider fully justified—is not my only nor necessarily my main motive in seeking non-nuclear futures; as I have pointed out elsewhere, even if nuclear power produced no toxic or explosive materials, it would still be unattractive. But in seeking an energy policy that makes sense, I must obviously consider as one essential element of making sense that the policy ensure the continued existence of people, without whom neither energy policy nor anything else is necessary.

11

BUTT vs. LOVINS

Critic Accuses Lovins of Gross Errors and Purposeful Falsehoods

One of Yulish's contributors is Sheldon Butt, President of the Solar Energy Industries Association and an executive of Olin Brass Company. He has a Master's degree in Chemical Engineering from the University of Louisville.

Butt: My fundamental reaction to Lovins' article, "Energy Strategy: The Road Not Taken?" is profound shock. It is shocking to me that Lovins, who has obviously benefited from an excellent education in the physical sciences, has allowed advocacy of his own concept of the future to lead him to support this position with a lengthy series of distortions and even misrepresentations of physical and scientific fact. This is doubly shocking because Lovins' training in the physical sciences should make him sufficiently aware of the need to consider facts objectively.

Lovins: As I shall show presently, Mr. Butt's accusation of "a lengthy series of distortions and even misrepresentations of physical and scientific fact" is unsupported by even one example, owing to the careful application of the physics education which he kindly ascribes to me.

Butt: In commenting upon Lovins' article, I will comment first upon the societal impacts of the "solution" which he advocates since I believe these to be of transcendent importance; and secondly, upon the realities of the development of solar energy as an alternative energy resource, as I see them from my point of view within the solar industry.

As I see it, the basic justification for our technological society, as it has evolved over a period of somewhat less than two centuries, lies in the exceptionally broad opportunity which it has provided to the citizens for individual expression and for the exercise of individual initiative. At no time in past history have more individuals had the opportunity to devote a large proportion of the product of their labor to those things which they individually elect as expressing their own desires, after first having satisfied essential needs. I think all of our recent history tells us that there is one further necessary ingredient. The maintenance of a society providing for broad individual election and for widespread exercise of individual initiatives depends upon individual rewards, which may be interpreted to mean increasing individual betterment and, particularly, betterment as defined by the individual's perceptions.

Lovins: While I must not be understood to advocate any sort of "return to primitive society," as some would see it, it is interesting to note that Mr. Butt's statement that "At no time in past history have more individuals had the opportunity to devote a large proportion of the product of their labor to those things which they individually elect after first having satisfied essential needs" is factually incorrect. There is a substantial anthropological literature documenting how "primitive" peoples, chiefly hunter-gatherers, spend their time and resources, and it generally turns out that such people have considerably more leisure and more surplus resources to devote to their highly developed arts than do typical members of the most affluent Western societies today. I note this parenthetically, not because it is directly relevant to either Mr. Butt's argument or mine, but rather to illustrate the dangers of assuming that our own culture has some unique superiority over those that have existed in some cases for thousands of years and that apparently offer ample scope for human happiness and fulfillment.

Butt: The history of our society has been that the vast majority of individuals have perceived individual betterment in material terms, whether this has meant the ownership of a dishwasher or the opportunity to travel. Even in totalitarian states, it has been amply demonstrated that it's necessary to provide considerable material incentive to the population.

My reason for this philosophical excursion is the attempt to establish some basis for evaluating Lovins' proposals in the context of their impact upon society. It may well be that there is a sufficient gap between the subsistence needs of the population of the United States and its economic productivity to accomplish a transition to the purely "soft" energy technology which Lovins advocates. However, it is implicit in the changes which he proposes that the population must sacrifice a great deal in terms of material well-being. If we are going to replace our existing electric power industry with some sort of vaguely defined, decentralized, "soft" energy technology, it would appear to follow that, somehow or other, these assets which have been paid for by the past "gap" between gross production and personal consumption must be replaced. The cost of replacement terms of reduced consumption must be deducted from the available product, thus diminishing that which is available within the system to provide rewards for the population. Indeed, Lovins proposes to replace a substantial portion of our energy production and conversion capability not once but twice within 50 years. (Lovins proposes that we first convert much of our energy industry to the direct burning of coal and, sequentially, replace this with "soft" energy technologies.)

Lovins: To return to Mr. Butt's thesis, he states without justification of any sort that a soft path implies "that the population must sacrifice a great deal in terms of material well-being." He has apparently overlooked my reference to "increasing comfort for modestly increasing numbers" who would be "significantly more affluent than today." If I had proposed scrapping the existing energy system and instituting a crash program to replace it with soft technologies, Mr. Butt's anxieties might be easier to understand. But in fact I have proposed gradually, smoothly replacing existing energy systems through normal attrition, and building transitional, soft, and conservation technologies instead of *new* power stations. Since the former cost less than the latter systems, consumer surplus would be increased, not decreased as Mr. Butt states.

As a glance at my Figure 3 will show, I do not propose converting "much of our energy industry to the direct burning of coal"—the part so replaced is indeed rather small, though critically important to oil and gas supplies—and do not propose to install large amounts of transitional equipment that is rapidly made obsolete by advancing soft technologies. On the contrary, I propose using transitional fossil-fuel technologies to stretch existing fluid-fuel reserves of relatively low extraction cost while building up infrastructures that can be adapted to soft technologies. This sort of re-use of investment is novel and has not been a major design criterion before.

Butt: This is by no means the end of Lovins' proposals to replace large blocks of our existing capital stock with new and different facilities. He proposes that energy consumption in the transportation industry be reduced by two-thirds. When we recognize that, within the next 50 years, we will have to provide transportation for a substantially larger population, and as we also consider that a substantial portion of our transportation needs are presently met by quite efficient modes of transportation, it would seem that reaching this goal requires much more than simply substituting small cars for large cars. It is reasonable to conclude that the reduction which he proposes would require very large-scale use of mass transit. Greatly increasing our use of mass transit implies much more than an investment in buses, rapid transit lines and the like. The pattern of our urban-suburban residential and industrial development has been such that the complexity of the travel patterns which have developed does not lend itself to cost-effective use of mass transit to the extent that would be needed to achieve the level of transportation energy efficiency which Lovins glibly represents as "a reasonable estimate for early in the next century." The sought for increase in transportation energy efficiency would appear to require a massive restructuring of our metropolitan areas; and with it, massive replacement of residential and commercial structures. Again, a process of forced abandonment of our capital stock and its replacement.

Lovins: It is true that population in the U.S. might increase 20–30% in the next 30-odd years, but it is not true that "a substantial portion of our transportation needs are presently met by quite efficient modes of transportation"—not unless "substantial" is construed in a strained and unnatural fashion. Of the 18.9 q used in U.S. transport in 1973, 9.8 went to cars and 3.9 to trucks (many well below easily obtainable efficiencies and with empty backhauls). A further 1.3 q went to aircraft—often the least energy-efficient but the fastest-growing freight mode. Pipelines, mainly to carry oil and gas, took 1.8 q, and ships a mere 0.3 q. Of the more efficient modes, rail took only 0.6 q and buses 0.2 q, with all others totalling only 1.1 q. Thus the energy impact of the more efficient modes is as slight as their relative use (though not in proportion, of course, as they *are* more energy efficient).

My analysis did not assume *any* change in settlement patterns—not even the shift to the countryside that has gone on steadily since about 1970—nor did it assume "very large-scale use of mass transit," still less "a massive restructuring of our metropolitan areas . . . and . . . massive replacement of residential and commercial structures." I assumed instead that a trebling of car efficiency—to levels readily attained in safe and comfortable foreign cars today—and less dramatic efficiency improvements elsewhere in the energy sector would suffice to treble the energy efficiency of the transport sector over 50 years despite the modest population growth expected over that period. The more detailed analyses of settlement patterns done by CONAES CLOP panel and others suggest that my threefold improvement leaves room for substantial increases in per capita air travel and driving despite only modest modal shifts that would affect freight more than passengers. The assumptions that Mr. Butt ascribes to my efficiency improvement do not stand up to quantitative analysis. [By 1978–9 it had become clear that a fivefold technical improvement in average U.S. car efficiency is rather straightforward to achieve.]

Butt: Granted that within the next 50 years we must substitute something else for liquid transportation fuels derived from petroleum, I fail to understand the logic of insisting that the alternative must be alcohol produced from farm crops. For some arcane reason, this method of providing transportation energy is looked upon by Lovins as a benign "soft" technology, while all the other alternates are evidently rejected as being "hard." The capital represented by the land area of the United States is one form of capital which we cannot expand or replace no matter how diligently we strive. Furthermore, our arable land area is now largely utilized for food production and increasing the extent of our crop land will inevitably require massive investment in land reclamation. Let us not forget that any crop which can be converted biologically into alcohol can be converted into food. Thus, alcohol production must inevitably compete with food production for the use of the present arable crop land and/or for the use of any additions to arable land area. The importance of this competition cannot be overlooked. Unpalatable though the prospect may be, there seems little doubt that, short of a succession of major catastrophes, the world population in 50 years will be much larger than it is today, with corresponding increases in the demand for foodstuffs from the United States.

Growing crops to produce alcohol may properly be regarded as one method of converting solar energy into a useful form. The conversion efficiency of plant life, measured on a whole plant basis, is on the order of 1% or 2%, under the best of conditions. There is a further loss of efficiency in converting plant tissue into alcohol and it is unlikely that the net conversion efficiency would exceed 1%, on the average. We may logically ask why we should consider this solar application as being desirable while dismissing as undesirable the use of solar generated electricity to power electric vehicles. After all, the conversion efficiency of solar electric devices is ten or twenty times higher and can make use of land area unsuitable for cultivation or rooftop and other areas unavailable for cultivation. Can it be that Lovins, in his wisdom, has decided for the rest of us that growing farm crops to produce alcohol is consistent with his model of "elegant frugality" for the American peasant 50 years hence, while investing in solar electric plants is not?

Lovins: As noted in my response to Dr. Lapp, crop wastes account for only about 39% of my assumed fuel alcohol production; and since these wastes are already being produced (and wasted for the most part), and often are already being harvested too, no additional land would be required and the process would not compete with food production. Indeed, it is quite possible that the shift away from highly intensive chemical agribusiness, which integrated farming and residue use implies, would increase long-term fertility and the resistance of our globally important cropland to ecological instabilities.

The comparison in efficiency between bioconversion and solar electricity to run vehicles ignores the likelihood that the latter would be far more costly than the former and would require precisely the costly changes in infrastructure (e.g. in our ways of making and repairing vehicles) that Mr. Butt decries. If really cheap photovoltaics become available (see *Soft*

Energy Paths at p. 143, n.28), this conclusion might be reassessed. More generally, Mr. Butt's comparison between solar-electric and bioconversion technologies wrongly assumes that the latter would require large additional land commitments. Neither would necessarily do so—solar-electric would not with photovoltaics on buildings, and bioconversion would not if, as I assume, the feedstocks came from existing farming and forestry wastes (and municipal wastes, which would indeed *save* land now used for dumping). Of course either solar-electric or bioconversion *could* be land-intensive if done in ways I do not assume.

I have not "decided" anything "for the rest of us," but rather encouraged public choice; and I do not envisage, save perhaps in the post-collapse phase of a hard energy path, what I presume Mr. Butt means by "the American peasant."

Butt: Lovins makes numerous claims that the capital cost of the proposed "soft" energy technologies is a great deal less than that of the "hard" technologies and therefore claims that the solution which he favors will have less impact upon capital needs than the "hard" course. In those solar areas where I have specific knowledge to bring to bear, I will discuss these allegations later. As we have seen from the examples discussed above, capital implications are far-reaching since his plan would require replacement of much of our existing capital stock.

Generating the capital required for the least capital intensive approach to resolving our energy problems must certainly be regarded as a difficult problem. Adopting a program which pursues a much less capital-efficient course and which, in addition, requires replacement of much of our present capital stock, simply magnifies the problem.

What does this mean? The process of capital formation is no mystery. Very simply, it is the difference between production and consumption. To the extent that we adopt plans which require accelerated capital formation, we must increase this difference. However, if we reduce the product available for consumption, we necessarily reduce or eliminate the material incentives available to meet the individual expectations of betterment which, as we have seen, are a key element in making a society based upon individual initiative and individual election work.

Recent history has shown us that nations can only briefly, and then only in times of great perceived danger, gain acceptance for policies which reduce consumption, without resorting to the most severe repression of individual liberties.

Chapter X of Lovins' article is most revealing. First of all, in Chapter X, Lovins gives the lie to his claim that the transition to the soft energy technology society can be effected while maintaining economic growth, by identifying the future society which he advocates as one of elegant frugality and one in which individual goals become entirely intellectual goals rather than materialistic goals. He proposes that the mass of the population can be induced to voluntarily accept such a change beginning in the immediate future. All of our experience, both recent and past, uniformly tells us that this will not happen and that a relatively rapid reduction in personal material incentives can be realized only through the mechanism of a severely repressive state. Lovins talks of a "substantial social movement, camouflaged by its very pervasiveness." This is a rather ridiculous statement since, historically, substantial social movements have never been invisible. I suppose Lovins' reasoning is that he accepts as an article of faith the presumption that his perception of the relatively near-term future is so "right" that it *must* be supported by a "substantial social movement." Therefore, if it is lamentably true that one cannot see such a movement, it must be invisible. This is the arrogance of the self-appointed Messiah.

Lovins: I do not claim that soft technologies are "a great deal less" capital-intensive than hard ones, but significantly less. A soft path would not "require replacement of much of our existing capital stock" in any sense in which a hard path would not also require the same replacement by attrition.

Happily—in view of Mr. Butt's rigid view of the materialistic foundations of American society—a soft energy path need involve no reduction in consumption as measured by delivered end-use goods and services, but rather envisages a substantial increase (e.g. a per capita trebling over 50 years) accompanied by a slight per-capita decrease in the energy required to yield those goods and services. Since delivered functions (like access gained by a car) are what people perceive, while energy required (like gasoline for the car) merely costs money, I do not see why such a policy should require "severe repression of individual liberties." Indeed, as a pluralist, I think people should be entitled to drive gas-guzzlers, live in uninsulated houses, etc.—and to pay the full social costs of doing so.

Contrary to Mr. Butt's statement, there is no inconsistency between the ideas in the previous paragraph and those in section X of my article. The purpose of that section was to raise contemporary questions that "can be neither answered nor ignored"—questions of personal values and public purpose that "are the beginning and end of any energy policy," and that must be raised because "Making values explicit is essential to preserving a society in which diversity of values can flourish." My intention in this section is "not . . . to resolve such questions—only to stress their relevance." Accordingly, I do not "advocate," "propose," or Messianically urge any of the things Mr. Butt says I do. Indeed, the things he puts in my mouth are a travesty of anything that could possibly be found there: for example, having "entirely intellectual goals" (where I would emphasize a concept of "enoughness" in the most affluent society on earth), having "a relatively rapid reduction in personal material incentives" (precisely the opposite of what I said), having an "invisible" but substantial social movement (what I said was "camouflaged," and I cited in n. 36 a book marshalling much of the available evidence on the subject: *An Incomplete Guide to the Future* by W. W. Harman). Mr. Butt's failure to perceive any signs of that social movement is not necessarily evidence that it does not exist. Indeed, I suspect that his obvious anxiety over what he mistakenly thinks I said may reflect his dislike for pervasive changes in values and perceptions which he *is* seeing in his own society.

Butt: Let us examine Lovins' "substantial social movement." Of what does it consist? We can and do recognize that a modest proportion of upper middle class individuals, having reached a substantially high level of personal affluence, have elected to establish personal goals which are no longer materialistic. They are joined by another small minority of individuals who are not affluent but who have elected to drop out of society. These are the dimensions of Lovins' substantial social movement. There have been other similar movements in the past, all of which have foundered as the missionary zeal of the proponents of the new revelation has been replaced by a more mundane concern for tangible rewards.

Lovins: Accounts such as Harman's may help Mr. Butt to find the evidence he seeks. Alternatively, he might reflect that successive Harris polls have tended to show by 3–1 (or larger) margins that most Americans feel inclined towards less materialistic values and more inclined towards those Harman describes—while also feeling they themselves are disproportionately affluent in a world of mass poverty and that this inequity may rightly cause poorer countries to turn against the U.S. The detailed Harris findings are remarkable.

Butt: Lovins claims that the tally of the ballot box supports his contention that a substantial—but invisible—social movement exists which supports his concept of an ideal, elegantly frugal society. I think most analysts of the last election will tell you that the central issue of the election was economic and that, generally, the winners were those who convinced the voters that their policies were the ones best suited to accelerating economic growth.

Lovins: Mr. Butt makes a more specific and tendentious case than I did. Of course I was writing before the November 1976 election, referring to the frequent successes of candidates who advocated environmental over material values (e.g. Governor Brown's "politics of lowered expectations"), but the same trends continued in 1976, including President Carter's election: he did, after all, say explicitly on several well-reported occasions during the campaign that if it ever came to a choice between the environment and economic growth, the environment would win.

I do not know what analyses Mr. Butt is referring to, but some prominent analysts felt the main issue, so far as one could be identified, was personal trust and competence. It also remains true that a candidate unfortunate enough to appear on Environmental Action's "Dirty Dozen List" is likely to lose his or her campaign—the effect of a very tangible social movement, as many diselected incumbents have found out the hard way.

Butt: In summary, Chapter X of Lovins' article truly identifies his goals as being those of creating a "new society" founded upon the concept of "elegant frugality"—in short, a society of peasants and craftsmen. The promises made earlier in the article that we can abandon much of our present energy capital stock and replace it with "soft" energy without giving up the material benefits which we have received from our present system are revealed by Chapter X as purposeful falsehoods.

Lovins: Mr. Butt's summary of my supposed "society of peasants and craftsmen" is a travesty of my views and a complete fabrication from section X, which merely raises salient social issues without stating what my views are. (Indeed, he has given altogether too much emphasis to section X, which is far less important than section IX.) If he wants to know what my views are, he should look on pp. 12–14 of *Soft Energy Paths*—but while these views obviously underlie the way I look at the world, they are not a blueprint I have any wish to force on other people: I should not like Mr. Butt to have to share my views any more than I want to share his. Meanwhile, I think it is important for him to ascribe to people who disagree with him the same honesty, competence, and good intentions that he would wish them to ascribe to him.

Butt: Let us turn now to some of the specifics and analyze some of the gross errors which Lovins makes.

Lovins advocates widespread use of solar heating and cooling. It is evident from studying the article that Lovins has considerable familiarity with the details of solar heating and cooling technology. So do I. Lovins states that solar heat is cost competitive with electric heat throughout most of the United States. This is true—but only if the solar system is limited in size so that it may operate efficiently. The fraction of the heating and hot water load which can be cost-effectively met by a solar installation varies significantly with climatic conditions, generally over a 40% to 60% range in the United States. Lovins states that the cost-effective "solar fraction" would be increased if the cost of electic energy as the alternative to solar were calculated upon the basis of the marginal cost of new electric facilities rather than upon typical electric rates which are based upon average cost, factoring in the cost of generation from less costly older facilities. Since inflation has, indeed, increased the cost of new electric generating facilities, this conclusion is, in some measure, true. To this point, Lovins adheres to the facts but, thereafter, departs from them with the statement, "On that basis, 100% solar heating, even with twice the capital cost of two-thirds or three-fourths solar heating, is almost always advantageous." The wording of this statement implies that a doubling of capital cost for a 100% solar system as compared to a 66% or 75% system is a conservative estimate. This is simply not true. Many careful analyses of the relationship between solar heating system size and the percentage of total heat requirements supplied by the system have been made. One of the more reliable such studies indicates in the case of Fresno, California, that a system intended to provide 85% of heating needs would be 2.8 times as large as one designed to carry 60% of the needs. In greater detail, system size for various percentages solar follows this pattern:

60%	1.0
70%	1.28
75%	1.65
80%	1.94
85%	2.80

As is typical of such studies, a 100% solar system size has not been calculated on the fundamental grounds that it would be too large to be of interest. Since there is a 44% increase in system size required to upgrade from 80% to 85%, while only an 18% increase in system size was needed to gain from 75% to 80%, it is obvious that any logical extrapolation of the data would show that the "100% solar" system was very large indeed. Climatically, Fresno is somewhat better for solar than the average for the United States.

Since the realities of climatic conditions will not change radically in 50 years, I must logically take the position that, although solar heating can make a major contribution to our national energy budget in the next 50 years, it cannot reasonably be a 100% source of heating energy. Since, unlike Lovins, I am also willing to accept the continuing need for the use of "conventional" energy sources to provide that portion of heating load which cannot be logically provided by solar energy, I can accept reality. Not so Lovins. He must claim that 100% solar heating is practical to support his thesis that we can dispense with conventional energy sources as a supplement to solar heating. Since the basic knowledge of solar technology which Lovins evidently possesses must tell him that his final conclusion is not only false but very substantially so, the only reasonable conclusion which I can reach is that his statement, as I have quoted it above, is not only untrue but deliberately so.

Lovins: A major point of disagreement here is the fraction of space and water heating load that can be satisfied by solar systems with storage systems (using sensible heat in water) having capacities of the order of 0.4–0.5 m^3 per m^2 of collector. I claim 100% under U.S. conditions (or, with about [0.7] m^3/m^2, under Danish conditions); Mr. Butt claims a much smaller value, typically 40–60%, with any economically acceptable storage volume. I have analyzed this disagreement at length in my earlier correspondence with Professor Bethe. Professor Bethe, accepting my analysis, calls for an assessment of the same problem in the U.S., since I had used mainly Danish and Canadian data; but I submit that if seasonal heat storage works in Denmark—a cloudy country at nearly the latitude of Anchorage—or in Canada, it must *a priori* work anywhere in the lower 49 U.S. states (and in much of Alaska too). [Excellent results have indeed been achieved, more recently, with passive solar buildings in Alaska.]

Mr. Butt's analysis of the increase in solar system size (meaning chiefly collector area) necessary for 100% solar heating rests on two unstated assumptions: that the standard of insulation and heat recovery in the house is far worse than is economically attractive at the margin (*e.g.* seven times worse than in the Danish house I analyze), and that covering a larger share of the heat load entails chiefly an expansion in collector area. The first of these assumptions not only increases the total heat load, but makes the storage problem much worse by doing nothing about the peak heat loads on cold winter days. This peaking greatly increases the need for both collection and storage. The second assumption, by constraining the design tradeoff between collector area and storage volume, requires emphasis on the former, even though an installed m^2 of collector costs some 4–6 times as much as an installed m^3 of storage.

What Mr. Butt's argument reveals, therefore, is a common and lamentable failure of solar specialists to work closely enough with specialists in the design of energy-efficient buildings. Seasonal heat storage, as Mr. Butt says, does not make economic or engineering sense in a house with as high and "peaky" a heat load as most houses have today. Neither solar nor conventional heating makes sense if you live in a sieve. But reducing and flattening that heat load is desirable and economically attractive no matter how the house is to be heated, and, if done before (or during) a solar retrofit, enormously expands the freedom of the solar designer—so far that seasonal storage, far from being impracticable, becomes cheaper than having backup. This is essentially because storage is water (or rocks, etc.) in concrete, whereas backup is a precious fuel plus copper, stainless steel, and other costly and precisely machined materials. I am confident that once Mr. Butt has had an opportunity to review my correspondence with Professor Bethe (and the references that it—like my original paper—cites), he will agree with us that 100% solar-heat retrofits do indeed work in conditions much less favorable than the U.S. The reasons he did not agree—including the "Pacific Regional Solar Heating Handbook" data which Professor Bethe also cited—lie outside his realm of expertise, in building design rather than solar hardware, making possible solar design philosophies that he probably has never considered before.

I am pleased to note that Mr. Butt feels solar heating "can make a major contribution to our national energy budget in the next 50 years." I think that once he has reexamined the tacit premises behind his erroneous conclusions about seasonal heat storage, he will agree that he has underestimated this contribution and so misjudged the accuracy of my own statements. There is nothing ideological about my rejection of solar-with-backup systems; rather, a straightforward economic and engineering argument that such systems introduce unnecessary cost and complexity to use a feature (backup) that is symptomatic of trying to add solar heat to an inadequately heat-conserving building, and that it is more cost-effective to get the building right first than to waste time, money, and fuel trying to substitute costly solar collectors and fuel for relatively cheap insulation and recuperators.

Butt: Page 82 in Lovins' article carries us one step further forward into the realm of fantasy. He now proposes that what we call "active" solar systems, using separate collectors and storage, can be replaced in new construction with entirely "passive" systems. It is true that, as a solar system is designed to carry an increasingly large percentage of the total heating load, thermal storage requirements increase exponentially. This is part of the reason for the exponential increase in cost of active systems designed to carry overly large percentages of the heating load. The problems of providing long-term heat storage in passive systems are much more severe than in active systems and, therefore, a passive system carrying 100% of the heating load becomes even more difficult to achieve than the "100%" active system.

Lovins: Mr. Butt's remarks on passive solar systems suffer from the same inappropriate assumptions about the thermal integrity of the building. I contend—and real houses exist to support my contention—that passive solar designs, in buildings with effective insulation and heat recovery today considered unusual, can carry the entire space heating load even in very severe climates. (Special collectors, though a small area of them, would still of course be needed for domestic hot water.) With care, this can be done in most climates by using the thermal capacity of the building itself: windows facing into properly designed rooms are very effective and rather selective solar collectors that capture substantial heat even on a cloudy winter day. In recalcitrant cases, storage in directly or convectively heated media (*e.g.* rockpiles heated by convective airflow) can be used to expand heat capacity virtually without limit, still within passive-solar criteria of not using forced circulation of a working fluid nor what are normally thought of as solar collectors.

Butt: It is quite true, as Lovins states, that the storage of low and medium temperature solar generated thermal energy is technically quite straightforward, using water tanks, rock beds or perhaps fusible salts. However, the economics of long-term storage of medium temperature thermal energy cannot be ignored. If one assumes a relatively wide operating temperature range for solar thermal storage, approximately 120 Btu per pounds of water can be stored. Fusible salt storage technology, which is still not fully developed, promises to increase storage capability per pound stored by a factor of at most six. On the other hand, storing coal provides 10,000 to 12,000 Btu of energy storage per pound. Even when this is discounted by two-thirds to reflect the loss of energy in electric power generation and transmission, we find that energy storage of coal at an electric power plant is substantially the more efficient mode. Thus, it is easy to see why the high capital cost of long-term, high volume, solar thermal energy storage is one of the reasons why "100% solar" is difficult and expensive to achieve. The optimization of solar energy storage has also been studied extensively and it appears that the optimum situation is to provide sufficient storage to assure that supplementary energy, as well as primary solar energy, can be stored in sufficient volume to make it possible to restrict demand for supplementary energy to off-peak periods.

Lovins: If coal were in infinite supply and cost nothing to use (and, though my economic analysis excludes externalities, if mining and burning it had no side-effects), it would be true that energy is more easily stored in the potential chemical energy of reduced carbon than as heat in water. But these conditions do not in fact obtain, and though storing heat for a house as hot water in an insulated tank is not cheap, it turns out to be cheaper than not doing it. It is possible that Mr. Butt has been misled by excessive assumed capital costs for storage tanks: there are several technologies available that are much cheaper than conventional steel tanks.

The optimization of solar systems has been studied fairly extensively, but unfortunately, the combination of cheap storage and very energy-efficient buildings that I analyze has not been much studied. This omission has led to the widespread use—by Mr. Butt and many others—of various rules of thumb that are not correct under different circumstances that can be realistically achieved. I therefore believe a thorough re-examination is called for, and I understand that the ERDA Solar Division and Los

Alamos are interested in doing one. [Later studies in several countries supported Lovins's contentions.]

Butt: We in the Solar Energy Industries Association are dedicated to aggressively pursuing the commercialization of solar utilization within the framework of our present society based upon individual election and individual initiative. We are dedicated to accomplishing this in a cost-effective manner, which means in a manner which allocates scarce capital resources to solar to the extent that they can be effectively employed. On this basis, an aggressive program, including substantial Governmental assistance but avoiding compulsive measures, can provide substantial contributions to our total energy budget. Somewhat over 20% of our energy budget is used as low temperature heat to heat domestic hot water, heat building space and to air-condition building space. We think we can provide about one-fifth of this as solar energy or 4% of our total budget in the intermediate term—by the early 1990s—and ultimately—by 2025—provide about 60% of those needs or 12% of our total budget with solar energy. Another large increment of the national energy budget is the category of industrial process heat. Solar thermal energy can provide a substantial contribution here as well, although, for technical reasons, the rate at which the solar contribution can be realized will be somewhat slower than in the case of heating and cooling. Less precisely, we estimate that 2% of the national energy budget may be solar process heat by the early 1990s and 10% by 2025. Although the percentages may seem modest, the magnitude of energy requirements is such that their importance is very great.

Lovins: Low-grade heating and cooling of buildings and of residential and commercial hot water accounts for 25% (26% with refrigeration) of U.S. delivered end-use enthalpy. If, like Professor Bethe, Mr. Butt accepts my analysis of 100% solar heating, then his estimate of market saturation by 2025 would imply meeting 25–26% of national energy needs with direct low-temperature solar collection by 2025 (assuming the end-use structure remains unchanged: in fact the proportion of all end-use needed as low-grade heat would decline). In the present temperature spectrum of industrial process heat, about 24% (including preheat) is needed at temperatures below 100°C, and approximately another 48% (including preheat) is needed at temperatures between 100 and 600°C, probably the practical limit for present solar technologies operable on fairly cloudy days (with highly selective surfaces). These process heat terms—documented in *Soft Energy Paths,* Chapter 4—would together account for a further 22% or so of all end-use enthalpy (assuming the present end-use structure; in fact the process heat spectrum may shift toward significantly lower average temperatures). Thus solar heat technologies usable in cloudy climates could asymptotically cover 48% of the present U.S. end-use structure, leaving a small fraction (8–10%) to be covered by heat resources yielding over about 600°C—precisely the category most likely to be reduced by industrial process improvements and shifts to more appropriate building materials. Such a roughly 50% contribution is, of course, even more important than the 22% Mr. Butt envisages for 2025.

Butt: The immediate problem which the nation faces is to develop replacements for petroleum and natural gas. The extent to which we depend upon these energy resources is several times greater than the potential solar contribution. Accordingly, we have concluded that our responsibilities as citizens dictate that we support a variety of other energy technologies which can also make significant contributions to the problem. Such information as we have been able to gather, relative to the potential contribution from other alternatives, convinces us that we must exploit not one but several. Among these, we include the "hard" technology of solar thermal electric generation and solar photovoltaic electric generation, as well as other "hard" non-solar technologies.

We feel that it is of transcendent importance that we maintain a society based upon individual election, individual initiative and individual incentive and, therefore, we support and will continue to support an appropriate mix of both the energy technologies which Lovins identifies as "soft" and as "hard." We perceive that any effort to limit the development of new energy resources to those somewhat arbitrarily classified as "soft" and, even beyond this, to abandon our existing "hard" technology assets, must inevitably require that we compel individuals to conform to Lovins' societal fantasies.

Lovins: Fortunately, the potential contribution from solar technologies is substantially larger than that from solar *heating* technologies alone. It includes, for example, the contribution of biomass conversion—enough, I believe, to run an efficient transport sector—and that of wind and other indirect solar technologies, as well as of potential high-temperature solar heat technologies, photovoltaics, etc. Like Mr. Butt, I support a wide variety of energy technologies to meet diverse needs in diverse regions. Unlike Mr. Butt, I have been unable to find a persuasive case for pursuing all technologies, since a program of improved end-use efficiency strong enough to avoid severe oil and gas shortages will make most of ERDA's favorite technologies unnecessary.

Like Mr. Butt, I "feel that it is of transcendent importance that we maintain a society based upon individual election, individual initiative and individual incentive." Unlike Mr. Butt, I believe that goal is consistent with a soft but not with a hard energy path, for the latter "must inevitably require that we compel individuals to conform" to the rigorous imperatives of a central-electric future. I do not feel that my proper role is to prescribe others' future or to declaim authoritatively on how American society must be run. It is puzzling that Mr. Butt—apparently speaking for the Solar Energy Industries Association, despite the disclaimer of organizational endorsement at the front of Mr. Yulish's booklet—should have diversified into this business. Perhaps the Association's talents might better be turned to promoting—and making sure it has not underestimated—the considerable market for its own products.

12

LAPP vs. LOVINS

Critic Alleges "At Least Two Consistent, Unjustified Practices"

One of Yulish's contributors is Dr. Ralph Lapp, an "energy/nuclear consultant" from whom we have already heard. Mr. Lovins calls Lapp's Yulish essay "a condensed and much more restrained version" of comments printed earlier in this book. There is some new material, however, to which Lovins addresses himself.

Lapp: ... I have also found at least two consistent, unjustified practices which Mr. Lovins employs throughout his article. First, a number of debatable assertive statements are presented as matters of fact with no documentation whatsoever. For example, continued use of coal is asserted to make an environmental catastrophe inevitable in the near future, through the greenhouse effect (carbon dioxide releases raising the earth's temperature and ultimately melting the icecaps)—even though Mr. Lovins advocates widespread use of coal in homes, businesses, and industry.

Lovins: Dr. Lapp notes "at least two consistent, unjustified practices" employed throughout my article. He states two. The first is that "a number of debatable assertive statements are presented as matters of fact with no documentation whatsoever." I would have expected Dr. Lapp to cite as his lone example of this "unjustified practice" some case where the Editor of *Foreign Affairs,* imposing space constraints on an article already half again as long as any previously published in his journal, deleted my proposed footnotes; for this often occurred, and could be remedied only by cross-reference to my Oak Ridge technical backup paper (and, later, to *Soft Energy Paths*). But instead, the only example of the "unjustified practice" which Dr. Lapp cites is:

> For example, continued use of coal is asserted to make an environmental catastrophe inevitable in the near future....

The passage to which Dr. Lapp refers actually reads as follows:

> The commitment to a long-term coal economy many times the scale of today's [in my illustrative hard energy path] makes the doubling of atmospheric carbon dioxide concentration early in the next century virtually unavoidable, with the prospect then or soon thereafter of sub-

> stantial and perhaps irreversible changes in global climate.[5] Only the exact date of such changes is in question.

Readers more careful than Dr. Lapp will note that (1) my statement is amply documented with survey articles by leading experts on climatic change, including references at the level appropriate for readers of *Foreign Affairs* (my Oak Ridge paper, n. 65, and *Soft Energy Paths,* p. 28, n. 5, contain fuller references for scientific readers); (2) these references fully support the statement I made (note particularly the analyses by Dr. Wolf Häfele's group at IIASA and by Dr. Alvin Weinberg's group at Oak Ridge); (3) I made no reference to "an environmental catastrophe" nor to its being "inevitable in the near future"; (4) the first-order effect of concern is not melting of the icecaps, which have a long time constant, but rather loss of Arctic sea-ice and hence of the semipermanent polar anticyclone, thus altering weather patterns in at least the Northern Hemisphere; (5) a possible cause of this alteration is a long-term commitment to a rapid rate of fossil-fuel combustion (presumably coal combustion owing to the scarcity of oil and gas during the period in question); (6) melting of the Arctic sea-ice, which is generally believed to be irreversible during periods of policy interest, would not alter sea-level, since the sea-ice is floating isostatically like an ice-cube in a glass; (7) the relationship between CO_2 release, atmospheric CO_2 concentration, and climatic change is subject to many uncertainties, properly indicated by such qualifications as "early," "virtually," "prospect," "or soon thereafter," "substantial," "perhaps," and "in question"; (8) my concern about the CO_2 problem, which rests on substantial knowledge of climatic change, is consistent with my advocacy of a soft-energy path, which minimizes the integrated fossil-fuel burn, getting us out of the fossil-fuel-burning business as quickly as possible. As I state on p. 88 of the *Foreign Affairs* article,

> The hard path entails serious environmental risks . . . . Perhaps the most awkward risk is that late in this century, when it is too late to do much about it, we may well find climatic constraints on coal combustion about to become acute in a few more decades: for it now takes us only that long, not centuries or millenia, to approach such outer limits. The soft path, by minimizing all fossil-fuel combustion, hedges our bets.

Thus the transitional use of coal—at a far lower level than advocated by many analysts, including Dr. Lapp—is a temporary stopgap to help

---

[Footnote 5 reads as follows:]

> [5]B. Bolin, "Energy and Climate," Secretariat for Future Studies (Fack, S-103 10 Stockholm); S. H. Schneider and R. D. Dennett, *Ambio* 4, 2:65–74 (1975); S. H. Schneider, *The Genesis Strategy*, New York: Plenum, 1976; W. W. Kellogg and S. H. Schneider, *Science 186*: 1163–72 (1974).

supplies of fluid fuels keep up with our needs pending improvements in end-use efficiency and deployment of soft technologies. The strategy of which this brief and sparing use of coal is an important part is designed in response to, not in ignorance of, such constraints as the CO_2 problem.

Lapp: Second, Mr. Lovins fails to discuss in any systematic way the environmental, economic, social, political, or lifestyle consequences of a radical conversion to a "soft" energy path, while dwelling at length on the difficulties he perceives with current energy strategies.

Lovins: Within the severe constraints of space in *FA*, I discuss all these items as fully as possible, with special emphasis on economics (sections IV–VII and a cited backup paper containing probably the most comprehensive economic analysis of a soft energy path published anywhere), sociopolitics (sections IX–X), and lifestyles (sections IV and X, the former making it clear that my analysis assumes a "pure technical fix" rather than a mixture of technical and social changes). These topics are also discussed more fully in the Oak Ridge paper and, with even less constraint on space, in *Soft Energy Paths*.

The "radical conversion" is Dr. Lapp's idea, not mine: what I said is that the soft path is a radically different approach to the energy problem, not that it entails any radical restructuring of society. Dr. Lapp, repeatedly echoing "radical" out of context, has fabricated the latter idea and implied that it is a fair representation of my proposals.

13

MEINELS vs. LOVINS

Soft Path Leads to a New Dark Age, Meinels Insist

Dr. Aden Meinel is Professor of Astronomy and Optical Sciences at the University of Arizona, where his wife Marjorie Meinel is a Research Associate. The Meinels are best known to the public for their advocacy of large-scale, centralized, solar-thermal electricity generating plants. Through Charles Yulish, editor of *Ten Critical Essays*, we have been informed that the Meinels refuse "to have any of their statements taken out of context." We take this to mean that they will not agree to our "dialogue" format, which couples Lovins's rebuttals as closely as possible to the material rebutted. Accordingly, we resort in this chapter to a parallel-column technique. Another condition insisted upon by the Meinels was that we use their article in its entirety. We have done so.

Meinels: We are chilled at the actions adovcated by this seductive and well-written article and appalled that a person claiming to be a physicist could resort to what appears to us as distortions of technical reality. The danger of such an article is that the appeal to a return to the "good old days"—ultimately to the Garden itself—plays upon a deep-rooted desire of the human species. Should this siren philosophy be heard and believed we can perceive the onset of a New Dark Age. To abandon high technology and large-scale cooperative organization is a step toward making mankind once more a slave to the dictates of a dispassionate environment and his own furies, with luxury for a few and depths of despair for the masses.

Lovins: It is astonishing that a soberly phrased article in an "establishment" journal should have moved two scientists to such flights of emotional rhetoric. It is neither customary nor seemly to frame professional controversies in such terms as "claiming to be a physicist," "appeal to a return . . . to the Garden itself," and "we can perceive the onset of a New Dark Age." I find it hard to respond substantively to such an outburst, but shall try to extract whatever matters of fact or logic I can find stated or implied.

The Meinels' first paragraph implies that I advocate abandonment of high technology, whereas in fact I propose its more intelligent application: the conservation, transitional, and soft technologies I advocate involve challenging technologies of all kinds, but direct technical sophistication into achieving simplicity rather than complexity. Some technical people under-

Meinels **Lovins**

stand what I mean by that and some don't.

The Meinels further imply that I want to see the abandonment of "large-scale cooperative organization." If by this they mean political chaos or a weakening of democratic institutions, no; if reversal of the current trend toward technocratic autarchy, yes. A soft energy path has nothing whatever to do with a penurious, uncertain, undisciplined, poverty-stricken, or inequitable society. How such a meaning could be extracted from my article is perhaps in the realm more of projective psychologists than of semanticists.

o

Lovins' characterization of the two choices as the "soft" road and the "hard" road is interesting. We would characterize them instead as "soft science" vs. "hard science." We object to his artful presentation method. He has meticulously taken and cited only favorable instances for his pet hypothesis and only the worst pertaining to the philosophy he wants to discredit. This is not good science. His citations deal with subjects where there is a proliferation of soft science articles and a scarcity of hard science articles, and the latter are often ignored by persons who don't want to listen to their results. The profusion of references appears to indicate scholarship, but unless some critical analysis is shown for both sides of the question, the citing of references is scarcely more than academic window dressing.

o

The Meinels do not justify the claim of "soft science"—possibly they overlooked my extensive references to rigorous and detailed work by reputable scientists—nor of an "artful" presentation method and asymmetrically loaded arguments. Citations of precisely which relevant references I have ignored or where I erred in fact or logic would have been particularly helpful, and are the elementary duty of a responsible critic to specify. To reject as "academic window dressing" the literature I cite, without saying why it is wrong, is unfairly to malign many authors more distinguished than myself.

o

When Lovins discusses solar and other exotic options, he is naive, accepting only what pleases him and ignoring very fundamental problems as though they did not exist. There are problems for both high technology and low technology solar options. We have traced the discouraging history of attempts to inject new low-level solar

o

The solar technologies I advocate are in no sense "exotic" and the charge of naivete is not supported.

The Meinels do not specify the "very fundamental problems" I have allegedly ignored any more than they specify the "distortions of technical reality." The failure of some solar projects (or, even more often, of high-technology indus-

Meinels

technology into developing countries and know that the facts of capital and economics are as inexorable in these societies as in advanced societies. To like only simple technology and eschew any technology that appears beyond the comprehension of the average individual is an illusion.

Lovins

trial projects of other kinds) in developing countries reflects in general a failure to appreciate the capital and social constraints of developing countries. The economic comparisons between soft and hard technologies are thoroughly treated in my article and its technical backup material.

I have not said that all technologies must be simple (or small), but rather that complex (or big) technologies have long ago saturated the limited applications for which they are appropriate. Moreover, "understandable" technology—a socially defined concept akin to Ivan Illich's "conviviality"—does not imply that an ordinary person can necessarily understand in detail how the device works, but rather that he or she can integrate its operation into everyday life, run the device without being run by it, and, in short, live with it as a tool rather than a machine. It is incidentally true—and psychologically important—that people living in a sensibly designed solar-heated house can generally gain, without special training, a good intuitive grasp of its general operation, and can anticipate its operating characteristics and diagnose most faults, even though they are unlikely to be able to design such a solar system themselves. But that is more a consequence of "understandable" design than a definition of it.

Meinels

o

Our work has taken us to much of the world, and we have seen that Lovins' type of simple society does exist in some developing countries. His proposals greatly resemble Mahatma Gandhi's "cottage industries," held by many educated Indians to have been a diversion from the effort to improve the lot of the masses. One of the greatest improvements in the welfare of the village inhabitants in India has been, contrary to Lovins' hypothesis, the electrification of the villages. This step parallels the dramatic changes caused by the Rural

Lovins

o

The "simple society," Gandhian cottage industries, and similarities to some developing countries that the Meinels claim to be similar to what I am advocating is indicative of their frame of mind but not of what I wrote.

Rural electrification, in India, the U.S., or Australia, was in many respects beneficial to those who received it, but in some cases may have gone too far into a region of diminishing marginal returns, and is generally felt by informed Indian analysts to have been a poor use of scarce capital: enormously

Meinels

Electrification Agency in the United States in the '30's, a period before the personal knowledge of most of the population today. His proposal also has echoes of Mao's "great leap forward," wherein technology was forced toward backyard industries, including even steel smelting! It, too, was a notable failure of an enticing dream. For the industrialized world to toy with such a set of ideas could be the beginning of a irreversible process leading downward, one from which recovery might be denied as problem piles upon problem and acrimony upon acrimony.

Lovins

greater welfare could have been won for the villages by, say, investment in biogas plants and clean water supplies, even though the latter programs would have cost far less. (See e.g. *Soft Energy Paths,* p. 7, n. 9, and p. 97, n. 29.) Rural electrification in India has tended to increase inequities within the villages and has been only about a tenth of the total electrification effort: most of the power goes to a small number of urban industries and rich households. Moreover, ill-considered electrification of such devices as irrigation pumps has led to absurd dependence of food supplies on scarce diesel fuel (interrupted during and since the 1973 oil embargo) even in areas where wind pumps offer a very attractive alternative. [See S. K. Tewari, *Science 202:* 481–6 (3 Nov. 1978)]

The Meinels nowhere specify why soft technologies matched to the end-use structure are not capable, as I claim, of meeting essentially all the energy needs of a mature industrial society such as the U.S.

o

Meinels

The series of Utopias that have lured many idealists have inevitably fallen short of their idealism when man attempted to translate them into the realities of the real world; although, as we in America know, bits here and there survived to enhance society and the human condition. Even Christianity, the bright hope for mankind, almost extinguished the light of reason and learning for a millenium due to the imperfections of man, who translated it into power politics and superstition of the worst kind.

Lovins

Abstract reflections on "the series of Utopias that have lured many idealists" do not help us with the concrete choice between a soft path—notable for its attitude of humility toward human frailties—and a hard path—notable for its Utopian emphasis on technological near-perfection. The Meinels seem to imply that my soft-path concepts are, or risk becoming, some sort of dogmatic theology. I hope not, and think they do no service by the type of attack that might encourage some impressionable people to make this error.

o

Meinels

Cheap energy is truly the basis upon which modern society is built and which frees us from the severity of a dispassionate environment. Agriculture and fire set the basis for cities, but since only slaves and animals were available

Lovins

Cheap energy in advanced industrial societies may be more a cost than a benefit (*Soft Energy Paths*, pp. 10–11), and only about 1.5% of all U.S. commercial energy flows are a substitute for human labor. I am not saying that

Meinels

for motive power and industry the lot of most humans was rather dismal. The harnessing of energy and inventions, and laws to control their exploitation, opened the way to freedom and dignity. To preserve these gains and extend them to all nations and peoples we need assured future supplies of inexpensive energy. This goal means that it will take the best of our technology to find the most economical mix as some present energy options become scarce and are reserved for higher priority uses. We see no way of avoiding nuclear power in that mix.

Lovins

energy is not indispensable; rather, that it is not at all obvious that energy should be unrealistically cheap (though I think it would be cheaper in a soft than a hard path), and almost certain that energy should not be disproportionately abundant to a species that has often demonstrated its indiscipline in managing large stocks of energy to the detriment of itself and its habitat. The Meinels' conclusion that nuclear power is essential is of some sociological interest, but, being completely unsupported by any sort of analysis (including consideration of quantitative arguments to the contrary), is judgmental and unpersuasive.

o

We personally do not fear nuclear power plants, but do feel that precautions must be taken to assure safety of these plants from local accidents or sabotage. Disposal of the wastes has gotten much recent attention and long-term solutions appear to be emerging. The fear-mongers, however, have so sensitized people that emotions may dominate any nuclear solutions, to the detriment of all of us. We would prefer other solutions than nuclear, but it presently is the only one having a long-ranged fuel supply within technical and economic grasp. Even the latter point, economic grasp, is being threatened, not by technical problems but by legislative and regulatory impact. In a sense we are creating our own problems concerning the economics of both nuclear and fossil energies by legislative overkill that in the perspective of future years may be deemed a curiosity characterizing our times.

o

While the Meinels, understandably, seem reluctant to diverge from their thesis far enough to engage in a debate on the nuclear issues that I mention only passingly and in outline, they take time for an innuendo about "fear-mongers" and "emotions." It is unfortunate and puzzling that the Meinels should concentrate so heavily on "a long-ranged fuel supply" when (1) few if any energy systems are likely to be resource-limited, and (2) they have devoted much of their professional lives to energy-income technologies and thus should appreciate better than most the fallacy of being fuel-oriented.

o

As we will expand upon later, the "every man for himself" approach for the future, which is what we read in Lovins' article—as contrasted to large-scale societal cooperation in maximizing the benefits for the entire

o

It is curious to contrast the Meinels' apparently Hamiltonian philosophy of central governance with Mr. Butt's philosophy of rugged individualism. One should be grateful for such a pluralistic and politically diverse soci-

Meinels

society—is curiously like the retreat into the castles, strong points and monasteries of the Middle Ages, when problems of society and safety forced continual withdrawal from communication and commerce with others. It is a lesson that history has for us if we but open our eyes to see.

Lovins

ety, and I am; but how the Meinels can construe my advocacy of major national programs in energy and other sectors as an abnegation of "large-scale societal cooperation in maximizing the benefits for the entire society" or as "retreat into the castles, strong points and monasteries of the Middle Ages" defies comprehension. Some—perhaps those of Mr. Butt's persuasion—might even interpret the Meinels' apparent rejection of individual initiative as a resounding call for dirigiste socialism. I do not want to read that into it; but I am also quite unable to fathom what the Meinels' political musings have to do with my thesis.

o

What are the real prospects for the "soft" solar technologies advocated by Mr. Lovins? Not nearly as bright as he portrays them to be. To be sure, there are simple applications of solar energy that have long been used, such as the drying of crops and production of salt. The dream of many is that the sun will become a major energy resource, eventually to supplant fossil fuels entirely. This dream has been frustrated in the past and could be once more, as the perpetual barrier of economics is faced. The enthusiasm of the chase appears to be obscuring the hard facts of economics as ideas for many solar devices for a variety of applications stream forth from imaginative people.

Mr. Lovins seems to believe that solar heating and cooling for homes and buildings is a proven and economically attractive alternative that can be used now. This just isn't the case. The fact is that solar heat is more expensive than heating by electrical resistance and far more expensive than by natural gas or oil. It is hoped that costs will come down in the future, but even the most optimistic advocates talk about competitive prices based upon "lifetime costs." Most people are more interested in "first costs." If you doubt this reality,

o

The Meinels' brief and unspecific mention of economic problems appears to overlook my arguments—set out in considerable quantitative detail in the Oak Ridge paper and even more fully in *Soft Energy Paths*—that though soft technologies, including solar heat, are costly, the hard technologies that are commonly proposed in their place (including the Meinels' nuclear power) are even more costly, in life-cycle cost and generally in capital cost too. The conclusion that solar heat is "more expensive than heating by electrical resistance" is hard to reconcile with even the highly pessimistic ERDA/MITRE study discussed in my response to Dr. Forbes's critique, and must be based on a limited number of pathological examples (possibly in Seattle with its historically cheap hydroelectricity). The reference to "even the most optimistic advocates" again ignores comparative capital costs of complete energy systems at the margin to do a given job (e.g. heat a house).

It is quite right that first costs loom large in a builder's or buyer's criteria—which is why I advocated, and why President Carter wants utilities to offer, capital transfer schemes that would ensure that capital is available for solar

Meinels

talk to any builder or manufacturer about whether the customer considers much more than first costs. However, Mr. Lovins is on the right track when he boosts the use of passive solar energy devices such as good architecture and good insulation and building materials, rather than expensive energy storage devices that would otherwise triple the cost of a solar heating system.

Mr. Lovins is certainly correct when he says that solar energy *can* be converted into heat. There are no serious technological barriers to be surmounted. No scientific breakthroughs are needed, except in the energy storage area. What defeats the effort is that cheap materials are not durable and durable materials are not cheap. Since solar energy is dilute, very large areas are needed to collect enough energy. The reason solar water pumping died in 1913 is clear when you realize that several acres of apparatus and machinery were required to yield 50 horsepower at the pump, whereas a 50-horsepower gasoline engine for a water pump can be lifted by two men in the back of a small truck.

The problems that impede the achievement of economic solar energy are the cost of collectors durable enough to survive the environment and the cost of energy storage. The operational problem is the need for solar and wind energy to have a backup energy supply, which adds to the overall cost through the existence of other expensive facilities not fully utilized during the year.

Lovins

and other energy investments on the same terms as it is available to build power stations. I appreciate the Meinels' advocacy of passive solar systems, but do not agree that storage systems would treble solar cost—unless storage is substituted for, rather than added to, energy-conserving building design. Likewise, I do not agree that "scientific breakthroughs are needed . . . in the energy storage area." Breakthroughs in latent-heat storage—in fusible salts, ammines, etc.—would be useful but are in no sense "needed" for seasonal storage to be widely and cheaply practiced.

The diluteness of solar energy is often overestimated—U.S. insolation averages a respectable 180 W/m^2—and may be a positive advantage in preventing us from accumulating amounts of energy that go beyond the realm of utility and into that of harm (to the earth or other people). Solar water pumping did not die in 1913; it was ahead of its time, and, with the cheap oil dream fading, is reviving. The gasoline engine the Meinels mention is an unfortunate example, for it relies on a fossil fuel made of primeval sunlight gathered over a very long period and a very large area of ancient swamp. As that legacy is used up, we are obliged to begin once more to live—more cleverly and sustainably—within our income, rather than spending our capital as it if were income.

The generic problems of collector cost and (depending on the application) perhaps storage are well known and have been carefully addressed in my analysis. [See also the Lovins-Weinberg exchange in the Appendix.] The Meinels have not shown any respect in which that analysis is unsatisfactory.

o

Meinels

Bioconversion may be the ultimate way of using sunlight, allowing nature to provide the storage capacity and continuity. But the problems here are very

Lovins

The efficiency of bioconversion is indeed low, but with the sort of bioconversion I have in mind—using current residues, not special crops—the

Meinels

serious. The basic efficiency of conversion is low, on the order of one to two percent at best, requiring vast areas of vegetation. This problem is complicated by the growing scarcity of arable land, water and fertilizer, which are already stressed to the limit by the earth's growing population. If bioconversion is the only economic answer, then we must look forward to a smaller population or a diminished usage of energy or both.

Lovins

marginal land requirement is nil and the low efficiency is irrelevant. It does not waste anything: there is lots more sunlight where that came from. The marginal requirement for water and fertilizer is also nil, and indeed fertility might be increased with clever management (see e.g. E. S. Lipinsky, *Science* 199:644–651 [1978]). Bioconversion is not "the only economic answer," but an important set of diverse technologies peculiarly well suited to producing portable fuels such as transport by conventional vehicles uniquely requires.

o

Meinels

There are a few possible ways of overcoming these obstacles to the use of solar energy on a large scale. One might be the technological analog of a plant, with small electrolysis cells powered by solar cells quietly accumulating hydrogen to be harvested occasionally for their stored energy. But the alternative which we prefer, and which Mr. Lovins would dismiss as a "hard" technology, is the production of electric power or hydrogen on large-scale solar farms located in the arid southwest on land not now in use.

Our advocacy of large-scale solar power farms is contrary to the stream of popular enthusiasm. Small-scale individual applications are the center of attention today. We feel this is contrary to the way society has gone for centuries, in fact, ever since the isolated castle of the Middle Ages gave way slowly before renewed commerce and order. There is no reason why each of us could not have our own gasoline, or diesel, powered generator and water well today—as Mr. Lovins advocates—except that it would be inconvenient, unreliable and costly. We have lived in a solar-heated house with a solar-heated swimming pool. We do not think many other persons, other than avid do-it-yourselfers, would enjoy it after the novelty wore off.

Lovins

I agree that photolysis or (with cheap photovoltaics) electrolysis of water offers an interesting range of opportunities; but I have never seen a respectable argument for centralizing either of these technologies, and doubt one can be made.

It is true that I am not enamored of the Meinels' favorite solar technology—centralized solar-electric farms in the southwestern deserts—partly because I do not see a need for such technology. I am surprised, though, that the Meinels should feel they are bucking the tide of interest in smaller technologies—quite the opposite of the story told by the ERDA budget. My impression is that the shift of interest and money away from the Meinels' approach and toward, for example, "power towers" was prompted not by ERDA's love for small technologies but rather by perceived technical weaknesses in the Meinels' proposals. (It is correct, however, that some analysts believe "power towers" make more sense on a smaller scale—a view that Otto Smith of Berkeley has been trying, with scant success, to impress upon the ERDA dispensers of R&D funding.)

If the Meinels' highly centralized and electrified approach is "contrary to the stream of popular enthusiasm," perhaps it is for the reasons mentioned in my

Meinels

Lovins

article. This does not imply, however, a return to "the isolated castle of the Middle Ages"; rather, a more percipient analysis of scale issues. Such analysis is not furthered by incorrect statements that I advocate individual "gasoline, or diesel, powered generator and water well" installations—a figment of the Meinels' imagination.

From my acquaintance with other dwellers in solar buildings, I do not think the Meinels' attitude toward their own solar house is widely shared. Possibly their house is not as thoughtfully designed as those whose owners I have talked to.

O

Solar energy may first come into use on people's rooftops, but we are certain that, as soon as possible, they will prefer to have it delivered as electrical energy and transportable fuels. People already use electrical energy for a vast array of needs and luxuries. Public and private utilities provide it for us and take care that the system supplies it with high reliability. American industry is geared to produce electricity-consuming devices. Solar electric power therefore meets the requirement of minimum perturbation of the socioeconomic system we live in, a requirement that must be met for change to be socially acceptable.

We think that the utilities have a long future of delivering energy and water to consumers at the lowest possible cost and maximum convenience, whether the energy be fossil, nuclear, geothermal, wind or solar. For everyone to abandon the utility and get his own energy system seems like a step back toward the Dark Ages, or like a person abandoning the ship to cling to his personal life jacket. We have confidence in the viability of the ship.

O

The Meinels' certainty that people will prefer premium products of solar energy to the direct delivery of energy forms matched to dominant end-use needs is unsupported by any analysis—particularly economic analysis, which I think would suggest the contrary. It is extraordinary that electricity's moving from a 13% to a dominant contribution to end-use energy should be regarded as a "minimum perturbation": it would entail electrification of most of the remaining 87% (space conditioning, industrial heat, and transport, for example), and would, according to my analysis in sections II, III, and IX of the *FA* article (which the Meinels nowhere contest), entail prohibitive economic and social costs. The Meinels oppose to this notion the straw man that redirecting our marginal effort away from further electrification implies "a step back toward the Dark Ages" with each person getting his or her own energy system.

By no stretch of a highly elastic imagination can the Meinels be said to be addressing the real issues of the appropriate long-term role of electrification when they assume that some mystical "preference" will dominate over the economic and structural issues whose importance they so lately stressed. It is

Meinels

Lovins

careful consideration of precisely those issues that has led me to the view that pervasive and indiscriminate electrification should not and indeed cannot occur, and that gigawatt scale does not make sense at the margin (any more than the Meinels' caricature of universally applied kilowatt scale would).

O

In conclusion, we would like to further disturb idealists such as Mr. Lovins by cautioning against emotional rejection of fission power as a true option for the future. The closest thing to economic reality with assured energy supplies for millenia could well be the fast breeder reactor. To foreclose this option, assuming the other options are just around the corner, could mean placing the destiny of civilization into the simple picture given by the old proverb: A bird in the hand is worth two in the bush. We are not even sure there are any birds in the exotic energy bush. We perceive the bush to be moving, but it could only be the gentle wind mocking us.

O

I am likewise bewildered by the Meinels' characterization of fast breeder reactors—which do not exist anywhere and may not in the future—as "a bird in the hand" and proven soft technologies, now in operation and with known costs and technical characteristics, as "exotic" and speculative. I fear, too, that the Meinels, apparently unfamiliar with my writings on the subject, are far more "emotional" in defending nuclear power than I have ever been in rejecting it—which is not to say that emotion is bad, but rather that my views about nuclear power were reversed in the early 1970s on the basis of reasoned and well-documented analysis published for peer review. I can only conclude that the Meinels' work on central-electric solar technologies, while it has led to some useful developments, has not left them enough time for an adequate study of energy or nuclear policy and has badly distorted their view of the relative merits of solar technologies which they and others have developed and advocated.

14

PICKERING vs. LOVINS

Ethicist Decries Localism as
"Center of Gravity in Lovins' Energy Policy"

Dr. Pickering has a Ph.D. in social ethics from the University of Chicago and is Associate Professor of Social Ethics at the University of Detroit.

Pickering: Whitehead said that "It is the business of the future to be dangerous," and while I agree with that, it has caused me to think that possibly this is one of the reasons why discussions of the future seem to triangulate between fearful anxiety, downright frivolity, and romantic nostalgia. As a case in point, we shall try to engage some of the issues which have been raised by Amory B. Lovins in his essay, "Energy Strategy: The Road Not Taken?" According to Lovins, the future is fulfilling Whitehead's criterion of danger, but that seems to be, from his point of view, the trouble with it. He sees a way in which the danger can be averted, the lights can be kept on, and the powers of the corporation restored to local communities in a non-nuclear future. Taken as a whole, his argument seems to rest more on a complaint about the present than on any persuasive vision of genuine alternatives for the future.

Lovins: Dr. Pickering's essay is more thoughtful, literate, and interesting than its companions in Mr. Yulish's volume, and seems out of place there. Focusing on issues of social philosophy (the sort of "soft science"—or worse—which some of the other contributors to the same booklet decry), it requires a more philosophical response. As will become clear, I do not believe Dr. Pickering's objections to my article have merit; and, as in numerical disagreements, readers will have to judge for themselves who has made the better case and whose case is more relevant to the article we are discussing.

Dr. Pickering's opening paragraph slides past future dangers—which are of a type and size to be not merely stimulating but mortal beyond human experience—and, unfortunately, rejects or disregards my attempt to frame a less dangerous and more resilient future. He does not agree that my article offers a "persuasive vision of genuine alternatives for the future"; yet it is precisely because it does so that Mr. Yulish's friends, fearing that the public finds that vision all too persuasive, were so anxious to assemble their booklet.

Pickering: By way of introduction, I should say that it is because there seems to be a foreseeable moment of reckoning over the next 50 years in the matter of the depletion of nonrenewable resources that Lovins believes we stand right now at a critical juncture in our decision-making on energy. There are two mutually exclusive paths, and he claims that we are going to have to choose between them very soon. "The first path resembles present federal policy and is essentially an extrapolation of the recent past. It relies on rapid expansion of centralized high technologies to increase supplies of energy, especially in the form of electricity." This, he calls "the hard path." "The second path combines a prompt and serious commitment to efficient use of energy, rapid development of renewable energy sources matched in scale and in energy quality to end-use needs, and special transitional fossil-fuel technologies." This he calls "the soft path." Admittedly it required an effort on my part to divest my mind of the sexual imagery suggested by such terms as "hard" and "soft," and to understand why one is better than the other. This problem, I understand, is entirely mine; but in having to disentangle the ways in which these terms were emotionally loaded for me, I was led, I think, to a better understanding of how they are emotionally loaded in this essay. For Lovins these two paths represent not only mutually exclusive uses of our non-renewable resources; they also represent mutually exclusive and irreversible directions for the development of our culture and institutions.

Lovins: As I noted at the start of section V of my article, I was unable to find a more satisfactory term than "soft" to summarize five specific characteristics of a class of technologies and the political structure of a class of societies; any term, including "soft," is bound to have unsatisfactory connotations for someone (though I must admit that its possible sexual imagery had not occurred to me). In retrospect I should perhaps have made up a nonsense word—if I could find one with no affective content. It was not my intent to choose an emotionally loaded term, and if it were, I would have failed, since it is evidently loaded in opposite directions for readers of different predilections: for every reader who thinks of "soft" as "benign" or "resilient" there seems to be another who thinks of "mushy" or "vague."

Be this as it may, Dr. Pickering is quite right to suppose that a hard path represents an irreversible and, as I see it, unwelcome direction "for the development of our culture and institutions." That direction excludes the diversity and personal freedom that could make a soft path work.

Pickering: It cannot be denied that we live in a time of relatively low confidence in our institutions and that the desire for alternative futures is altogether genuine. If only that were the whole truth, our problems would still be multiple and profound. When confidence declines, however, fear abounds and scare tactics take the place of reasonable arguments. There is no magic solution to this downward spiral in public understanding. Our best defense is the patient re-working of reasonable arguments to expose the scare tactics for what they are, to disentangle the fears from the on-going issues and thus, perhaps, by generating a genuine sense of alternatives, to reconstruct confidence, if not in our institutions, at least in our human capacities. Perhaps it is one of the structural anomalies of our time that an upward spiral of public attention to issues seems to go hand in hand with the downward spiral in public understanding of those issues, leaving the field open to strange mixtures of fear and fantasy at just those points where we had hoped to encounter a world of solid fact. This cycle appears to be pervasive in the arguments about energy, but it is certainly not unique to the energy issue.

Lovins: Dr. Pickering echoes my mention on p. 95 of the *FA* article of how "national purpose and trust in institutions diminish," and the consequent risk of increased centrism and technocracy. Dr. Pickering adds, however, an innuendo about "fear" and "scare tactics" which, if intended to refer to my article, he would be hard put to justify, and if not, he should have omitted.

I share Dr. Pickering's view of the great social importance of coherent positive visions around which to mobilize informed discussion and a renascence of hope, and it was largely for this reason that I prepared my article.

Pickering: It is not my intention to put the whole burden of this downward spiral on Lovins. Similar problems appear on "the other side" of the energy issue. It is not uncommon among the defenders of the status quo to threaten us with a new Dark Age if the engines of the modern corporation are not given complete possession of our future, their every wish our command, their every appetite satisfied just as they, in their bureaucratic wisdom, perceive their needs. Lovins' argument deserves our critical attention because he is trying to generate a sense of alternatives, and because he does pose, however backhandedly, the problems of social change which are necessarily entailed in any serious discussion of the future—whether it be the future of our energy needs or the future of democratic institutions. Unfortunately, these issues are posed too indirectly—too much by implication and, alternately, by assertion—to yield sufficient clarity about the range of real choices which are before us.

Lovins: It is unworthy of an ethicist, presumably trained in logical thought and attuned to the sympathetic understanding and refinement of diverse views, to let such innuendoes as "strange mixtures of fear and fantasy" slither into his prose—and then to try to excuse them by a symmetrical innuendo applied to "the engines of the modern corporation." If Dr. Pickering means to argue that I am a huckster of fear or fantasy, let him say so in a declarative sentence and cite evidence for this view—evidence I think he will be unable to adduce. If he means rather to indulge in oblique, polite character assassination through the niceties of rhetoric, let him then present himself as a rhetorician, not an ethicist, and abandon any implied claim to objectivity or scholarship. He is trying to have it both ways: to gain the persuasive power of the rhetorician without the responsibility of the scholar.

The same objection applies to Dr. Pickering's implication that the "sense of alternatives" I seek to generate is sense without substance; his obscure but plainly pejorative use of "backhandedly"; and his inexplicable assertion that I pose social issues "too indirectly—too much by implication . . ." Again, no doubt it would be desirable to expand such expositions as section IX of my article to many volumes (I do indeed amplify my overcondensed account in the Oak Ridge paper and, more fully, in Chapters 9 and 10 of *Soft Energy Paths*); but it was my impression that it was the very directness and clarity of my description of social issues, among other things, that gave my article its impact and that therefore aroused such concern among Mr. Yulish's clients and colleagues. If my paper were as opaque and muddled as Dr. Pickering makes out, it could simply be ignored without fear of its influencing anyone.

Pickering: As Lovins conceives them, we have basically only two choices, and they are mutually exclusive. If this were true, it would be an amazing historic first. Part of the reason Lovins insists on the idea of an excluded middle is because he wants to present the possibility of a non-nuclear future. His conception of the "hard path," therefore, is an intensification of the recent past, leading inevitably to the exhaustion of nonrenewable resources, thus creating an inescapable dependence on nuclear power. His conception of the second or "soft path" offers, he says, "many social, economic and geopolitical advantages, including virtual elimination of nuclear proliferation from the world."

Lovins: Exclusivity would not be "an amazing historic first"; it is rather a moral that can be drawn from almost any branch of human history. My "excluded middle" is excluded carefully and gently: "Though each path is only illustrative and embraces an infinite spectrum of variations on a theme, there is a deep structural and conceptual dichotomy between them." The reason the middle is excluded is that not to do so would lead to a category mistake. Possibly Dr. Pickering has misconstrued exclusivity as meaning technical incompatibility, despite my careful development in the article of the concepts of resource competition, cultural exclusivity, and institutional antagonism.

Dr. Pickering's tortured logic to try to invent a rationale for exclusivity by means of nuclear power is a much less natural line than the reasoning I actually used: that a hard path is one whose polity is dominated by certain structural issues (autarchy, centrism, technocracy, vulnerability, etc.), that a mixture of hard and soft technologies (to the extent that such a mixture could in practice be constructed) would have these same problems in significant degree, and that it is thus a category mistake to imagine that a path can be hard and soft simultaneously, as if its softness were a function only of choices of hardware. It cannot be overemphasized that the distinguishing feature of hard and soft energy paths is sociopolitical, not technical, and that the sociopolitical conditions—the social architecture—that characterize the two paths cannot, logically, exist in the same society at the same time.

Pickering: I find this a truly startling claim [that the soft path would promote "virtual elimination of nuclear proliferation"], and to give the reader some idea of what it takes to startle this writer, let me say that I consider the possibility of a world without war to be a politically meaningful goal! But it has never occurred to me that the elimination of nuclear energy was one of the preconditions of that goal. Therefore, as a layman in this technical field, I would like to know whether that is a real possibility or a scare tactic designed to advance a choice which must actually be grounded on other considerations. My first inclination is to disbelieve the possibility—to think that, from here on out, our problem is to find ways to govern nuclear proliferation, not to eliminate it.

Nuclear power has made its appearance in human history, and while we certainly want to govern its development and uses, it seems unlikely that we either can or should look forward to its disappearance. To become more and more aware of its demonic possibilities is one thing which any prudent policy should accomplish; to foster a belief in secular exorcism seems like a retreat from reality, a flight into other-worldly fantasy. All the more so, since the author wants us "to recognize that the two paths are mutually exclusive." That may or may not be the case; but if it is, it is incumbent upon the author to make an argument demonstrating that it is the case, or at least why it is plausible to think so. If his other readers are anything like me, this is not something that he should ask us to "recognize." I lack the power of immediate intuition into these complex matters; so his point amounts to an appeal for faith, and that happens to be just what I am not willing to yield in this matter—not to the advocates and not to the opponents of nuclear energy. The gravity of the issue demands a "show me" attitude on the part of all of us, and the assumption that we are capable of following the arguments, indeed rather eager to hear them, if someone will just be so kind as to offer them.

Lovins: Dr. Pickering might not have quite the level of sophistication in nonproliferation problems and geopolitics that is common among the rather select readers of *Foreign Affairs*, and it is understandable that in a field so far removed from his own professional work he might have found the condensed argument on nonproliferation in my article's section VIII difficult or unconvincing. But I think the fuller account of a rather subtle and complex techno-political argument in Chapter 11 of *Soft Energy Paths* will offer him ample grounds for hope that "secular exorcism" of the "demonic possibilities" of nuclear energy is indeed "a politically meaningful goal" and a realistic basis for real policies. I think he will also find in Chapter 11 of *SEP* an arguable—and, according to the reaction of some very sophisticated analysts, a convincing—case resolving his doubts about whether elimination of nuclear proliferation is a "real possibility."

I am aware that fatalism is currently fashionable, and that many respected students of these matters believe that proliferation can only be moderately slowed, not stopped or reversed. But I am encouraged that several European experts recently exposed to the Chapter 11 argument have reversed their long-held views and now feel that it is indeed possible to approach, and perhaps to attain, general and complete denuclearization in both the quasi-civilian and the military sphere.

It is a pity that Dr. Pickering has muddied the waters of the exclusivity argument by mixing in the nonproliferation argument—whose only relation to exclusivity is that with continued proliferation, both soft and hard paths become academic. It is likewise unfortunate that he has not thought a little harder about resource competition and about cultural and institutional exclusivity. As I pointed out in response to Mr. Yulish's introduction, "Where we are today is a fine example of the sort of exclusivity I mean: decades of lopsided commitment to one sort of technology have left us with a limited range of perceptions and technical options and a rigid set of institutions."

Of course I cannot *prove* exclusivity in the same way that I can prove a Euclidean theorem or measure the velocity of light; it is more akin to many of the empirically unverified but historically plausible predictions that scientists (hard or soft) make every day on the basis of their experience and understanding. That is why the proposition of exclusivity must be—and is, both in my article and elsewhere—supported by reasoned argument rather than by some sort of pseudomathematical proof.

Accordingly, if Dr. Pickering will kindly refer to the last paragraph on p. 86 of my article, the sociopolitical and social-perception arguments of sections IX and X, and the timing, resource-competition, and institutional arguments of section XI, he will find the allegedly omitted discussion of exclusivity which he seeks. If he reads it as carefully as I wrote it, I am confident that he is indeed capable of following the argument. But he must not attempt to read into it interfering arguments that bear it no relation.

Pickering: In my reflections on this matter, I have no intention of getting into any of the arguments based on technical knowledge about energy sources, systems, or uses. I leave those matters to my colleague, Margaret Maxey. That may sound irretrievably irresponsible since, ostensibly, that is what the argument is about. But I wish to excuse myself from that level of the argument (a) because of my total incompetence and (b) because I believe that there are some other dimensions of this argument which are being obscured by their absorption into the morass of technical details, and (c) because I think that it may aid the entire discussion if the ethical and political dimensions can be identified for what they are and can be "recognized" for what they are when they appear in what otherwise may seem to be technological choices.

As far as I can see, there are three kinds of ethical and political questions to be made in the question of "energy policies" for the present and the future:

(1) There is the question of the prudent use of existing but scarce energy sources.

(2) There is the question of the alternative energy sources, including possibly renewable ones, and the general effort to eliminate physical scarcity caused by approaching resource exhaustion.

(3) There is the question of the institutional matrix required for the generation and distribution of energy, how this institutional matrix relates to the rest of the community and, ultimately, the question of what kind of community or communities we actually hope to live in. This is a question of the technological limitations on our vision of community. For those of a more up-beat mind, it is a question of the technological possibilities for alternative visions of community. In either case, the technology is not the essence, unless one envisions a technology-serving community. The essence lies in the relationships, activities, and institutions to be sustained with a structured community life—including the institutional means required for the generation, distribution, and governance of energy in the life of the larger community.

[Ms. Maxey, when approached about participation in this book, refused permission for her work to be used.—Ed.]

Lovins: I welcome this passage by Dr. Pickering and agree with him that technical and numerical arguments, though important, are not the whole story nor even the decisive arguments about the broad outlines of energy policy—a point I emphasized at the end of section X. I agree also that his three classes of proposed questions, though not exhaustive, are relevant—though in (2) I would delete the "possibly" as smacking of prejudgment and would not confine my definition of scarcity to the purely physical, as economic, geopolitical, or other kinds of scarcity are generally more proximate.

I did not use the analytic structure proposed by Dr. Pickering because it does not seem to me particularly useful. My failure to use it hardly makes my thesis more obscure—more likely it helped me to avoid the sort of obscurity (at least for a *Foreign Affairs* audience) into which Dr. Pickering is about to lead us. But there are, without doubt, infinitely many ways to slice into a problem, and it is not for me to deny Dr. Pickering his pastimes. As will become clear, his intent is to dismantle my carefully constructed heuristic device—which obviously has some power, else he would not be writing about it—in order to insinuate his own and (even more) Dr. Maxey's bizarre view of what the energy problem is and why, according to that view, my thesis leads to undesirable results. I regret that Dr. Pickering is not as willing to analyze my arguments within my own paradigm as I was to analyze hard-path arguments within the terms, criteria, and data normally used by hard-path proponents. Without such intellectual flexibility, discourse disintegrates and people talk past each other.

Pickering: Mr. Lovins has presented us with an argument in which these three dimensions of the question are not very well distinguished analytically. It is therefore difficult to evaluate just what his argument (a) is really based on and (b) is actually proposing. It may help to clarify the matter if we review his argument and try at least to distribute his points under our three headings: (1) What is he saying about the prudent uses of existing resources, (2) what is he saying about alternatives and (3) what is he saying about the institutional matrix. For reasons which will become clear below, we shall take up these dimensions in reverse order, that is, institutions first. I will address myself to these considerations. Margaret Maxey will take up the question of alternative energy systems and prudence in the use of existing resources.

Since, in any case, the prospects of continuity and change, consolidation and diversification, centralization and poly-centralization and the like are *policy* questions, the issue of energy begins and ends with assessments of the institutional matrix—that which exists and that which may be desired. One of the disciplines which the topic imposes is the limitation of the discourse to real possibilities, and that means to institutionally available alternatives. There is no point in posing abstract possibilities against an existing state of affairs as if the two had practical quality. This may seem like an unimaginative principle, but there are two things to be said about this as a general methodological procedure. On the one hand, the abstract always has a certain advantage over the concrete because it retains its integrity of intention and lacks the inevitable ambiguity of embodiment. On the other hand, this procedure forces us to look at the existing state of affairs itself, not merely as a mess already made, but as a set of real possibilities, as the potential locus of real novelty—if there is to be any. It is not the conservative methodology which it might, at first blush, seem to be. If there is any general relationship between analytical methodologies and the prospects for meaningful change, it would seem to be this: that practical conservatism goes hand in hand with abstract radicalism, because meaningful novelty is effectively postponed by both parties—the "radicals" awaiting some very new vision while existing arrangements continue to roll ahead with all the advantages of "realism" on their side, no matter how crazy they may be. If change is desired, then effort should be focused on uncovering and, if possible, expanding the range of real possibility. It is simply not enough to construct an argument which has the form: since the existing state of affairs is "impossible," some other arrangement is more desirable. Unfortunately that seems to be the structure of Lovins' argument, and that structure does a disservice to some of the valid points which he may have in his presentation. When, in his final section, Mr. Lovins says that the choice between his "hard path" and his "soft path" "may seem abstract," he is only half right. The abstract choice is the one he prefers, and it is a weakness of his presentation that the possibilities to which he would have us commit our hopes and resources remain abstract, mere possibilities which have not yet taken the shape of real choices that can effectively be made. His two paths leave us with more of a sense of opposition than of real alternatives.

Lovins: This passage seems a long-winded way of saying that policy exploration should take into account the attributes of existing institutions and the plausible means of adapting them to new tasks. This I have done. I am less worried than many analysts about the antagonism of existing institutions to soft-path concepts and implementation, both because present policies are obviously not working and because the benefits of a soft path are large enough so that they can be distributed in a way that gives each actor an adequate incentive. In several places—for example, in discussing the role of utilities in capital-transfer schemes—I indicated how potentially antagonistic institutions could be coopted into the transitional project, to their own advantage.

That said, I did not and do not think the proper role of the article was to attempt a detailed policy blueprint with draft legislation, microscopically described administrative actions, dates, etc. My purpose was rather to define the nature and comparative merits of a possible class of approaches to the energy problem, giving enough (and, for reasons of space and audience, only enough) policy and technical backup to show that these approaches fall within the universe of "real possibilities" to which Dr. Pickering wishes us to confine ourselves. To the extent that my proposals are "abstract," others' proposals—including the proposals of hard-path proponents—are equally "abstract." Indeed, as Herman Daly points out (*Soft Energy Paths*, p. 15), low-energy advanced societies exist today, whereas "it is the high demand, hard technology scenarios that have never before been experienced and are completely hypothetical. Yet our 'crackpot realists' all treat the hypothetical high energy projection as if it were empirically verified, and the empirically verified low energy scenario as if it were the flimsiest conjecture!"

The "abstract" label cuts both ways: a hard energy path, not as practiced today but as projected into the future, is arguably more conjectural than a soft path at similar dates. It is not enough to point to the momentum of established institutions: one must consider the present or incipient failure of these institutions (for example, utility finance, nuclear regulation, or synthetic-fuel finance) and the drastic, highly speculative institutional innovations now proposed to bail out those failed institutions. In contrast, I propose briefly in the *FA* article, and at the greater length available in *SEP*, soft-path policies that move closer to, not further from, the principles of the market and social processes that hard-path proponents tend to espouse: private and individual enterprise, democracy, and, with light regulation (such as antitrust) for equity, a quasi-free market. Accordingly, Dr. Pickering's attack on the soft path's "abstraction" seems to miss the point.

Pickering: Let me show what I mean in saying this.

In his world of two mutually exclusive choices, Lovins holds the "hard path" constant, while attributing to the "soft path" all the possibilities of what he considers desirable social change. Indeed he tells us that the distinction between his two paths "rests not on how much energy is used, but on the technical and sociopolitical *structure* of the energy system," thus focusing our attention on consequent and crucial political differences. What makes these two paths mutually exclusive is, he tells us, their "logistical competition and cultural incompatibility." As I read the essay, cultural incompatibility is the more basic factor. In fact, it seems to be for "cultural" reasons that Lovins wants to exclude nuclear energy and indeed any continued or increasing dependence on other forms of high technology. The underlying reason for this seems to be Lovins' perception that high technology requires large-scale corporate enterprises—which is surely true. The "softer" technologies appeal to him because they seem to offer a way of addressing our difficulties and having a decent future, as he says, "unimpeded by centralized bureaucracies." We need to examine this claim and the configuration of values in terms of which it is made, because it would appear that the "soft path" is put forth not only as a policy to define and meet our needs for energy, but also—and maybe even primarily—to use energy policy as a lever for social change.

Lovins: Dr. Pickering omits from his list of forms or grounds of exclusivity the very argument he has just been using—institutional exclusivity or antagonism (illustrated by the difficulty of adapting today's ERDA staff, tax laws, R&D practices, and—in part—technical community to understanding or exploring soft-path concepts). He then guesses—wrongly—that it is for "cultural" reasons that I oppose nuclear fission technology (though its sociopolitcal effects would still make it unpleasant even if it generated none of the toxic or explosive materials that are my main source of concern). He then states—also wrongly—that I oppose "any continued . . . dependence on other forms of high technology." In fact I advocate increased dependence on some forms of high technology such as "chip" microprocessors and advanced materials, where they are appropriate, and see a continued need for limited amounts of large-scale energy systems (such as many of the existing hydroelectric dams) into the indefinite future.

What Dr. Pickering seems to be trying to say is that I worry about the sociopolitical effects of big, complex, highly centralized energy technologies used on an unnecessarily large scale. That is correct, but for many reasons beyond the single reason he picks out (encouraging oligopoly). Many other cogent reasons are cited throughout my article, especially in section IX.

Pickering: I do not complain about this. Simple common sense says that energy policy or the lack of one is going to be an important determinant of social change in coming years, just as it has been for the past century or more. Insofar as possible, therefore, any energy policy should be called to account for its social and political presuppositions as well as for its foreseeable social and political consequences. Unfortunately in Lovins' case, his presuppositions seem to prevent a candid assessment of probable consequences. He seems to assume that by making the "right" technological choices we can solve certain long-standing social and political problems. Lovins appears to be looking for a technology which will free the people from domination by corporate interests, bureaucratic barriers to participation, and the expropriation of their own judgement by oligarchic experts. His assumption, therefore, is that genuine powers need to be restored to local communities—something like neighborhoods—if we are going to have anything like a democratic future; and to be really meaningful powers, they must be such as to render technology familiar, intelligible, and controllable at the local level. This, it seems to me, is the underlying cultural drive in Lovins' argument for the soft path.

The problems are real, but the solution will not hold. And some of the facts, I fear, are trimmed to favor the solution. Over the years many attempts have been made to cure the social ills of corporate industrialism through the repristination of local neighborhoods, the "restoration" of decision-making powers to the presumed locus of face-to-face relations, the residential neighborhood. Even John Dewey, the philosophic and generally insightful guide for perplexed democrats, looked to a pluralistic localism as the solution to the existential problems of a corporately organized world. It is worth recalling his pronouncements in this matter, because they represent an enduring strand in American social thought and they reappear, it seems to me, in Lovins' argument:

> In its deepest and richest sense [Dewey proposed] a community must always remain a matter of face to face intercourse. This is why the family and neighborhood, with all their deficiencies, have always been the chief agencies of nurture, the means by which dispositions are stably formed and ideas acquired which laid hold on the roots of character . . .
>
> Democracy must begin at home, and its home is the neighborly community . . . If there is anything in human psychology to be counted upon, it may be urged that when man is satiated with restless seeking for the remote which yields no enduring satisfaction, the human spirit will return to seek calm and order within itself. This, we repeat, can be found only in the vital, steady, and deep relationships which are present only in an immediate community . . . The local is the ultimate universal, and as near an absolute as exists . . . Unless local communal life can be restored, the public cannot adequately resolve its most urgent problem: to find and identify itself.

I do not quote Dewey as a way of putting Lovins down. On the contrary, he stands within a very distinguished company and his vision of the future represents one of the enduring themes in American democratic social thought, namely, that the vitality of democratic values requires the revitalization of local institutions.

Lovins: A hard energy path may not be intended as "a lever for social change," but I see this as one of its primary and unavoidable effects. A soft energy path is intended not as such a lever, but on the contrary, if anything, as a way to *prevent* social change in the hard-path direction (centrism, autarchy, technocracy, vulnerability, etc.). Obviously a soft path cannot do this in every sphere of life, but primarily—perhaps only—in the energy sector, and thus cannot be expected to "solve certain long-standing social and political problems." At least it should prevent those problems, in the context of energy policy, from getting worse. If it returns them to the lesser prominence and intractability of a decade or two ago, so much the better. To do much more than that—to achieve the genuine political decentralization that Dr. Pickering describes—would require a general political reform far beyond the scope or purpose of what I proposed. A soft energy path would be compatible with such reform, and might even be a necessary condition for it, but it is certainly not a sufficient condition. It was patently *not* my intent to concoct a recipe for the general overhaul of American political institutions. Such an overhaul may, as I suggest in section X, be on the way through the normal pluralistic functioning of our democracy, but is the business neither of my article nor of energy policy.

For this reason it is facile and misleading of Dr. Pickering to criticize my article, either directly or obliquely (" . . . some of the facts, I fear, are trimmed to favor the solution"), for not doing thoroughly and convincingly what it never set out to do—namely, provide an explicit set of necessary *and sufficient* conditions for reforming and revitalizing the entire sociopolitical life of this country. My task is rather more modest: to suggest a promising approach to the *energy* problem which avoids the prohibitive costs (including political costs) of the way we have been going lately.

Quotations of Dewey's general social philosophy, though interesting, are beside the point.

Pickering: Over the past century we have seen a number of interesting and socially innovative attempts to act on this perception of the democratic prospect—from Jane Addams and the settlement house movement, to Saul Alinsky and the community organization movement, to Milton Kotler and the community corporation movement. Perhaps now we should add Amory Lovins and the community energy movement. In candor, however, we have to say that the experience of these other movements does not augur well for the actual prospects for Lovins' vision. The settlement houses have been absorbed into the web of professionalized social services. The community organizations, while they have had some success in commanding political attention, have not been influential in the organization of urban political power. Community corporations have remained negligible as contributors to social change. Maybe there are reasons to suppose that "energy" is the missing link in the chain of powers needed to restore neighborhood dignity, but I doubt it. As I view the history of these other movements, the problem is not that they have achieved nothing. It is that they have not achieved their larger aims. In the end, they have been supplemental to the social order, not levers for changing any of its basic arrangements. That seems to be the fate of localistic orientations under our social conditions.

Lovins: It is idle for Dr. Pickering to criticize past examples of various social reform movements (not always justly, according to my limited knowledge of these examples) as a way of suggesting that a soft energy path will fail, as they have failed, to achieve its "larger aims." Since my largest aim in the *FA* article is to propose reforms of energy policy which will *avoid* the various nasty side-effects (fiscal, environmental, military, social, etc.) of an unsustainable policy, my failure to achieve an imaginary "larger aim" of general political transformation is a fault not of mine but of Dr. Pickering's understanding.

Pickering: If local autonomy and neighborhood values are indeed the citadel of democratic values, then we are without hope and we had better face it. I doubt very much that this is the case. While this may have been a plausible analysis of the early encounter between democratic institutions and the world of the corporation, I am now inclined to think that its continued repetition is a case of aggrieved nostalgia, mistakenly identifed as idealism. I am prepared to say that the prospect for democratic values and democratic institutions lies in our ability to work out reasonable forms of governance for large scale corporate enterprises by creating new channels for both popular and professional participation in public policy formation, not in the attempt to restore long lost "local autonomy."

Lovins: Though Dr. Pickering's criticism of my article for failing to provide precisely what many other critics accuse me of providing—a sort of Manifesto for general and complete revolution—is a fatuous diversion from the real political issues I raise, all of which he ignores, I cannot resist pursuing him along the detour far enough to remind him that local autonomy is alive and well and is living in New England town meetings, Wyoming ranches, New York block associations, Midwestern villages, and urban neighborhoods around the world, from London to Melbourne to Tokyo to Nairobi. It is—as "developers" of all kinds can attest—a potent political force.

Dr. Pickering's argument about "effective forms of governance for large scale corporate enterprises" would be more persuasive if he could suggest at least one such form, or if he could show that political institutions as centralized as those we now have, or even more so, are really manageable and functional.

Pickering: Lovins argues that his soft technologies are "ideally suited for rural villages and urban poor alike." And he thinks that they "do not carry with them inappropriate cultural patterns or values . . . ; they can often be made locally from local materials and do not require a technical elite to maintain them; they resist technological dependence and commercial monopoly." Nonsense. Any technology wears out, and if the soft technologies are as commercially viable as Lovins maintains, they will attract investment and they will become absorbed in the web of commercial corporations competing in the manufacture, distribution, marketing and servicing of them. They will remain "local" only to the degree that they are not generally marketable and, therefore, remain marginal to the supply of energy.

Lovins: I fear Dr. Pickering has again been reading too hastily. I said, correctly, that soft technologies "resist technological dependence and commercial monopoly": resist, not prevent. The point is that even in overdeveloped countries, soft technologies allow mass production by large corporations, but give those corporations no natural monopoly. Small enterprises and even individuals also have their own competitive niche, based on low overheads, greater innovation, and more exact adaptation to local needs. These small enterprises may well assemble mass produced materials or components to their own designs. I expect a mixed large-and-small-scale-production economy of this sort to develop quickly, with both its parts making important contributions.

In developing countries, Dr. Pickering misses the mark even more widely: soft technologies would be based entirely, or almost entirely, on locally available skills or materials, and commercial enterprises could not penetrate the equipment market because there is no such market, owing to the lack of purchasing power. There are precedents—for example, bamboo tubewells in parts of India—for the very rapid spontaneous spread of simple new technologies, capital-saving and labor-intensive, on an exceedingly large scale and with results highly successful for local welfare.

Pickering: What is more, Lovins does not seem to be aware that "localism" is itself a cultural value whose appropriateness needs to be assessed. He seems to think that this orientation is by nature less coercive, more participatory and, somehow or other, more equitable. And that seems to be why he is opposed to nuclear power.

Lovins: Dr. Pickering gratuitously reminds me that "localism is itself a cultural value whose appropriateness needs to be assessed." I agree, and trust that readers have been assessing it. My own political philosophy, being Jeffersonian in essence, tends to favor making political decisions at the lowest possible level, and I note that this is the custom both of most of the world's people (now and previously) and of the most durable cultures now observable: the sort of centrism permitted by modern instruments of rapid communication is a very recent development, yet seems already to have outstripped our ability to manage it and to digest the incoming information. (Kenneth Boulding describes a hierarchy as "an ordered arrangement of wastebaskets"—a device for *preventing* information from reaching the executive.) But my personal political views are neither intended nor needed to justify the arguments in my article; nor are they, as Dr. Pickering again supposes, why I have criticized nuclear power.

Pickering: There may, however, be another, although related, reason for

Lovins' opposition to nuclear energy, and I would like to divert from my main line of argument for a moment to consider it. Lovins would have us "consider the impact of three prompt, clear U.S. statements": (1) to phase out our own nuclear power program and our support of others; (2) to commit our resources to soft technology at home and abroad; (3) to treat nonproliferation and nuclear power as "interrelated parts of the same problem." Such a policy, he believes, "would be politically irresistible" even though "nobody can be certain that such a package . . . would work." What interests me here is Lovins' assumption that there is no meaningful distinction between civilian and military technology in the nuclear field. This he considers "the hypocrisy that has stalled arms control." I think he sees any increasing dependence on nuclear energy as an increasing dependence on the military. He says, for instance, ". . . by no longer artificially divorcing civilian from military nuclear technology, we would recognize officially the real driving forces behind proliferation." Apparently Lovins believes that the United States is developing nuclear energy at home and exporting the idea and the technology abroad for military purposes. I find the logic odd. If nuclear energy were basically a military matter, it is difficult to conceive of any strategic advantage to be gained by spreading it around. Indeed, its diffusion would seem to neutralize any advantage while increasing exponentially the military risk. This is a matter on which I need more light, and perhaps Mr. Lovins does too.

Lovins: Dr. Pickering admits the possibility that my concern about fission technology might also be related to concern about nuclear weapons proliferation. This is correct: such proliferation to governments and to subgovernmental groups is the leading item on my list of nuclear worries (as is the case with most informed analysts). Where Dr. Pickering first goes astray in summarizing my feelings about this problem is where he refers to U.S. exports of nuclear technology "for military purposes." It is *others'* military purposes that are the most important of the several reasons that such exports are wrong (which is not to say that U.S. military purposes in continuing to brandish our nuclear arsenal are any more moral or proper). I agree that the logic of such exports is odd: they do indeed greatly increase the risk of nuclear war while destroying the basis of the strategic-deterrence doctrine on which the U.S. relies. I urge Dr. Pickering to consult Chapter 11 of *Soft Energy Paths*—and any of the excellent books on the proliferation problem that it cites—for the "more light" that he seeks.

Pickering: At any rate, coercion—civilian or military—is not a good thing in Lovins' lexicon, and as with so many other things, Lovins would rather find some way to eliminate it rather than accept it as a permanent factor to be brought under discipline and governance. He seems to think that an emphasis on localism will be both more consensual and more diverse or pluralistic, whereas "large-scale projects requiring a major social commitment under centralized management" may require "quasi-warpowers legislation." I make no argument for the use of state powers to serve corporate interests when I say that this way of posing the problem of coercion is not only irrelevant but obscurantist to the democratic prospect. I happen to think that non-violence is a decent, underdeveloped democratic principle with many more institutional ramifications and possibilities than we have even begun to entertain. But non-coercion is a social fantasy, a political fraud, and therefore not a moral ideal.

I see no way whatsoever to get serious about actually governing the behavior of corporations or solving the problems of distributional equity if we have already taken the pledge never to drink the tainted potion of coercion. I realize that Lovins believes that localism will take care of that. Soft technology, he says, is "an understandable neighborhood technology run by people you know who are at your own social level," whereas hard technology is "an alien, remote, and perhaps humiliatingly uncontrollable technology run by a faraway, bureaucratized, technical elite who have probably never heard of you."

This is a classic example of posing the abstract against an existing state of affairs as if all one had to do was to choose in order to supplant the one with the other. The grievances and abuses to which bureaucracy and red tape are subject are all too well known, but they do not seem to vary simply with the scale of the enterprise, nor even with its ideology. In our current organizational environment, local, small scale institutions are as likely as multinational corporations to adopt the impersonal bureaucratic style. It is true, I believe, that we need to experiment with alternative models of administrative order; but simply localizing powers has greater potential for leading in the opposite direction. The creation of another level of localized institutions could simply provide an additional screen behind which really decisive powers continue to operate, hidden from public view. No matter how local operations may be or become, they will still have to participate in a wide web of organizational inter-connections. Domesticating the alien and the remote is an important institutional problem; but neighborhood organizations are not likely to solve it.

Lovins: The first paragraph begins Dr. Pickering's excursion into the role of coercion in political structures and theory. While this is neither my field nor his, I agree that governance, even in a tribal village, seldom consists wholly of self-discipline, and that mutually agreed checks, balances, rights, obligations, and sanctions are necessary to provide the needed balance between majoritarianism and justice or between equity and power. The nature of such checks, etc. is a respectable problem that has occupied political philosophers for some millenia. I happen to agree with Churchill that democracy is the worst system of government—except for all the rest; and that in particular, the approach embodied in the U.S. Constitution is probably the most sophisticated, flexible, and potentially durable so far developed for a large, diverse, and powerful country.

The "coercion" to which I object, and to which I think most Americans also object when they encounter it, is that of State or private power (corporate, union, Mafia, or other, collective or individual) which, by the test of our democratic principles, is imperfectly restrained, so infringing or abrogating Constitutionally protected rights. I have nowhere put upon "coercion" the infinitely broad meaning that Dr. Pickering criticizes. I agree with his implication that in a society consisting entirely of Christs, Gandhis, Buddhas, and the like, no "coercion" would be necessary, but no such society is now observable.

Thus in criticizing my soft-path exposition for not erasing the problems of political power and individual rights that have plagued our nascent democracies for several thousand years, Dr. Pickering is again shying a red herring at the previously introduced straw man. To point out, as I do, that hard technologies incur nasty political problems which soft technologies, thoughtfully deployed, need not entail, is not to suggest that "localism" (or soft technology) "will take care of" such problems as "governing the behavior of corporations or solving the problems of distributional equity."

It is precisely because soft technologies *could* be coercively deployed, because what Dr. Pickering calls "localism" *could* be repressive, because small bureaucracies *can* be just as petty as big ones, that a soft path is defined not merely as one that relies upon soft technologies (though it must use them), but rather as one whose political order is not dominated by hard-path problems (centrism, vulnerability, etc.). It is only because hard technologies *necessarily imply* a hard-path political structure that a hard path can be characterized by its hardware. Because soft technologies do not make a soft path unless properly applied, the definition is asymmetrical. This asymmetry reflects the difference between "hard technology *implies* a politically hard path" and "soft technology *permits* a politically soft path." This difference is faithfully preserved in the language of my FA article, though I might have given it more emphasis had I known that the article was to be subjected to detailed textual analysis by people anxious to put on it a meaning it will not bear.

Pickering: I do not wish to belabor this critique of localism, but it does seem to be the center of gravity in Lovins' energy policy, and it is a continuing theme among the advocates of democratic social change in America. Furthermore, it seems to be the point at which our powers of constructive engagement are enjoined right now: can we give concrete meaning to the concept of democratic social change under conditions of truly large-scale corporate enterprises? I do not wish to join Lovins in saying "No" to that question.

Yet socially at least, that seems to be his trump card against nuclear power. It is alien, remote, subject to large-scale malfunctions in which "one may have to choose between turning off a country and persisting in potentially unsafe operations," running the risk of a garrison state to protect the technology and to guard the wastes, depending on an "elitist technocracy," leading ultimately to "social engineering," etc. "For all these reasons," Lovins maintains, "if nuclear power were clean, safe, economic, assured of ample fuel, and socially benign per se, it would still be unattractive because of the political implications of the kind of energy economy it would lock us into."

Lovins: At the risk of disappointing Dr. Pickering in his assiduous search for still more straw men, I must deny having answered "No," or anything else, to the question "Can we give concrete meaning to the concept of democratic social change under conditions of truly large-scale corporate enterprises?" Had I asked such a question, I would have pointed out that democratic social change has been trying to cope with this problem for a long time. This issue is addressed peripherally in *Soft Energy Paths*: e.g. in my reference (p. 150, n. 6) to Mr. Justice Brandeis's dissent in *Liggett Co.* vs. *Lee* 288 U.S. 517 ff (1933). My residual concerns with fission technology—those that would come to be dominant if the problems arising from its toxic and explosive materials somehow disappeared—go far beyond the control of oligopolies. Those concerns include, for example, the vulnerability of electric grids, the difficulty of making political decisions about large plants with correspondingly large impacts (even "conventional" impacts), and the social-control implications of these problems.

Pickering: Lovins seems to believe that if we will just make the right technological choices, we can deliver our grandchildren, if not ourselves, into a fraternal future in which local, face-to-face relations are structurally supported, coercion is minimal and, apparently, distributional questions will not loom large. "Where we used to accept unquestioningly," he says, "the facile (and often self-serving) argument that traditional economic growth and distributional equity are inseparable, new moral and humane stirrings now are nudging us." I cannot find a single sentence in Lovins' argument to support this claim.

I can neither see nor foresee any *acceptable* "path" into any future at all which does not face the distributional problem squarely and give an account of its probable distributional consequences. Neither can I subscribe to any "path" which does not at least face the problem of growth. It may be "facile"; I do not see how it is "self-serving"; but I continue to think that the prospects for growth and the problems of equitable distribution are inherently related. The discussion of this relationship, however, has become enormously tangled. In the main, it seems that there are three basic or general problems with growth which should be distinguished from each other.

(1) There are serious problems in the present and for the future if we try to prolong indefinitely the patterns of economic development and growth which have been characteristic of the last century. If this is what Lovins has in mind with his phrase, "traditional economic growth," then there is something to his assertion. It is this pattern which endangers the very biosphere, if continued, and which has brought the distributional question so much to the fore.

(2) One of the critiques of this "traditional economic growth" is that it is wasteful and costly in every sense of the word—humanly, financially, environmentally, institutionally; and its stubborn continuation promises to increase all these costs exponentially in the foreseeable future. The search is therefore underway for ways and means of "doing more with less." Lovins devotes an entire section of his paper to ideas in this range and concludes that we could double the uses of existing energy levels through various "technical fixes." All ideas in this range should be welcomed and tested to see if their promise can be fulfilled; but it should be remembered that this is a strategy *for* economic growth, not an alternative to it. The emphasis is on doing *more* with less, not on *doing less*. Conservation can be a strategy of growth.

(3) Finally, there is the fact that there are more of us every day. Unless we can sustain economic growth at least of the same magnitude as population growth, then, de facto, we are falling into a distributional deficit. That is the inherent relationship between economic growth and distributional equity. It is true that economic growth does not automatically translate into materially equitable relations; but that does not negate the fact that growth is the economic condition for egalitarian politics.

Lovins: Dr. Pickering picks out an isolated sentence from section X, omits the subsequent sentence that amplified it, and says he "cannot find a single sentence . . . to support" it. But the two sentences I devoted to the subject of economic growth and distributional equity were (1) peripheral to the thesis of the article, (2) self-explanatory, (3) part of a section heavily condensed by the editor, and (4) amply supplemented elsewhere, *e.g.* in Chapter 10 of *Soft Energy Paths.*

I agree that any acceptable path must squarely face the distributional problem; and, as my several references to the problem make clear, I think a soft path is likely to contribute much more than a hard path to resolving this problem. (See also Chapters 1 and 10 in *SEP*.) Unfortunately, Dr. Pickering's digression into the alleged necessity of economic growth to improve distributional equity makes an already confused subject more confused. His point (1) is correct, thus unnecessary. His point (2) ignores my statement (*FA* article p. 76) that doubling end-use efficiency can be a recipe for maintaining present economic activity with half as much energy, rather than for doubling economic activity with present energy; it can also, for example, mean halved activity with quartered energy. I assume for the purposes of my analysis that the composition and growth rate of economic activity in the soft path are no different from those in the hard (save that ultimate economic growth can be much *higher* in the soft path—see *SEP*, p. 47, Fig. 2–3); but there is nothing about the soft-path argument that requires such growth. It is an exogenous *assumption*.

Finally, Dr. Pickering's point (3), through the loose use of pronouns such as "we," falls into an elementary confusion between a deficit of per-capita quantity and one of distribution. Growth says *nothing* about distribution; that depends on what is growing and who gets what (both from the growth and from the previous level). Distribution can change with or without growth of anything. Conversely, economic growth can and often does make distribution *more* skewed (as in the world's rich-poor gap since 1950 or before).

Pickering: There are genuine political problems on this globe. There are existing states of advantage and disadvantage. There are protective power relations covering those advantages. Natural, organizational, and cultural resources are not evenly distributed; and no one policy is going to make all those rough places plain. If we are constrained, either by nature or by our own imaginations, to a policy of no economic growth, it will make all those rough places rougher. It is too easy for the affluent to proclaim that we have enough; but who is this "we"? As far as I can see, the idea that "small is beautiful" only applies in affluent societies. It is rather like voluntary poverty. It may be an acceptable and even desirable personal life commitment; but is hardly the basis for social policy. Politically, economically, and socially, I cannot see that it solves any known problem in our common life, except possibly middle-class boredom and alienation. And frankly, I do not believe that it solves that problem very well. The retreat into communities of people just like ourselves is hardly the path into higher levels of social responsibility and wider embodiments of political justice. There is a world out there with hundreds of millions of people in it whose very subsistence depends on our collective ability to sustain at least moderate levels of economic growth *and* on our ability to govern our institutions and to control their behavior both at home and abroad.

Lovins: This is not the place to pit Dr. Pickering's superficial arguments in detail against those of such authors as Daly, Boulding, Mishan, Ward, and Galbraith; but by ignoring *what* is growing, *where*, and *who* reaps the costs and benefits of that growth, Dr. Pickering is evading the very compositional and distributional issues he claims to be addressing. He further confuses arguments about the nature, composition, rate, and distributional aspects of economic growth with scale issues ("small is beautiful"). It should be sufficient reply to point out that his statement that "There is a world out there with hundreds of millions of people in it whose very subsistence depends on our collective ability to sustain at least moderate levels of economic growth" is just as much an arguable personal opinion as my statement that "We can now ask whether we are not already so wealthy that further growth, far from being essential to addressing our equity problems is instead an excuse not to mobilize the compassion and commitment that could solve the same problems with or without the growth." If Dr. Pickering could adduce evidence that traditional economic growth has significantly narrowed the gap between rich and poor Americans, or rich and poor nations, I presume he would have done so. He has not. Perhaps the evidence does not exist.

Pickering: Neither growth nor distribution, it seems, can be delegated to laissez-faire localism. Lovins is concerned that any system which "pits central authority against local autonomy" is bound to be coercive. I am concerned that any system which does *not* pit central authorities against local "autonomies"—including corporate autonomies—will force us to relinquish any meaningful pursuit of distributional justice. As matters stand now, it is privilege and deprivation which are aggregated in self-proclaimed local autonomies. Up to now, the only major modern public institution which we in the United States have turned over to the ideology of "local autonomy" is the system of schools; and they are a distributional disaster, subject to every kind of economic, social, and racial discrimination, messed over with politics of protecting presumed privileges and in urgent need of having legally regulated, equity-oriented, centralized authorities pitted against the veil of local autonomy under which their mischief is done. Difficult as it may be to achieve, the maintenance and the strengthening of centralized authorities capable of wielding legitimate coercive power is the necessary condition for the people as a whole to be able to protect the environment, to cope with their domination by corporate powers, and to be able to expand their opportunities in the face of localized bastions of privilege. I doubt very much that the democratic prospect, either at home or abroad, will be enhanced by institutionalizing more privilege at the local level. And that, I believe, is what Lovins' proposal would inevitably come to, even though it is not his intention.

Lovins: Dr. Pickering once more puts words in my mouth—that "any system which 'pits central authority against local autonomy' is bound to be coercive." What I said about such systems is that in the particular context of centralized, hard-technology energy projects they tend to yield "an increasingly divisive and wasteful form of centrifugal politics that is already proving one of the most potent constraints on expansion"—a point Dr. Pickering does not deny. Nowhere did I advocate abandonment of the Constitutional sanctions and checks which prevent abuses of power, whether by the State or by a locality, by a corporation or by a person. To the extent that local autonomy has abridged Constitutionally guaranteed rights, the courts have rightly intervened, and the battle for justice continues. (That said, it is a bit unfair to criticize local schools for not providing what an inadequate local tax base often does not let them afford.)

It is as important for central authority to intervene to protect equity as for local autonomy to intervene to prevent excesses of central authority. The balance between those forces is the essence of federalist theory and practice. Attaining that balance implies neither that all decisions must be made locally nor that all must be made centrally, but rather that all must be made at the several levels that are administratively convenient, politically appropriate, and equitably efficacious. Jeffersonians and Hamiltonians differ, of course, in how they construe these flexible guidelines.

I fear, too, that Dr. Pickering has missed the point about hard-path centrism: the reason it incurs a political cost (in the view of all but the most dedicated autocrats) is that it upsets the politically acceptable balance by wielding central power *il*legitimately (as perceived by many people and courts), thus clouding, not "enhancing," the "democratic prospect."

Pickering: I have no doubt but that many of Lovins' "soft technologies" can be developed in ways which can contribute to the general welfare without our having to adopt the social philosophy which he believes they necessarily entail. I wish I could be equally confident that we could be done with the idea that the future of democratic values and institutions is all tied up with small town institutions and values. We live in a world of large-scale institutions which both have and generate large-scale problems which will need large-scale resolutions. The interpersonal values of friendliness and neighborliness are important to us all, and they need to be safeguarded; but they are not sufficient criteria for general social policies, and their value can only be strained to the point of absurdity in trying to make them so.

Lovins: In his concluding paragraph, Dr. Pickering admirably summarized the fallacy of his own position. The "social philosophy" which, he says, I believe soft technologies "necessarily entail" is one that they *permit* and *support* but *do not entail*; it is Dr. Pickering who supposes they entail it, not I. The idea that "the future of democratic values and institutions is all tied up with small town institutions and values" to the exclusion of "large-scale problems which will need large-scale resolution" is likewise his, not mine. The notion that the "interpersonal values of friendliness and neighborliness are important to us all, and they need to be safeguarded," with which I heartily agree, does not entail that those values are "sufficient criteria for general social policies," which is again Dr. Pickering's fabrication, not my statement. These values are, as I state, important elements of a sustainable social order, and may be enjoying a renascence, but are hardly a sole basis for "general social policies" even were it my role or intent to devise such policies. And those values, as Dr. Pickering has amply demonstrated, "can only be strained to the point of absurdity" by trying to make them play a role which I never made them play save in Dr. Pickering's impressive, but sadly inaccurate, imagination.

15

SAFER vs. LOVINS

Soft Path Could Lead to Disruption and Instability, Economist Claims

The final volley in Yulish's fusillade against Lovins is fired by Dr. Arnold Safer, a Vice President of Irving Trust Company and formerly Professor of Economics at Long Island University. A graduate of Brandeis, he holds advanced degrees from the University of Rochester.

Safer: The principal point of Mr. Lovins' paper is that there are two paths leading to a solution of the Energy Crisis, each mutually exclusive of the other. The first is a "hard path," dominated by large scale technology, centralized energy production, and increasingly pervasive control over our daily lives. The second and, according to Lovins, the preferable choice is the "soft path," idyllically flowered with many simple energy conserving processes, undertaken by local communities, and thereby preserving the essentials of Jeffersonian democracy. That Lovins' "soft path" will *by itself* solve the energy problems of the next quarter century and beyond is a vivid example of how sound technical ideas can lead to highly oversimplified socio-economic generalizations.

There is much to recommend [in] many of Mr. Lovins' specific suggestions regarding energy conservation and the more efficient use of existing sources of energy. Making do with less energy is a commendable goal. How to do that in a market-oriented economy without the incentives of the price mechanism is, of course, the underlying bone of political contention. Mr. Lovins does not address himself to this aspect of the problem, and thereby reveals a major gap in his understanding of how invention becomes transformed into innovation within the context of a market-oriented economy. The avoidance of energy waste through better thermal insulation, improved furnace efficiencies, and the recovery of waste heat are all sensible suggestions to be encouraged by government policy. Yet government policy, in the mistaken attempt to alleviate the pain of transition to reduced energy use, is deliberately holding down the energy cost incentives needed to achieve many of Mr. Lovins' prescriptions. While it is true that the U.S. economy has a host of local regulations, labor-imposed barriers, and conservative business practices, which all serve to constrain the path to energy conservation, the process is nonetheless going inexorably forward, and would be rapidly speeded up by permitting the energy markets to clear at considerably higher prices. The current natural gas crisis is testament to the fact that the speed of adjustment is insufficient. To the extent that higher energy prices might imply unacceptably high profits for any one group these could be taxed away. And to the extent that higher energy prices might cause undue hardships for lower income groups, the tax revenue so generated could be utilized in the form of income supplements (e.g., fuel stamps).

It is, therefore, not the specifics of Mr. Lovins' suggestions which bear criticism; it is the naive proposition that significant conservation can be carried out under some sort of new ethical standard which will make us all more sensitive to both the environmental and energy costs of our own actions. When it comes to jobs and profits, energy and environment take a back seat, unless the appropriate pricing policies are in place. Government could, of course, mandate by fiat many of these changes, but unless they are economic, they are likely to be severely resisted.

Lovins: Dr. Safer's first paragraph errs both in supposing that both hard and soft paths can lead "to a solution of the Energy Crisis" (though both are so intended by their proponents) and in caricaturing the soft path as essentially (shorn of rhetoric) energy conservation. Of course conservation cannot *by itself* solve the energy problems of the next quarter century or even of the next 50 years; but conservation by itself is not what I proposed. Nor did I ignore, as Dr. Safer states in his second paragraph, the problem of price signals, some of which I proposed on *FA* p. 75 and others of which are mentioned throughout my article. (For further details, see *SEP*, Chapter 1.5.)

Accordingly, if Dr. Safer rereads the passages just cited, he will find there a concise statement which in essence echoes his own views on energy prices, and indeed improves on them by a severance-royalty proposal that is much easier to implement than mere deregulation and windfall-profits taxes. I have nowhere proposed reliance only on "some sort of new ethical standard," but rather on the better use of existing market mechanisms. If ethical standards change, so much the better, but I have not assumed that they will change. The question whether the conservation measures I describe "are economic" is long ago settled, so the market resistance Dr. Safer mentions would not occur.

Safer: In addition to exploring the possibilities of reducing our present extravagant energy consumption, Mr. Lovins claims that these [new energy] sources are widely available, relatively simple to tap without dependence on high technology, and matched in scale, geography, and quality to end-use needs. Above all, they are renewable, implying that they do not deplete the non-renewable gifts of nature. As Mr. Lovins puts it, we can live on our energy income instead of on our energy capital. He is, of course, suggesting a widely increased use of solar, wind and biological energy sources. There is no question that *on the margin* all of these technologies should be encouraged, both in their application and in their future technical development. But to suggest that society will be able to utilize these technologies to the degree required to cause a substantial decline in our use of fossil fuels and electrical generation is again a naivete of the highest caliber. The capital alone required to modify both our housing stock and our industrial plants to these new technologies would dwarf the amounts needed for conventional fuels and power, not to mention the resulting disruption of existing economic interdependencies.

There is a further critical point which Mr. Lovins sorely misses. Just as oil in the ground is useless without drilling rigs and refineries, so are the rays of the sun without solar collectors and distributors. All of these require enormous amounts of capital, which must be spread out over a wide volume of usage to become economical. Traditional energy technology is large scale because it is economical, therefore, relatively inexpensive to the final user. So far, at least, few of these newer technologies offer that prospect.

Lovins: Dr. Safer seems to have misunderstood me to say that our whole society should be rapidly retrofitted at vast cost. I suggested instead a smooth 50-year transition that would involve essentially no writing-off of unamortized capital stocks. Replacement is mainly at the margin by normal attrition or secular growth. Retrofitting is modest and is indeed less than would otherwise be needed for massive electrification in a hard path.

Dr. Safer states that the capital requirements for these soft-path actions "would dwarf the amounts needed for conventional fuels and power," but unless my economic analysis is wrong—and he nowhere contests it—the opposite is true on a case-by-case basis, and therefore presumably for totals too. Dr. Safer has not justified his charge of "naivete of the highest caliber," but rather implied that his own thinking about scale issues is superficial. Unless he can point out some error of fact or logic which invalidates my economic analysis, I can only assume he has not read it and therefore does not appreciate why his orthodoxy is out-of-date.

Safer: The relationship between energy and employment demonstrates the single overriding fact that it takes more and more energy to supply a job at the level of productivity we have come to expect from the U.S. economy. Energy is an important segment of the increasing capital intensity of the U.S. employment structure. Increased energy intensity makes for increased labor productivity and thus for gains in real income which we call prosperity. Because we sustain a high per capita energy consumption, we are prosperous. In terms of output, this means we require large amounts of energy to produce those goods and services which give us a high standard of living.

Postwar economic history bears out this important relationship. While the amount of energy per dollar of real output steadily declined over the 1947-66 period, the number of jobs required to produce a unit of output declined even faster. As a result, it took more and more energy to sustain the same levels of employment and output. This trend accelerated over the 1967-73 period, as energy use grew by over 30%, while employment increased by only around 15%. With the 1974-75 recession, both the use of energy and the level of employment declined. However, because of the high price of energy relative to other inputs, energy use declined more than employment, so that the energy/labor ratio declined from its peak 1972 level.

In 1976, the upward trend returned; energy grew by 4.8% and employment grew by around 3%. At the same time, real GNP grew by an estimated 6.2%. Due to the forces of cyclical recovery from the depressed levels of 1975, employment grew by more than its long-term trend in relation to real GNP growth. Energy, on the other hand, grew by somewhat less than its long-term trend relation to GNP, primarily due to the initial conservation programs brought about by the dramatically higher price of energy. The outlook for 1977 and beyond, however, suggests that cyclical recovery may be giving way to secular expansion.

Lovins: Dr. Safer's energy-employment argument suffers from two serious flaws: failure to take account of the scope for improved end-use efficiency, and identification of labor productivity (which he loosely calls "productivity") with "gains in real income which we call prosperity." The income accrues only to people who have jobs. People who have been disemployed by energy- and capital-intensive black boxes do not benefit from the increase in productivity of the remaining workers, and their own real income is likely to *de*crease. The limiting case of Dr. Safer's thesis is presumably that labor inputs to the U.S. economy are virtually or entirely replaced by other factor inputs such as capital; labor productivity thus becomes astronomical but everyone is out of work.

It is better economics, I suggest, to seek an optimal balance between the several factor inputs rather than seeking to maximize the ratio of output to any one factor input without regard to the effect on the others. Indeed, many of our current ills appear to arise from treating a capital-short, labor-surplus economy as if it were the reverse. (I should also note that for reasons which seem persuasive to economic formalists—though not always to physicists—energy is traditionally treated in production functions as an intermediate good rather than a primary factor input, and therefore "washes" in computations relating final output to primary factor inputs.)

In short, the historical behavior which Dr. Safer cites reflects a period in which, for reasons which no doubt seemed good at the time, cheap energy was substituted for costly people. This behavior does not tell us what is rational for aaaaaaaaaaa period when many people lack jobs and energy will no longer be cheap. Nor does short-term fluctuation (such as the behavior of the energy-economic system in the past few years) tell us much about long-term elasticities, since, as Dr. Safer would doubtless be the first to point out, major changes in the energy system take a long time. What little it does tell us, though, is consistent with the view that since marginal energy costs became larger than historic costs, around 1970, the economy has begun—with a short perceptual lag—to seek a new equilibrium with a mixture richer in labor and leaner in energy. Of course, this use of economic metaphors begs many important questions, such as the merits and composition of economic growth, the quality of employment, the concept of work, and the dubious relationship between gross economic activity and net welfare; but my point is that even within Dr. Safer's paradigm, he has not made a sound argument that my thesis is wrong.

Safer: The secular question here relates to the structure of longterm growth in the economy. Can the growth in energy consumption be restrained to a rate no greater than the increase in employment? Both energy and employment are tied to the growth of output. And output must grow faster than employment to generate the increased productivity and the real per capita income gains which we call prosperity. But if output grows faster than employment, energy use will also grow faster than employment. Both the historical and technological evidence point to this conclusion. Despite the dramatic shift in employment away from manufacturing and capital-intense industries generally, the consumption of energy per productive job has continually risen. It is an inherent part of the process of capital formation.

Lovins: Dr. Safer cites no evidence for his statement that "Both energy and employment are tied to the growth of output." If he is saying that output cannot increase unless primary energy use increases, he is plainly wrong, as end-use efficiency can be at least doubled, at present world oil prices, by purely technical measures. Further, since welfare derives from economic stocks whereas Dr. Safer is measuring it by economic flows, the two measures can radically diverge if the goods made become more durable. And increased real income can come from sources other than increased labor productivity: for example, from increased capital productivity or decreased transaction costs. Nor does any physical law or other empirically verifiable principle require that "if output grows faster than employment, energy use will also grow faster than employment"; whether this occurred in the past does not tell us whether it need occur in the future under very different circumstances. I suspect that Dr. Safer is taking Cobb-Douglas production functions more seriously than their shaky foundations warrant.

Safer: These relationships also hold true in a cyclical sense. The often-used rule of thumb that it takes 3% incremental growth in real output to generate a 1% decline in the unemployment rate implies that energy use is likely to grow more rapidly than employment. As a corollary to that rule, we would estimate that energy will grow at about two-thirds the rate of growth of incremental output, thereby generating an increase of around 2% in energy use per 1% decline in the unemployment rate. As a result, it will take an increasing amount of energy to sustain the same number of jobs in the economy, if these jobs are to generate a higher level of real income to working men and women.

In assessing the outlook for 1977, we would expect to see the ratio of energy to employment increase, once again, as continued economic expansion stimulates an increased use of energy. We [Irving Trust] are forecasting a 5.5% rate of real GNP growth for 1977, coupled with a 4.0%—4.5% increase in energy consumption. We expect conservation programs to gradually reduce the energy/GNP ratio, but these efforts will become tougher and more costly as time goes on. We estimate a lower limit of around .67 BTU's in the incremental amount of energy required to sustain a dollar increase in real GNP.

Looking to the longer run, if we assume that a target of government economic policy is to reduce the unemployment rate to a 4½–5% range by 1980, it will require around a 2% per annum increase in employment. This employment growth in turn is tied to at least a 6% per year growth in real GNP. If these new jobs are to yield the productivity gains necessary for non-inflationary growth, energy use will likely grow by around 4% per year over the same period. That is, if the U.S. economy is to create 8 million new jobs between 1977 and 1980, based upon an assumed growth of around 6% per annum in real GNP, there must be around a 2% per annum growth in the level of employment. Productivity gains of about 4% per year make up the difference, and for the most part, that productivity gain is intimately linked with an even more energy-intensive employment structure.

Lovins: Considerable skepticism should be applied to historic rules of thumb. What we *do* know from ample physical analysis, with due attention to its economic aspects, is that if we had chosen some years ago to build a different sort of capital stock, we could today have just the same amount and composition of economic activity that we do now but with 40% less energy (a conclusion shared by, among others, Dr. Schlesinger [*Bus. Wk.*, 25 April 1977]. The extensive work of the CONAES economics panel has shown that the same flexibility in energy/GNP ratios extends to at least a factor of two in the U.S. These findings are not consistent with Dr. Safer's views, which appear to rest only on historic precedents. I am accordingly inclined not to attach much weight to the energy forecasts of the Irving Trust Company over the next three years—let alone the next fifty. [Later analyses suggest technical scope for a technical improvement in primary energy efficiency of a factor 6 in Britain and about 8 or more in the U.S., all cost-effective at long-run marginal cost, e.g. against new power stations.]

Dr. Safer again *assumes*, on the basis of how people behaved when energy was cheap, that increased labor productivity implies increased energy intensity. Precisely the opposite is more plausible: that human skill will be re-substituted for black boxes, natural materials for inappropriate synthetics, cheaper resources for dearer ones.

A further respect in which Dr. Safer's analysis lacks sophistication is his tendency to treat only first-order effects. I have argued (*e.g.* in *SEP*, p. 10) that second-order effects collectively dominate first-order effects, making cheap energy more a cost than a benefit. A simplistic analysis of first-order correlations tells us little about the complex causalities that really make the energy-economic system work.

Safer: We are faced with the dual problems of unemployment and excessive dependence on foreign sources of energy. Although we have idle manpower resources and the need for a greater domestic supply of energy, we have been unable to weld together an effective employment and energy policy. Solutions will not come easily or quickly. But the longer the delay, the greater the vulnerability of many American jobs to foreign economic and political pressures. Between now and 1980, it is likely that the U.S. dependence on foreign oil will increase, regardless of what we do in the next few years. The advent of the Alaskan pipeline will only arrest the decline in U.S. oil production, while delays in increased coal production and nuclear power will have to be compensated for through increased oil imports. While we have imposed restrictions on rising domestic energy costs, we must still pay the price internationally. What had been costing us $7 billion to $8 billion annually for imported oil now costs us $35 billion. And the longer we delay in resolving the domestic energy supply program, the more we will have to import from OPEC—at what will likely be ever-rising prices, given OPEC's near monopoly power over the world oil market.

In 1976, the U.S. spent about $35 billion for imported oil—almost equal to our capital investment in domestic energy supply. Compare that to 1962, when the U.S. invested around $10 billion in domestic energy while paying $1.8 billion for oil imports. This means our cost of importing oil has increased from less than 20% of domestic energy investments in 1962 to almost 100% today.

It would seem possible that a proportion of those resources devoted to importing oil could be fruitfully invested in increasing the supply of U. S. energy and in creating productive employment. A recent study by the Economics Department at U.C.L.A. suggests that replacing U.S. oil imports of around 2.0 million barrels per day with an equivalent amount of domestic energy would ultimately generate as much as 800,000 productive new jobs in the U.S., depending upon the particular policies adopted. According to that study, about 25% of these new jobs could come from additional domestic energy production and the construction of new plant facilities. The balance of the growth in employment would be derived from non-energy sectors which would produce more goods and services, both to support the energy-producing industries and to supply what would become a generally faster-growing economy. Although the policy alternatives to achieve these improved employment opportunities may differ, the overall implication for the economy is the same: idle manpower can be put to work in the implementation of a policy of greater energy self-sufficiency.

Lovins: The ideas here are unexceptionable—I have been saying much the same for several years—but I would lay more stress on the capital-saving and job-creating benefits of both improved efficiency and soft technologies compared to hard-technology increases in supply.

Safer: The history of industrial progress demonstrates an evolution of technological change, rather than a revolutionary uprooting of existing practices. To be sure, the core discovery or application is often revolutionary in its intellectual approach, and in comparison to what is being done at the time. But the widespread adoption of the new technology always requires a long period of gradual infusion into the economic system. This is the antithesis of the course of new energy technology suggested by Mr. Lovins. There is no reason why centralized power stations, with their advantages of scale but disadvantages of transmission and storage, cannot easily exist side by side with solar or wind devices, with their concomitant costs and benefits. Competition and consumer demand will gradually select the more efficient and beneficial. The companies authorized by government to supply these energy needs will still be there, regardless of which technologies ultimately prove more reliable in this higher energy cost era.

Lovins: These ideas echo my own in the article except for the sentence "This is the antithesis of the course of new energy technology suggested by Mr. Lovins"—which seems to reflect a serious misunderstanding of what I actually suggested. I fully agree with the second and third sentences, which therefore cannot contradict what I wrote. I envisaged a 50-year transition, not a revolutionary transformation, and I pointed out that in a technical sense, power stations and solar or wind devices can coexist (which is not to say that more power stations make sense *at the margin*).

Safer: Probably the most serious criticism of Mr. Lovins' article stems from his repeated insistence upon the mutually exclusive aspects of his "hard" and "soft" paths of energy production and consumption. This appears to be the unique point in Mr. Lovins' approach. In his conclusion, he states:

> A hard path can make the attainment of a soft path prohibitively difficult, both by starving its components into garbled and incoherent fragments and by changing social structures and values in a way that makes the innovations of a soft path more painful to envisage and achieve. As a nation, therefore, we must choose one path before they diverge much further.

This kind of broad generalization and sweeping conclusion is frequently repeated in the article. As a result, the less informed reader may begin to believe it, while the more informed may become increasingly irritated. Since Mr. Lovins is insisting upon an "either/or" choice, the burden of proof must be upon him. Society cannot bet its energy future on a host of unproven technologies, nor radically alter its present social structure to reach some distant utopian future.

Lovins: Dr. Safer's diffuse irritation with the exclusivity argument does not substitute for some explanation of why it is wrong. I have addressed the rationale of exclusivity at length, both in my article and elsewhere in these responses, and can hardly offer a better example than the present state of the US energy system. If Dr. Safer will not address the arguments I presented, then I fear I can think of no better way to satisfy his need for "proof." His underlying views seem to emerge, however, where he speaks of "unproven technologies" and the need to "radically alter . . . present social structures." Readers more careful than Dr. Safer will appreciate that I have advocated neither of these things.

Safer: It seems to me that the implementation of Mr. Lovins' "soft" path must stand the test of *gradual* adoption by both energy users and suppliers. Government policy can be utilized to further advance these new approaches, but they must remain marginal changes until their full implications are seen. Tax policy, regulation, and research and development funding can all be used to "get our feet wet," so to speak, but the step-by-step process of technological change is a fundamental ingredient of the multiple decision-making structure upon which our economy rests. For government to force radical change, of the kind Mr. Lovins advocates, would require a major uprooting of the socioeconomic framework within which the United States has prospered. Although his goals may be noble, Mr. Lovins' "soft" path sounds like a phony fork in the road, which could lead to serious economic disruption and social instability.

Lovins: I am content to introduce a soft path gradually at the margin: indeed, that is what I proposed. I have not proposed "a major uprooting of the socioeconomic framework within which the United States has prospered," but rather an evolutionary change resting foremost on the more efficient use of the market mechanisms that Dr. Safer advocates. It is hard to see how even the most orthodox economist can fault my proposals for long-run marginal-cost pricing, capital-transfer schemes, replacement-cost accounting, desubsidization, life-cycle costing, internalization of externalities, whole-system costing, antitrust enforcement, and other touchstones of responsible and conservative fiscal management. Having overlooked these proposals and wrongly suggested that I have ignored the economic aspects of energy policy, Dr. Safer tries instead to argue that my proposals "could lead to serious economic disruption and social instability." Such a charge could be made, and I have made it in detail, about a hard path; but any attempt to make it as credibly about a soft path requires a modicum of specificity, documentation, explicitness, and logic. These requirements place only a modest burden on a critic with Dr. Safer's resources, and I regret that he has chosen not to bear that burden.

16

CRAIG/NATHANS vs. LOVINS

Lovins's Analysis Is Called Deficient and Seriously Flawed

A lengthy critique of Amory Lovins's *Foreign Affairs* article was prepared at ERDA's request by Paul Craig, Department of Applied Science and Energy Resources Council, University of California, Davis, and Robert Nathans, W. Averill Harriman School of Urban and Policy Science, State University of New York, Stony Brook. Of the Craig/Nathans critique, Amory Lovins said: "It contains much correct and commendable material on which I shall not comment in detail." Since the major purpose of this book is to show how well Lovins' thesis holds up under adverse (and often hostile) criticism, we shall omit major portions of the critique that Lovins finds unobjectionable. In fairness to the authors, however, we retain their complete "Introduction" and "Overview" before beginning to excerpt selectively.

Craig/Nathans: Last fall Amory Lovins published in *Foreign Affairs* magazine a highly provocative article describing an alternative energy strategy for the U.S. based upon renewable resources, distributed technologies, and energy conservation. This paper, together with a technical analysis presented at the Oak Ridge Associated Universities Conference, "Future Strategies for Energy Development," held October 20–21, 1976, have led to extensive debate within the U.S. and elsewhere throughout the world on alternative energy futures (recently the material has been expanded and published in book form). The debates have focused on a variety of types of issues—including ethical questions, technical feasibility, economics, environmental issues, and many others.

In this paper we endeavor to develop a broad perspective on issues raised by Lovins. Our approach has been to dissect the Lovins arguments according to type and thereby to provide improved understanding of the implications of his thinking upon policy, research, and general discussions about energy futures. While there are many issues of a technical nature which need to be analyzed, we have not attempted technical analysis here. Indeed, we do not believe that answers to the most fundamental issues are to be sought in numbers, for the state of technology and economics is such that many issues perceived as important simply cannot be answered with assurance at this time. The paper offers instead a framework for thinking about the issues and a set of questions for further analysis.

We conclude that Lovins has brought to a focus a broad set of widely held concerns, and that while some of Lovins' conclusions may be questionable, the issues he has raised are germane. Changing social objectives, limits on technology, and other factors fully justify—if indeed they do not mandate—exploration of futures significantly different from "business as usual."

Attitudes toward energy have been changing with remarkable rapidity in the past few years. Lovins' ideas are extraordinarily controversial today. Yet one can see in them many concepts which might well become tomorrow's dogma. A careful analysis of his work has yet to be accomplished. We hope that this situation will quickly change.

Overview

"Lovins" energy scenarios are attracting increasing attention among disparate groups in the U.S. and other industrialized countries. Based on the writings of Amory B. Lovins and others, such scenarios encompass a wide set of objectives. On one level they portray the technical/engineering feasibility of employing an alternative set of energy technologies to satisfy future energy demands. On another level, they question current R&D strategies of the developed nations. And finally, they introduce hitherto unconsidered societal considerations into the determination of national energy goals and priorities.

At the most fundamental level, however, Lovins' work reassesses the structural parameters and criteria for long-term development of the U.S. system of production, distribution, and end-use consumption of energy. In effect, he offers a conceptual framework to guide this development, priorities for its decision-structure, and criteria for the choice of technologies.

While many terms have been suggested to characterize Lovins' work—"decentralization," "distributed," "small-scale," "soft"—none are sufficiently suggestive of either the radical nature of the blueprint or the enlarged societal domain into which energy systems decisions are placed in its formulation. To quote Lovins:

> The second path combines a prompt and serious commitment to efficient use of energy, rapid development of renewable energy sources matched in scale and in energy quality to end-use needs, and special transitional fossil-fuel technologies. *This path, a whole greater than the sum of its parts, diverges radically from incremental past practices to pursue long-term goals.* [Emphasis added.]

Lovins provokes a reexamination of our view of the role of energy in its relationship to society, and thereby urges exploration of new means for evaluating possible modes for national energy system development. Its purpose is to encourage the process (going on in this paper among others) of trying to define new models. Such models can gain acceptability—but only if they are more successful than their competitors in offering acceptable approaches to solving some of the central energy problems of the country.

This more fundamental, and perhaps more grandiose, interpretation of Lovins' work has been somewhat obscured by the tenor of the discussion and commentary elicited in reaction to Lovins' writings. In part, this reaction stems from the manner in which Lovins has decided to couch his arguments and theses. Faced with the difficult task of attracting the attention of both his policy analyst peers and policy-makers in government and industry, Lovins has (as is customary for all those who wish to cast themselves in the role of heretics) pushed the conventional model of "centralized high technologies" to its limit in order to call attention to the points at which it may break down. Lovins sharpens the discussion by arguing that there is a mutual exclusivity between the conventional path or model and his own, thereby establishing the entire issue as one requiring immediate attention.

It is the thesis of this report that the proper arena for starting critical discussions of Lovins' work lies in exploration and critical evaluation of his conceptual framework for energy system development. Two areas deserve particular attention. One is an examination of the assumptions built into Lovins' conceptual model and the conditions under which it is to be applied. Unless we understand the sensitivity of the model to the character of the energy demand which is assumed, or the importance of the geographical scale, we will have difficulty clarifying its basic tenets. The second is a delineation of the term "soft technologies." The task of classifying technologies can serve to clarify the meaning of "soft." Only after these two aspects of Lovins' conceptual model are better understood, can one proceed to interpreting features of the spectrum of soft alternative futures, their compatibility with more conventional energy supply and distribution systems and the process by which such futures may be realized.

Special comments are needed on Lovins' views on several specific points. On nuclear power, it is clear from much of Lovins' writing that he is intrinsically and irrevocably opposed to nuclear systems. Indeed, one can even argue that his emphasis on soft technologies derives in large measure from the (to him) unacceptability of nuclear alternatives. We believe that for many policy purposes the question of whether society should choose to exclude all energy derived from nuclear systems is fundamentally separable from the generic issues raised by Lovins. Moreover, however correct the arguments Lovins and others have put forth against large-scale, high technology energy development, they constitute only the basis of a rationale for exploring other paths, not a case for adopting any particular path.

With respect to Lovins' attempt to pose the issues in terms of a confrontation between two opposing paths, we believe this places an unfair burden on the new entry, for it seeks to compare a model of energy system development about which we know a great deal with one which exists as only a "qualitative vehicle for ideas." The intensity of the reactions demonstrates the risks of trying to move too quickly from a new conceptual framework for thinking about energy system development to a specific "solution" strategy. As with any major change in paradigms affecting important areas of societal concern, an initial period of assimilation and exposition is required.

On Lovins' assertion that we must choose now between two well-defined paths with no hybridization of hard and soft paths either possible or desirable: leaving aside the question of whether such a distinct separation is conceptually plausible, it dismisses *a priori* the promise of what could be a new and interesting combination of hard and soft technologies—a combination which may be extremely effective in allowing a soft technology path to function effectively on a national scale.

In the body of this paper we endeavor to characterize these issues and raise questions. We do not attempt to provide answers.

Lovins: This 87-page report, a revision of a May 1977 draft, was prepared for ERDA (Nuclear) and presents the authors', not ERDA's, views. It contains much correct and commendable material on which I shall not comment in detail. I welcome, for example, its emphasis on social criteria in energy decision-making, on modes of analysis other than the numerical, and on certain structural issues. I agree with its assessment that the fundamental character of my analysis "has been somewhat obscured by the tenor of the . . . commentary elicited." My soft-path arguments are indeed separable from my criticisms of the uniquely nuclear features of nuclear power. (Saying, though, that "for many policy purposes" the acceptability of nuclear power is "fundamentally separable from the generic issues" that I raise goes further, and is hard to accept unless I construe "separable" and "generic issues" quite restrictively. Elsewhere in their paper, too, Drs. Craig and Nathans insist that technological decisions are *not* separable from social, environmental, and other implications. I agree.)

Several points are less satisfactorily handled. Such statements as "some of Lovins's conclusions may be questionable" and [in a later passage] "somewhat exaggerated" are vague and unjustified. The discussion of exclusivity seems to confuse soft technologies with soft paths (defined sociopolitically) and technical incompatibility with the logistical, institutional, and cultural exclusivity of paths. The transition I envisage does entail having hard, soft, and transitional technologies coexisiting for decades.

Craig/Nathans: Lovins correctly points out the weight accorded in most current analyses to maintaining the current private sector and utility control structure of energy production, distribution and end-use. Given the magnitude of the energy problems facing the nation and the world, however, new approaches may be essential. For just as we find it necessary to consider a broader set of social criteria with respect to technical energy R&D decisions, there is a need to examine new infrastructure arrangements which recognize the possibility for much more extensive change than is normally considered. Basic to those considerations are the organizational arrangements making up the energy industry and its relationship with decision-making agencies within government. If current arrangements are expected to remain in effect, there is probably little innovation or interest in considering radically different energy options to be expected in "official" circles.

Lovins: The discussion ignores the possibility that "official" circles may be looking for an escape from current policy failures that does not require new institutions. This is largely what I try to provide.

Craig/Nathans: In both Lovins' analysis and those adopted by the federal government, a view of the future is taken which assumes that the country will have the time to proceed in a somewhat orderly and systematic fashion from its current heavy dependence on oil and natural gas to one of dependence on essentially inexhaustible resources. Lovins' article in *Foreign Affairs* deals with a transition strategy which moves away from dependence on the use of coal to his targeted soft technology future. Coal is similarly viewed in current federal plans as the primary transition fuel with breeder fission, fusion, and solar-electric seen as the ultimate inexhaustibles. In both scenarios, however, less attention is paid to the details of the transition than to the form of the targeted outcomes. Indeed, the ability of either approach to accommodate itself to relatively sudden and drastic localized and national shortages of oil and natural gas during this transition period is not considered to any extent. This perspective is at odds with the concern that the nation will not be able to respond to such sudden oil shortages in the 1985–2000 period without major economic and political disruption.

Lovins: I did in fact consider how soft vs. hard paths respond to "relatively sudden and drastic localized and national shortages of oil and natural gas"; I concluded that the soft path is far more resilient, owing to its more efficient end-use, greater supply diversity, reduced needs for oil and gas, geographic dispersion, and more flexible infrastructure (multi-fuelled boilers, cogeneration, district heating, etc.). These are major design criteria precisely because such contingencies are likely no matter what we do.

Craig/Nathans: Lovins gives insufficient attention to the technical problems and the economics of supporting a centralized energy electrical system whose peak and baseload can fluctuate widely if it is widely utilized to adjust the loadings of a wide-spread network of dispersed systems.

Lovins: This subject is neglected because my analysis assumes zero-backup solar [heating] systems. It is true that integrating wind machines, photovoltaics, microhydroelectrics, etc. into a grid presents interesting technical problems (on which I have been trying to stimulate research), but these problems can hardly be less tractable than those of integrating many huge and unreliable reactors into a grid. The integration problem exists and deserves study, but I do not think it looks excessively difficult. [1978 studies by Ed Kahn (Lawrence Berkeley Lab.) and others have tended to confirm this.]

Craig/Nathans: More recently energy decisions have become a focus of a new set of interest groups. Because the source of interest of the groups emanates from the espousal of a broad set of social goals and values applicable to a wide range of individual-societal interactions, the attacks of these groups on the existing energy system in the U.S. are much more pervasive. Decisions and choices affecting what is or is not included in such energy system must follow, these groups maintain, from their relationship to certain societal outcomes. Energy must serve to support the desired social structure and not vice versa.

Much of the underlying rationale for Lovins' approach is based on this perspective. Lovins' view is that the characterization of projected energy end-use must follow from the distribution and level of social amenities to be accorded each individual. Moreover, the notion that the design of the energy production-delivery system must not violate these amenities is consistent with this philosophical position of the machine serving man.

Lovins: The discussion is obscure. The "interest groups" are not specified. While in my perspective energy is indeed a means, not an end, I do not seek to specify desirable ends. Carefully distinguishing my personal preferences from my analytic assumptions, I assume, for purposes of argument, the same (or higher) end-use needs for goods and services that advocates of hard paths assume.

Craig/Nathans: Expressions of organized labor opposition in the automobile and petrochemical industry to elements in the National Energy Plan suggest that both the industry and its unions will present a united voice in this regard. This is not to convey the impression that the arguments of these entrenched groups are any less valid or subject to suspicion because of their source. No one doubts the short-term economic disruption of adopting a massive shift from a vertically integrated energy system to one that employs new technologies, new fuels, and new end-use devices in a horizontally integrated fashion. Rather, it is raised to acknowledge the presence of strong coalitions which will be working to defend the present arrangement. These groups will insist on their right, just as environmentalists have, to have their own special interests included in energy policy deliberations. All of this raises an interesting question. As we add more and more actors into the decision structure affecting energy technology choices, each of whom has his special set of interests, will an obstacle result for management—be it centralized or decentralized—to deliver what everyone admits is an essential commodity to all?

Lovins: The argument that proliferation of actors yields a universal veto power and paralysis correctly describes the hard path's result. With good information, this outcome is unlikely for a policy which, like a soft path, benefits virtually all actors simultaneously, including corporations and unions.

Craig/Nathans: The future characterized by Lovins in his *Foreign Affairs* article and in the ORAU presentation assumes or implies a particular set of technologies and a reasonably well-defined set of attitudes and societal orientations. These technologies and attitudes represent only one choice out of an unlimited spectrum. In this section we review the major assumptions which characterize the long-term future depicted by Lovins.

The criteria explicitly stated by Lovins for selection of technologies include five elements:

1. The technologies will rely upon renewable energy flows rather than depletable resources. That is, flux sources of energy will be utilized rather than energy forms stored in one era and released rapidly during another.
2. They are diverse. That is, many technological approaches are used simultaneously, none of them large.
3. They are flexible and based upon low technology, requiring limited technical skill to utilize.
4. They are matched in scale and geographic distribution to end-use needs. They take advantage of the free distribution of natural energy flows.
5. They are matched in energy quality to end-use needs.

Lovins: The third criterion should deal as it does with ordinary "skill to utilize" but should not imply limited skill to make or design, as these activities may well require great sophistication to achieve operational simplicity, just as a solar pond uses simple materials to achieve complex optical results. Using it does not require *specialized* skills.

Craig/Nathans: Diversity as used by Lovins seems to refer primarily to energy supply, not energy use. How large must a technology be before it is excluded under this criterion? Individually-operated units, such as home solar heating systems or family-sized windmills, would clearly be included. As the unit size becomes larger, the dividing line becomes difficult. Approaching the issue from the other side, one may try to identify units so large as to be clearly disqualified under this criterion. The answer depends perhaps upon the total energy needs of a particular region. In an underdeveloped nation using, say 100 megawatts of electricity, a single 100 megawatt unit would not be diverse. But what about 10 units of 10 megawatts? Or 100 units of 1 megawatt? A 1 megawatt unit is quite small by present U.S. utility standards. On the other hand, it may be large in an underdeveloped nation, or large by the standard of an individual home (which typically requires about 10 kilowatts under peak load conditions). A unit in the size range 1-10 megawatts would be just right for a U.S. shopping center which could in turn make use of the waste heat, thereby increasing overall efficiency (cogeneration). In the U.S. today there are installed about 400,000 megawatts of electrical capacity. With such a large capacity, may it not be argued that under these conditions a 1000 megawatt unit is diverse, making only a modest contribution compared to the total need. Note that this definition depends on the size of the power grid.

Lovins: The discussion of "Diversity" deals with end-use matching and allied issues, not diversity. Diversity is clearly a property of an aggregate of various soft technologies; seeking diversity in single objects is a category mistake akin to asking about the first digits of numbers less than ten.

Craig/Nathans: Lovins is ambiguous on the issue of technological complexity. On the one hand he states that energy systems must not be arcane; on the other, they need not be unsophisticated.

Lovins: My discussion of complexity may have been overcondensed, but was hardly ambiguous. I proposed technologies that are simple to decide about, use, diagnose, repair—simple from the end-user's point of view—but that can embody sophisticated technological understandings and (as in microelectronics) manufacturing processes.

Craig/Nathans: Matching of technologies in scale is to some degree allied with matching them in diversity. The orientation of this criterion appears to be toward minimization of energy distribution. There are many consequences. A transportation network which endeavors to move persons and goods over long distances must be supplied with energy forms of uniform character on a temporal and geographic scale appropriate to the transport system. Thus energy would appear to have to be moved extensively.

Reliance upon solar energy as a renewable resource, combined with minimum energy transmission and distribution, constrains reserve-poor regions of the nation in their energy potential. Regions with low insolation or extreme variations in cloud cover can never have the energy base of those located in the sun belt. Further, solar fluxes are high in regions of the nation with low population density. Non-energy resources constitute an issue in this criteria element, just as they do in several others. Construction of complex goods requires many resources—iron, copper, aluminum, energy, and others. These minerals are, in general, not colocated. Indeed, many minerals exist only in low concentration in the U.S. and have traditionally been imported. A decision to minimize movement of energy resources carries with it an implication of minimization of transportation of other resources. The result of such a policy would certainly be to decrease national homogeneity and to circumscribe many types of choice presently available.

Lovins: The discussion confuses scale with geographic dispersion and invents the meaningless "matching . . . in diversity." The aim of end-use scale matching is to reduce as far as possible the costs (including the social costs) of distributing energy. I do not agree that energy must be moved about the country to avoid inequity, or that the U.S. is or should be homogeneous; I prefer pluralism and diversity. Californians may enjoy not having New England winters, but some New Englanders enjoy their peculiar autumn foliage, maple syrup, snowshoeing, having less sunlight than Arizona, etc. There is no more imperative to give Wisconsin the isolation of Florida than to give Nevada the water of Mississippi, Kansas the mountains of Oregon, Vermont the cotton-fields of Alabama, or Connecticut the prairies of Wyoming—though doubtless Bechtel could do all these things for a price. Diversity conflicts with the mentality that spawned the California and Central Arizona Water Projects, but I do not see why energy, any more than any other resource or natural feature, must be uniformly distributed everywhere. The important thing is that all parts of the country do in fact have enough sunlight for their needs with relatively small differences in cost.

Craig/Nathans: Matching of energy quality is a criterion based upon laws of thermodynamics. By energy quality matching it is possible to utilize more effectively the full thermodynamically available energy (free energy). Irrational energy pricing policies have occasionally led to situations in which energy input to certain energy production processes actually exceeds energy produced (just after the 1973 OPEC embargo this occurred in stripper oil wells where cheap regulated gas was used to operate pumps which produced unregulated oil, which sold at a high price). In principle, a rational energy pricing policy might lead to incentives toward high thermodynamic efficiency and thereby to achievement of this criterion.

In practice, many other forces act. A much discussed current example of improved thermodynamic matching arises in industrial cogeneration of electricity and process steam. Yet even when economics appears to support such projects, there can be legitimate concern arising from "locking in" of particular production processes. A characteristic of systems with high thermodynamic efficiency is that they are carefully designed, and the elements are tightly interconnected. Under these circumstances it becomes exceedingly difficult to redesign or change in any major way any individual portion of a plant. The entire system must be handled together. An advantage of inefficiency is that it can provide flexibility—both of process design and of process and plant location.

One example of this occurred in the 1950's in the phonograph industry. The decision to standardize record speeds at 78 rpm facilitated the establishment of a large industry. Yet at the same time this standardization delayed a shift to lower speeds when reproduction technology improved. The situation finally changed, but only after a major battle between RCA and Columbia, who introduced 33 rpm and 45 rpm. A similar situation occurred in the technology for color television. Columbia's system was compatible with existing black and white transmission, but it was mechanically complex. The RCA system required changes in transmission but used electronic rather than mechanical techniques. In this case technical change carried the day, and transmission techniques were modified.

A major contributor to growth of per capita GNP in the U.S. has been the consistent rapid growth in industrial output per unit of (labor dollar, etc.) input. The ability to replace existing plants (even though still operational) with new more efficient plants or plant subunits, using new or improved processes has contributed to this growth. It is presently unknown (and is a good subject for research) to what degree thermodynamic matching of energy sources to loads might impede industrial output growth, assuming such continued increase is desirable.

Lovins: The pathological cases of net-energy loss have nothing to do with quality matching, which is designed to reduce as far as possible the costs (including the social costs) of energy conversion. High Second Law efficiency is a means to ends such as frugality with resources (not mentioned in the Craig & Nathans criteria), not an end in itself. End-use quality matching is not quite the same as high Second Law efficiency, though they are related in the formalism. Cogeneration hardware, like any other kind, can have flexibility designed into it, at a small penalty in cost or performance or both—but still with a vast advantage over duplicated utility and factory steam-raising systems. I do not see the relevance of the phonograph and television examples to this question of flexibility. Nor do I see how end-use quality matching might "impede industrial output growth."

Craig/Nathans: The use of coal as a bridge to renewable energy forms is assumed by most energy analysts today. Expansion of coal production and use is a key element of the President's Energy Plan. Lovins has proposed the use of small-scale coal conversion systems as a part of a transition to soft technologies (he has not argued that coal is a "soft" energy form). There are many issues associated with the technical possibilities for such systems. At the end-use point these include the capability of building small-scale coal units capable of reliable operation and capable of meeting air and water pollution regulations. In principle, however, such systems can be built (although they may prove quite expensive).

Lovins: The statement about small-scale coal end-use ignores historic experience, and the speculation that it "may prove quite expensive" is incorrect.

Craig/Nathans: The bulk of the criticisms raised thus far against Lovins' concepts tend to be numerical. One reason for this is that Lovins has drawn extensively upon well-known numerical data bases in order to argue for his ideas. For example, many of his data are taken from the Bechtel Energy Supply Planning Model (Carasso, 1975). Critics have charged that Lovins' numbers tend to be too optimistic regarding renewable resources and must be too pessimistic regarding conventional energy forms and nuclear power. This category of criticism turns upon the specific values which one group or another believes characterize a particular technology at a particular time. Examples of the types of numbers which are discussed in technical criticism are costs of technologies, reliability, performance characterization in a particular environment, etc.

A warning needs to be raised about undue emphasis upon numerical analysis in considering alternative futures. On a time scale of a few years, or for investment decisions which are being made today, economic intercomparisons are justified and offer a valuable decision criterion. On longer time scales economics become very shaky indeed, and uncertainties loom large. Engineering numbers similarly lose significance as one moves further into the future. While there is a temptation to emphasize numerical analysis because of the feeling of solidarity which appears to obtain, it is our belief that factors which are difficult or impossible to quantify are likely to prove substantially more important in decision making vis-a-vis soft energy paths.

For long-range energy system evolution—which is the focus of Lovins' analysis—economic factors are not and cannot be controlling. Costs are inevitably dominated by uncertainty about technical feasibility, environmental impact, reliability and vulnerability of the new systems.

Long-range decisions do not fundamentally turn on economics. Whether or not one agrees with this point of view, it certainly establishes the role of societal preferences as a legitimate factor in considering long-range energy options.

Lovins: While I agree that numerical arguments are not dispositive, Chapters 6-8 of *SEP* make a *prima facie* case that economic arguments tend to favor soft technologies by a larger margin than that normally used today to make hard-technology investment decisions.

Craig/Nathans: Lovins' suggested coal use in 2000 is deserving of special attention. His projections call for an increase in coal use from a present 15 quads (about 650 million tons per year) to 29 quads (about 1.3 billion tons per year). This use rate is slightly lower than that called for in the ERDA-48 baseline scenario, and about the same as that called for in the ERDA-48 intensive electrification scenario (ERDA 1975). It is also about the same as would be produced by 1985 under President Carter's April 20 [1977] plan. Production of 1.3 billion tons of coal per year would require extensive increases in coal mining, with the attendant environmental problems from mining, combustion, etc. It is by no means clear that such expansion is consistent with the philosophy of a "soft technology" future. Technologies for coal production at this level are known which would be massive in size, centralized in management, expensive and require extensive new transportation networks.

Lovins: My assumed coal use in 2000 (and in 1990) is 25 q, not 29 (it is 23 q in 1985), and need not entail significant Western stripping. It obviously involves some non-soft technologies, but they are part of the transitional fossil-fuel economy, which is not bound by the same criteria even though it shares the same goals. Coal-mining is brief, sparing, and held to high environmental standards.

Craig/Nathans: The only sector of the economy which is critically dependent upon liquid fuels is transportation. In the absence of major progress in electrical energy storage systems, liquid fuels will be required in the future if our transportation system is not to be drastically altered. Lovins' work says almost nothing about this problem. To the extent that it is addressed, bioconversion schemes are used which would appear to require extremely high levels of coordination, and probably centralization. Federal programs which deal with liquid fuels have to date received low emphasis, despite the key role these energy forms play.

Lovins: I explicitly discuss the liquid-fuels problem in several places, notably the section dealing with bioconversion. The kind of bioconversion I propose does not "require extremely high levels of coordination, and probably of centralization.'

Craig/Nathans: A major concern about large-scale nuclear power plants is their potential vulnerability of sabotage or to natural disaster. Indeed, it seems likely that the primary barrier to further deployment of nuclear systems may result from public concern about the vulnerability of these systems. While there has been extensive (albeit inadequate to date) analysis of nuclear systems from a safeguards point of view, only a few studies have been carried out on the vulnerability of other energy systems. It may well turn out that vulnerability of complex technological systems is exceedingly great no matter how carefully they are designed, and that this vulnerability may constitute a major argument in favor of dispersed or decentralized energy systems.

The Pacific Gas & Electric Company has on a number of occasions had its power transmission system attacked by terrorists intent on damaging the system. Since transmission line towers are necessarily often located remotely, security of the electricity distribution system appears to be impossible. Further, the likelihood of apprehending individuals who destroy transmission lines appears slight.

There is, of course, no way to estimate the likelihood that technically skilled saboteurs will undertake widespread operation within the United States. On the other hand, however, the incidents of sabotage on a nation-wide basis are certainly increasing. Thus, it is plausible that at some time in the future technically sophisticated attacks on centralized systems of all sorts might be undertaken. Decentralized energy systems cannot, of course, be defended as intensely as major (and capital-intensive) centralized systems. On the other hand, the potential for societal disruption with decentralized systems is very much less and the publicity associated with a successful sabotage endeavor correspondingly reduced. Individual elements of the system can be put out of operation, but the bulk of the system would not fail.

Lovins: Vulnerability to disruption is a major design criterion favoring soft paths [as confirmed theoretically by Ed Kahn of Lawrence Berkeley Lab. in 1978].

Craig/Nathans: Lovins has succeeded in bringing together a broad set of materials so as to construct a picture highly appealing to many groups in the United States and the world. His study is deficient in a number of areas, some of which we have discussed. These deficiencies have been used by critics to argue that his entire approach is so far removed from reality as to be irrelevant to public policy ("irresponsible" is the term used by Ralph Lapp). At the other extreme, David Brower, founder of Friends of the Earth, introduced a reprint of Lovins' paper by saying " . . . if the Nobel people will send us an application for the Peace Prize, we think we know whose name we will submit."

Lovins' work is and will remain controversial. We believe it is important. This importance stems only in part from the actual analyses presented. Lovins' analyses are seriously flawed, and there is a strong temptation to miss the importance of his work through preoccupation with details. What is of vastly more importance than the numerical detail is the fact that Lovins has identified major problems with existing approaches to energy, and that he has proposed alternative approaches which are plausible and which appeal to diverse—and probably extensive—constituencies.

Many of the technical difficulties with Lovins' technical analyses can be answered through extensive analysis. Lovins operates as an individual. It is only to be expected that his work will will not stand up in every detail under close scrutiny by assembled expert opinion. Should his work continue to capture attention—as now appears virtually certain—then more extensive analysis will be undertaken both by his supporters and his critics. And from that analysis will certainly emerge "Lovins-type" futures worked out in careful detail, some of them fully defensible against all but the most sophisticated attack. As this analytic process continues our picture of what constitutes a "Lovins" future will certainly change. Some of the weaker concepts will be modified or replaced. The concept of "mutually exclusive" paths will gradually be replaced by a less clear but more practical continuum with a gradual divergence of a Lovins future from a "business as usual," or centralized future. To concentrate too much upon the process by which Lovins' ideas may be refined and modified, and upon the analytic problems, omissions and factual errors of his work misses the essential point. Lovins has captured the discontent of many individuals and groups with "business as usual" approaches to energy.

It is our conclusion that the forces he has catalyzed will not simply fade away but rather will become increasingly important as time goes on. The government will have to deal with these forces one way or another.

Lovins: Neither the Craig & Nathans paper nor any other so far published justifies the statement that my analysis is "deficient in a number of areas" and is "seriously flawed," or that "it is only to be expected that . . . [my] work will not stand up in every detail under close scrutiny by assembled expert opinion." Such conclusions, in the light of this record, seem premature, and should be either supported in specific detail or deleted. As they stand, they seem inconsistent with the findings of the reviewer (Dr. Gene Rochlin) whom Dr. Craig asked to form a judgment of the numerical controversies surrounding my article. Similar reservations apply to "Some of the weaker concepts," "more practical," and "analytic problems, omissions and factual errors." These phrases should be justified or omitted.

[G. I. Rochlin is with the Institute for Governmental Studies, University of California, Berkeley. Excerpts from the paper referred to by Mr. Lovins follow.]

Rochlin: Many of the critiques of the Lovins soft path thesis have focused on the various numbers calculated to some extent in the *Foreign Affairs (FA)* article. To a great extent, the documentation for these is not in the *FA* article, but in the Oak Ridge paper (ORAU) and the footnotes therein. If there is a single generic statement to be made about the numerical responses in the critiques of the *FA* article, it is that the critics have, by and large, either failed to read the ORAU paper and its references, or have read them incorrectly. Both Lapp and Bethe, in particular, have made serious errors in their public responses that arise primarily from shooting from the hip instead of examining the documentation.

It is critically important to recall that Lovins himself characterizes his work as a synthesis and not, by and large, a rederivation of numbers and figures. As I will explain below, these numbers can be presented in a way that obscures their derivation or their meaning. But they are not off the cuff or *ad hoc* estimates; by and large they are based on the careful work of others. That this is so cavalierly ignored by many of the critiques is in itself revealing of a not untypical attitude that environmentalists and critics of present programs are sloppy, careless, or ill informed. . . .

Bethe has now conceded (see letter to Lovins of April 12, 1977) that Lovins is correct in his statement of the cost of solar home heating and storage. [Lovins points out that Professor Bethe did not accept all his cost estimates, but only his arguments about the feasibility of seasonal heat storage with relatively small storage volumes.] Bethe and Lapp made several conceptual errors in differing with Lovins' estimates.

His other cost estimates, as well as estimated prices for delivered energy, need to be examined. One needs to determine whether such firms as, e.g., Stal-Laval are quoting production line costs and prices, or using the first few units as loss-leaders to bring in business. As Lovins almost always gets his numbers from primary sources, and usually cross-checks them as well, I would be surprised if they were off by large factors. . . .

As I have pointed out before, comparing Lovins' figures with those of his critics is more than an apples and oranges problem. It is comparing apples and horned toads. The two sides have not yet agreed on which is edible, if either. However, Bethe's last letter shows that, however grudgingly, his critics must agree that only a whole-system comprehensive model is a fair one for comparing decentralized and centralized energy technologies. While Lovins' basis—in bbl of oil equivalent—is interesting and suggestive, it would be better to develop a more complete and comprehensive set of indicators. Perhaps none are any more useful, perhaps they are. Certainly, a great deal of work needs to be done in this area.

In the meanwhile, Lovins' numbers are, with a few exceptions (most notably the extravagant statement on nuclear temperatures and his diesel electric statement) standing up in good form to the assaults.

[It is the editor's understanding that Dr. Rochlin later agreed that when correctly interpreted, the statements concerning nuclear temperatures and diesels were *not* exceptions to the general rule that Lovins's numbers are standing up in good form.]

17

SOUTHERN CALIFORNIA EDISON Vs. LOVINS

Utility Executive Disparages Fluidized-Bed Technology

Southern California Edison Company, a major utility, reacted strongly to publication of Lovins's article in *Foreign Affairs.* An anonymously authored 20-page paper containing "factual clarifications" was who-knows-how-widely distributed to "Dear Colleague." All points raised by this rather ineffectual document are dealt with by Mr. Lovins elsewhere in this book, so we didn't ask SCE for permission to use it. As part of its package of anti-Lovins material, however, SCE distributed "An Analysis of the Potential of Fluidized Bed Combustion Technology for Generating Electricity from Coal," a paper by Alexander Weir, Jr., Ph.D., Manager, Chemical Systems Research and Development, SCE. If this paper is to be believed, Lovins is wrong about a rather important point: the availability and suitability of fluidized-bed combustion as a transitional technology to help bridge the gap between where we are and where we need to be. The issue isn't quite joined on an eyeball-to-eyeball level because Weir is talking about (what else?) huge central station power plants while Lovins is talking about comparatively small installations, used in the main for purposes other than the generation of electricity. Still, the confrontation is pretty direct and we thought it would be of interest to you.

Assistant Counsel John Wentworth Evans of SCE's Law Department answered our request to reprint the Weir paper. Dr. Weir "heartily agrees," it seems, that public discussion of the issues addressed in "the Paper" would be good, but he maintains that the "editing technique" we employed resulted in "distortion of his ideas."

Assistant Counsel Evans alleges that by "deleting entire passages of the Paper, you have taken Dr. Weir's ideas out of context and generally misrepresented his thoughts." Not remembering any deletions that might reasonably be complained of, we checked. The omitted material was as follows:

1. A 28-line "Summary" (which, obviously, contained nothing that was not in the body of the text).
2. A heading on page two reading "The Fluidization Process."
3. A heading on page three reading "Electric Power Research Institute

'Fluidized Bed Combustor Development Issues.' "

4. A heading on page six reading "Potential Problems With Fluidized Bed Combustion."

5. A heading on page nine reading "Other Applications Of Of [sic!] Fluidized Bed Concept."

6. A heading on page eleven reading "Conclusion."

7. Footnotes, eleven in all, on pages 2, 3, 4, 6, 7, 8, and 10.

We also altered Dr. Weir's text by restoring a line that he had inadvertently omitted from a direct quotation of Lovins.

Southern California Edison did not refuse outright to let us use Dr. Weir's paper, but it imposed conditions:

1. We could use a complete, unedited version, but critical comment by Lovins could only appear before or after "the Paper"; it could not be interspersed. (The material had already appeared in that form in the hearing record, and in that form, we found it less illuminating than it might be.)

2. We would be required to publish a rebuttal of Mr. Lovins's comments by Dr. Weir, without any counter-rebuttal or comment by Lovins. (Interestingly enough, we were told that "This rebuttal may be interspersed . . . at the proper location in Mr. Lovins' comments. . . ." It might be amusing to hear John Wentworth Evans explain why interspersing is naughty in one case and nice in the other.)

3. Conditions were placed upon excerpting of material in promotion copy, magazine serializations, and so on. (These conditions were reasonable.)

4. Friends of the Earth, a non-profit organization that has never known what it's like to be totally free of debt, was to donate $1,000 to the 1978 SCE Employees United Way Campaign.

SCE's assistant counsel ends his letter with a not very heavily veiled threat that although Dr. Weir "believes discussion of these ideas is vital," he reserves the right "to institute whatever legal action he feels is appropriate to protect his name."

In order to bring to readers what SCE and Dr. Weir purport to believe is a vital discussion, we would have agreed to restore the omitted summary, headings, and footnotes. We would have agreed not to intersperse Lovins rebuttals among Weir comments, resorting instead to the second-best parallel-column format. We would certainly have agreed to reasonable conditions concerning excerpting. Could it be that the demand for a $1,000 donation was SCE's way of ensuring that its terms would be unacceptable? We had offered Dr. Weir, as we offered all the contributors, a smaller reprint fee.

The safe and sane thing, no doubt, would have been to drop the whole thing. But Lovins has interesting and important things to say about

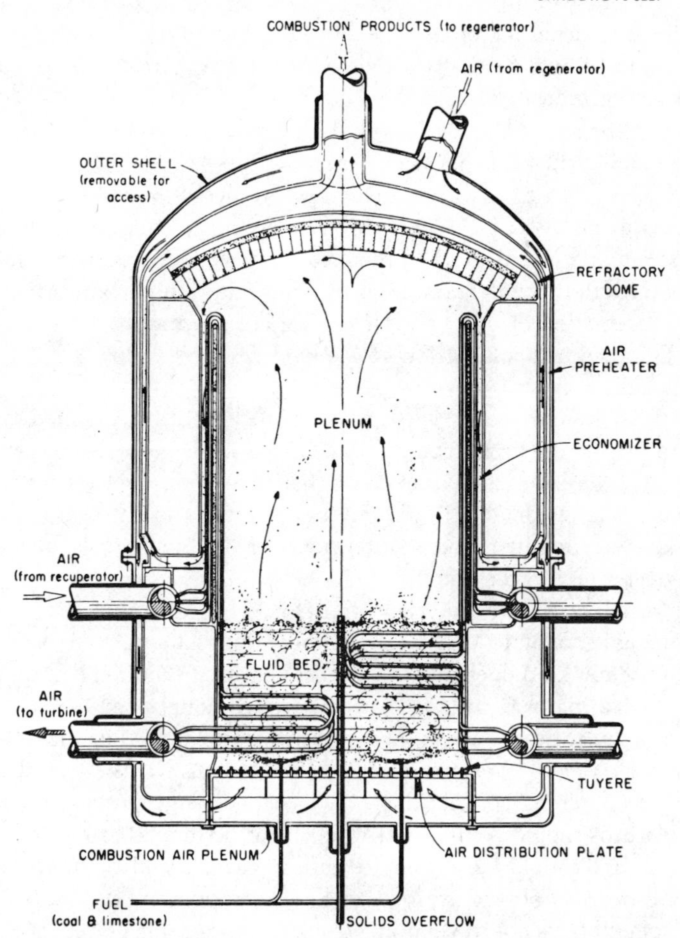

SCHEMATIC DIAGRAM SHOWING FLUIDIZED-BED COAL-COMBUSTION SYSTEM DESIGNED TO SERVE AS A HEATER FOR A CLOSED-CYCLE GAS TURBINE

ORNL Reference Design Combustor

fluidized-bed technology. And since the things he says are couched in the form of responses to someone else's comments, we reluctantly adopted the third-best expedient of paraphrasing the verbatim comments that Lovins was responding to in the hearing record.

Paraphrase of Weir: U.S. coal reserves are plentiful compared to oil and gas. An environmentally acceptable method of burning coal to generate electricity would be desirable. Environmental considerations require that part of the sulfur and other inorganic matter contained in coal be removed before it can escape to the atmosphere. This can be done in various ways, listed here in declining order of technological maturity:

1. Conventional combustion followed by stack gas treatment to remove fly ash and gaseous sulfur oxide.
2. Solvent refining of coal to produce a solid fuel from which most of the ash and some sulfur has been removed.
3. Gasification of coal to produce low-Btu gas.
4. Gasification of coal (in a reaction involving steam and oxygen rather than steam and air, as in 3) to produce intermediate or high-Btu gas.
5. Production of methanol from intermediate or high-Btu gas.
6. Liquefaction of coal by means of various processes.
7. Fluidized-bed combustion of the kind described below.

Lovins: Dr. Weir's paper is narrowly framed in utility-oriented terms: how to use fluidized-bed combustion on a very large scale to generate electricity. Maslow once pointed out that if the only tool you have is a hammer, it is remarkable how everything starts to look like a nail. Dr. Weir has the opposite problem: interested only in driving nails, he tries to make everything into a hammer.

My article, seeking the right tool for each job, argued (on thermodynamic, economic, and social grounds) that building more big power stations of any kind does not make sense, but that fluidized-bed combustion is a sensible way to meet particular transitional needs—mainly for industrial direct heat—with coal at a scale ranging from the order of a few kW to the order of 10 or perhaps 100 MW(t). Thus Dr. Weir and I begin with very different conceptions of what the energy problem is, what fluidized beds should be used for, who should be interested in them, and how big they should be. Of course, utilities could well be interested in fluidized-bed technology for industrial cogeneration and district heating; but if something is not worth doing (such as building big power stations at the margin), it is not worth doing well, so Dr. Weir's comments on the problems of replacing monstrous coal-fired power stations with equally monstrous fluidized-bed versions thereof are pervaded by a certain irrelevance.

Dr. Weir unaccountably *assumes*, without addressing any of my arguments to the contrary, that replacing oil and gas by coal, even transitionally, requires an increased supply of electricity. Neglecting boiler backfits (e.g. for cogeneration) as outside his brief, and ignoring some classes of coal-electric technologies (notably fuel cells), he then provides an incomplete list of coal-fired fluidized-bed electrical systems. His omissions include combustors with separate heat exchangers, either in the radiant zone in the freeboard region above the bed or in a separate heat-recovery section (Elliott's rotating fluidized beds, for example, achieved power densities over 100 kW per liter, confined by 15g accelerations, and had no heat-transfer surfaces in the bed); open-cycle gas turbines; mixed-cycle gas turbines such as the Stal-Laval system that I cite; closed-cycle gas turbines; combinations of gas turbines with steam cycles; and, for completeness, the use of sulfur acceptors other than limestone. (At elevated pressures, dolomite is approximately twice as effective a sulfur acceptor per Ca atom as limestone, since magnesia can capture sulfur too, yielding sulfur capture over 90% with a Ca/S atom ratio of 1.0–1.5.)

Paraphrase of Weir: In fluidized-bed combustion, particles of coal and limestone are suspended in an upward-flowing stream of air. A heat exchanger in the bed of coal converts water to steam. Temperature of the combustor is controlled by the removal of heat by water and steam. Gas leaving the main combustion chamber passes through a cyclone separator, which removes about three-fourths of the solids in the gas. The solids, including unburned coal, ash, limestone, and the calcium sulfite and calcium sulfate produced by the "burning" of limestone, are fed to a second fluidized-bed combustor known as a carbon burn-up cell, which also contains a heat exchanger. Unburned solids are removed from the main combustion chamber and the carbon burn-up cell by some continuous process not yet fully developed. Solids in the gas escaping from the cyclone may be removed by electrostatic precipitators, baghouses, or scrubbers.

Operation of a fluidized-bed combustion system appears to be more difficult than operation of conventional coal boilers, even if the latter are equipped with scrubbers.

Fluidized-bed technology is in a "relatively early" stage of development. The most advanced demonstration project is an ERDA-sponsored 30 MW plant at Rivesville, West Virginia. The system was started up in January 1977. Part of the boiler was operated for 40 hours, but was shut down because of difficulties with the materials-handling system. According to the Electric Power Research Institute (EPRI), which manages the program, the Rivesville project has limited design flexibility and built-in operating problems, so its operation cannot be optimized.

Lovins: The discussion tacitly assumes a relatively high fluidizing velocity. This carries over ("elutriates") large amounts of particulates from the bed, requiring a separate carbon burn-up cell. Dr. Weir seems to regard this requirement as normal and inevitable. In fact, the more sophisticated designers of fluidized-bed combustors, especially in Europe, have found empirically that elutriation of unburned carbon fines from the bed and the need for a carbon burn-up cell (or recycling of carbon fines) are a symptom of poor design and can be eliminated by proper design. This often involves moderate fluidizing velocities (typically 0.6–2.5 m/s) and the careful use of bed circulation patterns designed to maximize residence time for complete burn-up even of large lumps of coal (up to 2–5 cm). Secondary heat exchangers and waste-heat boilers can still be and often are used to recover a third to a half of the total heat. Ash and sulfated sorbent are normally removed by the simple and well demonstrated method of overflow over weirs. With proper design, no precipitator is needed to capture particulates (see below), and certainly no scrubber. Since the headaches (and costs) of scrubbers are avoided, it is generally not true that "operation of a fluidized bed combustion system is . . . probably more difficult" than for "conventional coal boilers" of equivalent thermal and air-quality performance.

The 30-MW(e) Rivesville project is years late and was obsolete before it was started up. It may be "the most advanced (today) demonstration" in ERDA's jurisdiction, but is many years behind commercial European experience that I cite in my Oak Ridge and *FA* papers and in *Soft Energy Paths* (pp 118–21). [This was broadly confirmed in a 1978 European fact-finding tour by Sam Biondo of the U.S. Department of Energy.] The Rivesville boiler is lamentably misdirected and has no bearing on the approach I advocate. European experience ranges from the Stal-Laval and Enköping systems I cite to the 18-MW(t) Renfrew boiler of Babcock & Wilcox (UK) Ltd and the commercial 5-MW(t) Cheshire boiler used for crop drying at Widnes, to the extensive commercial experience of Stone-Platt Fluidfire—to say nothing of working fluidized beds in Czechoslovakia (the Duklafluid two-stage gasifying bed), India, etc. It is significant that the National Coal Board (UK) expects atmospheric-pressure fluidized-bed combustors of about 25 MW(e) capacity to be on commercial offer in the UK by 1978, and that Babcock & Wilcox (UK) Ltd is reportedly prepared to build and warrant fluidized beds up to 100–150 MW(t) right now. [Several U.S. and U.K. vendors began offering large warranted fluidized beds during 1978–9.]

Not surprisingly, EPRI and ERDA are not very well acquainted with these developments. Last September (1976), the people purportedly in charge of ERDA's international coal programs had never heard of either Stal-Laval or Fluidfire—neither, after all, is in the public sector—even though the former had been prepared for nearly two years to build a coal-fired fluidized-bed gas turbine of about 70 MW(e) on normal commercial tender, and the latter probably has more operational experience of fluidized-bed combustion than the rest of the world put together, having sold large numbers of metallurgical furnaces, recuperators, and combustors (typically with capacities of a few MW[t]). EPRI's large-scale, electrified, high-technology design philosophy and ERDA's poor access to foreign (especially foreign academic and industrial) information automatically exclude the fluidized-bed developments that could save them both a decade's reinventing of the wheel.

Fortunately, as shown by TVA, American Electric Power Company and AB Enköpings Värmeverket, it is possible to pursue technically exciting fluidized-bed technologies more aggressively than EPRI or ERDA have done so far (on their 1977 fluidized-bed budgets of about $5M and $43M respectively). The commercial vendors I cite—Kymmene-Mustad and their four international competitors for the Enköping contract, Stal-Laval Turbin AB, Bobcock & Wilcox (UK) Ltd, and Stone-Platt Fluidfire—have confidence in their systems and back them up with normal commercial warranties. The grounds for this confidence are set out in my citations and notes. If Southern California Edison chooses not to read those references or not to buy hardware offered turnkey with full warranties by reputable vendors, that is not my fault.

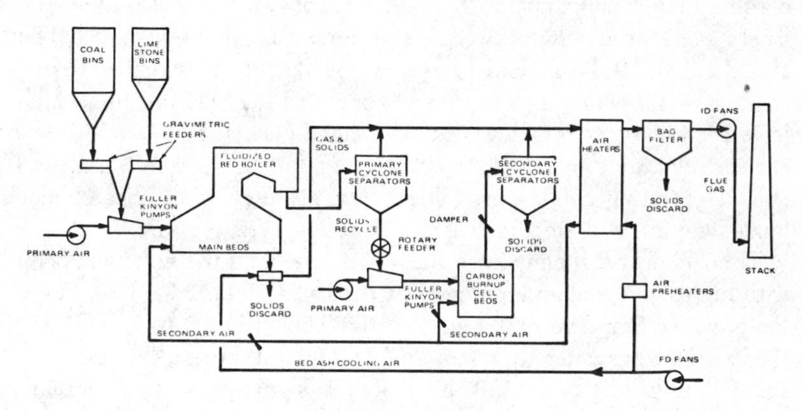

System Flow Schematic

Paraphrase of Weir: Partly because of difficulties encountered at Rivesville, EPRI is conducting research on a number of "Fluidized Bed Combustion Development Issues." The first of these is "High Sorbent Consumption." Sulfur is removed by the reaction of limestone and sulfur to form calcium sulfite and calcium sulfate. Current fluidized-bed systems require a Ca/S ratio of 3, i.e., three atoms of calcium are needed for each atom of sulfur removed. EPRI wants to reduce the Ca/S ratio to 1.5.

At Rivesville, the Ca/S ratio is expected to be between 3 and 4. This compares unfavorably with ratios of between 1.0040 and 1.0081 obtained by SCE on a 170-MW scale in scrubber studies at its Mohave Generating Station. Even if fluidized-bed systems reach their goal of a 1.5 Ca/S ratio, their reagent requirements will still be 50 percent greater than those of conventional plants with scrubbers. And it's doubtful that the goal can be reached. The idea is for sulfur dioxide to react with solid particles of limestone carried in the gas. One would not expect this to be as efficient as passing the gas through a bed of lime or limestone, let alone as efficient as treating the gas with a liquid or slurry reagent as is done in scrubbers. Larger amounts of reagent will be required for the same amount of sulfur removal.

Lovins: The first technical issue Dr. Weir raises is the allegedly excessive Ca/S ratio for effective sulfur capture in fluidized beds. His data are out-of-date. BCURA tests cited by Squires (*SEP*, n. 3, p. 118) show SO_2 reduction of about 85–98% as Ca/S atom ratios vary from 1.0 to 2.0 in a pressurized bed (not, incidentally, requiring a carbon burn-up cell). The Kymmene-Mustad system is expected, based on thorough pilot tests and warranted specifications, to remove over 90% and over 80% respectively of sulfur in high-sulfur coal with a Ca/S ratio of 1.5. It is thus warranted to meet not only the current Swedish emission standard of 600 mg S/MJ but also the post-1981 standard of 240 mg S/MJ with no scrubber (only a baghouse—for which a panel-bed or horizontal fluidized-bed filter might have been substituted had it been built slightly later). Stal-Laval likewise expects sulfur capture over 90% with a Ca/S atom ratio of 1.5-2.5, depending on operating conditions. (The Enköping result is perhaps of more immediate interest, since that bed is at atmospheric pressure.)

The 240 mg S/MJ emission standard warranted for the Enköping plant, with high-sulfur coal and no scrubber, equals 93% of the US New Source Performance Standard of 1.2 lb $SO_2/10^6$ BTU. (Emissions of SO_3 from fluidized beds are generally absent.) With similar strictness, the particulate standard to be met by the Enköping plant is 36 mg/MJ, which is 38% of a typical USEPA standard of 0.1 lb/10^6 BTU for large units. Emissions of NO_x will undoubtedly be far below the large-unit New Source Performance Standard of 0.7 lb/10^6 BTU, owing to combustion temperatures roughly half those of a conventional boiler. (The small Fluidfire boilers typically achieve NO_x emissions of order 1–3 parts per million—two orders of magnitude below a normal gas furnace.) The Enköping plant will control oxidation of nitrogen in the fuel by controlling recirculation of flue gas. For fluidized beds in general, low combustion temperatures also minimize volatization of heavy metals in the fuel.

Dr. Weir's discussion of reagent requirements is misleading, since reagents are not a major cost component for fluidized beds as they are for scrubbers, and since the dry aggregates produced by sulfur capture in fluidized beds can be regenerated by roasting in a suitable gas atmosphere (or, since they have a low leaching rate, can be used as construction aggregates, substituting for materials now mined for that purpose). Further, Dr. Weir neglects to mention that desulfurization produces a smaller mass of waste if done in a fluidized bed than if done by a scrubber on a normal boiler. For example, a rather unsophisticated fluidized-bed boiler producing 100,000 lb/hr steam by atmospheric combustion of Illinois coal and capturing sulfur with limestone might produce the following wastes compared to an equivalent conventional boiler with flue gas desulfurization:

fluidized bed		**conventional boiler** *(spreader stoker)*
55 ston/day of dry granules consisting of:		66 ston/day of wet sludge
$CaSO_4$	33%	—
CaO	36%	—
$CaSO_3$	—	24%
$CaCO_3$	—	9%
inerts	10% (from limestone)	3%
ash	21%	18%
H_2O	—	46%

(Source: M H Farmer et al., "Application of Fluidized-Bed Technology to Industrial Boilers," FEA/ERDA/EPA Interagency Task Force report, January 1977.)

It is theoretically correct that under laboratory conditions, such reactions as sulfur capture can proceed faster in wet than in dry systems. But dry (especially fast) fluidized beds are in fact widely used as chemical reactors in the chemical industry because their turbulent flow and very large particle surface area give very respectable reaction rates and material throughputs without the intractable chemical and materials-handling problems associated with two-phase materials such as scrubber sludge.

Paraphrase of Weir: The second problem EPRI is researching is low combustion efficiency. Current fluidized-bed combustion efficiency (without a carbon burn-up cell) is 85 percent, according to EPRI, which seeks to increase this figure to 99.8 percent. Even with a cyclone (which is only 70 to 80 percent efficient) and carbon burn-up cell, the maximum efficiency that could be expected is 96 percent. Compare this with combustion efficiencies measured at Four Corners units 4 and 5: 99.5 percent in Test 1 and 99.7 in Tests 2–5. Fluidized-bed systems are inferior to conventional coal-fired boilers on a combustion-efficiency basis. Both kinds of systems lose energy because of heat lost in stack gases and Carnot-cycle losses in steam turbines, but the overall energy efficiency of fluidized-bed systems "is believed to be" lower than that of a conventional system (including the energy requirement for flue gas desulfurization).

Lovins: Dr. Weir's second generic issue is low combustion efficiency. He cites 85% without a carbon burn-up cell, and an upper limit of 96% with one owing to cyclone losses (since he says cyclones "are only 70–80% efficient"; in fact a high-efficiency cyclone captures about 98% of particles above 1½ microns, and in a well designed fluidized bed the total mass of elutriated carbon even below this size is negligible). Unfortunately, Dr. Weir's data, based on unspecified "current AFBC systems" [atmospheric fluidized-bed combustion systems], are simply wrong if one follows the European principle of eliminating elutriation of carbon and the need for a carbon burn-up cell through proper design, both by using modest fluidizing velocities and by ensuring adequate residence time in the bed. The Enköping plant, typically, has the following design characteristics (4000 h/y operation, normal conditions): 5.0% of the net system efficiency lost to

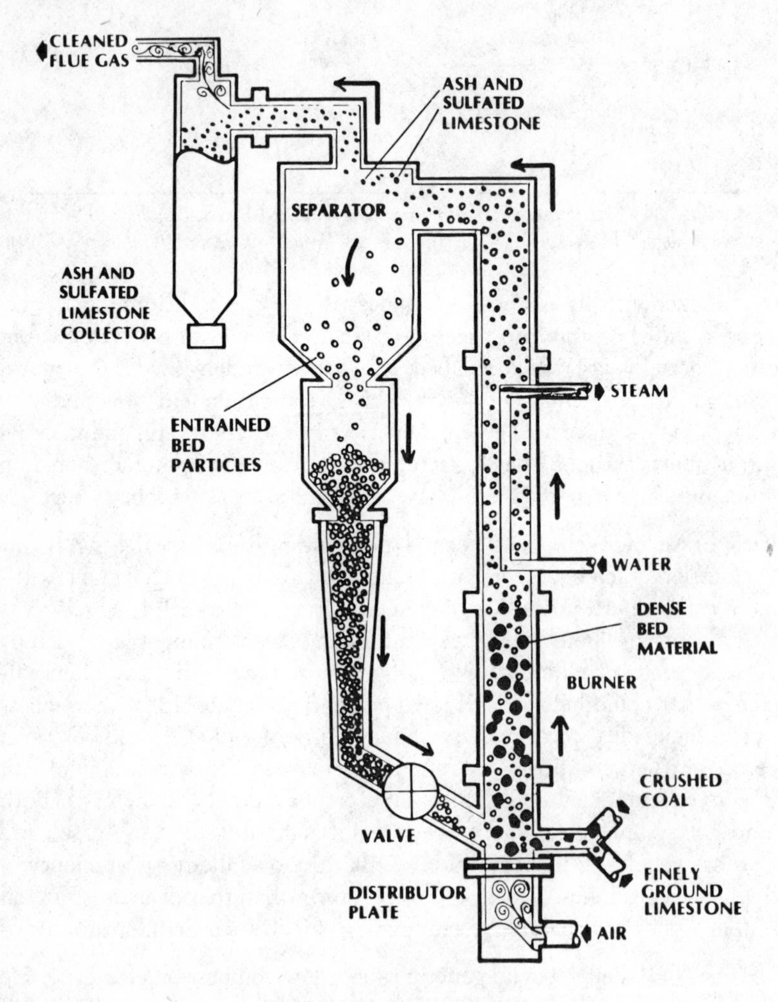

Battelle Multisolid Fluidized Bed

dry flue gas, 0.6% to moisture in flue gas, 0.5% to radiation, nil (much less than 0.1%) to unburned carbon, and 1.5% tolerance, yielding combustion efficiency over 99.9% and thermal efficiency of 92.4%. Likewise in the Fluidfire systems, carbon and sand carryover are negligible, and thermal efficiencies are generally about 84% without a recuperator (94% with) for the less efficient units, 92% (over 95%) for the more efficient units. Combustion efficiency in all cases is over 99.9%, not 85% as Dr. Weir claims. (The Fluidfire industrial units of about 3 MW(t) are at least 82% efficient, with 99.9+% complete combustion, at bed temperatures around 850–900°C, which, with (Ca + Mg)/S atom ratios around 2.5, will capture over 90% of sulfur, meeting the New Source Performance Standard with coal up to nearly 8% sulfur.)

It is also noteworthy that the remarkable heat-transfer coefficients obtained in fluidized beds reduce the amount of fuel burned, hence the amount of emissions, per unit of useful heat obtained. Heat-transfer coefficients about 4–8x better than in conventional boilers have been measured in atmospheric fluidized beds; about 4–30x better in pressurized ones. Volumetric heat-release rates are also typically 10–25x higher in shallow fluidized beds than in ordinary boilers, reflecting a capital savings, and stacking factors of 1000 m^2 of heat-transfer surface per m^3 of bed volume can be obtained in shallow beds. Thus properly designed fluidized beds save fuel (and money) compared to normal boilers, not only by higher combustion efficiency but also by higher net thermal efficiency. This is contrary to Dr. Weir's conclusion, but consistent with several other studies comparing large, utility-type fluidized beds with normal pulverized-fuel/flue-gas-desulfurization power stations. For example, the Energy Conversion Alternatives Study (more formally, "Study of Advanced Energy Conversion Techniques for Utility Application Using Coal or Coal Derived Fuels," contract NAS3–19406, 1976–7) yielded the following General Electric data (1975 $, five-year construction time, $1/$10^6$ BTU fuel, all emissions with NSPS):

characteristic	conventional	atmospheric fl. bed	pressurized fl. bed
net output (MW(e))	760	748	751
net thermal efficiency (%)	32	36	39
capital cost (1975 $/kW(e)	840	608	694
busbar cost (1975 m$/ kW(e) – h output)	39.4	30.9	33.1

The interagency study cited earlier in this response likewise found a significant cost advantage of fluidized-bed over conventional boilers for raising industrial steam.

Paraphrase of Weir: Corrosion and erosion of materials is the third problem addressed by EPRI, which says that excessive heat exchange tube erosion (caused by fluidization of coarse particles at high velocities) and corrosion (caused by the "presence of both oxidizing and reducing zones") occurs. It is believed that these problems will always be more severe in fluidized-bed systems, whose heat exchange surfaces are scoured by burning coal and limestone. The possibility of plaster-of-paris type deposits, such as those observed in boilers operated by TVA and Union Electric, where coal and limestone were mixed (in a fixed, not a fluidized bed), "cannot be discounted." SCE tests at Mohave show that a mixture of coal ash, lime, calcium sulfite, and calcium sulfate does form such a deposit. A build-up of deposits would not only create a corrosion problem, but would also reduce heat transfer. Deposits on boiler tubes would complicate boiler turn-up or turn-down, which depend in a fluidized-bed boiler on the rate of steam production. With respect to erosion and corrosion, conventional coal combustion is better than fluidized-bed combustion.

Lovins: The third issue raised by Dr. Weir is corrosion, erosion, and deposition on boiler tubes and other fluidized-bed materials. These problems have not been observed in the European beds (including those with vertical boiler tubes), even though extensive tests have included limestone injection, caking coals, and the use of coals with up to 70% ash content—essentially bituminous sandstone. The deposits cited by Dr. Weir have never been reported in a fluidized bed and would not be expected at its lower temperatures (below the sintering temperature of ash). As a precaution, however, Fluidfire generally separates combustion from heat-transfer zones in its beds.

Paraphrase of Weir: Before fluidized-bed combustion can be used by the electric utility industry, other "potential problems," not all addressed by EPRI or ERDA research, must be solved. First, we'll discuss boiler turndown and load following.

"It is proposed in the fluidized bed concept" that combustion will be controlled by changing the amount of heat-exchange surface exposed to the fluidized bed and by changing the rate at which coal and air are fed to the combustor. Keeping these three factors in balance will be far from easy, creating difficult control problems. Fluidization takes place within a rather narrow range of gas velocities. SCE tested a scrubber using the fluidization concept; about 1.6 million ping pong balls were fluidized by upward-flowing flue gas. At low gas velocities, the bed wasn't properly fluidized; at high velocities, the ping pong balls went out the top. This is not, of course, an exact analogy to a fluidized-bed boiler, but ping pong balls have about the same mass as particles in a fluidized-bed combustor (5 grams) and gas velocities are almost the same (12.5 ft/sec in the scrubber and 12 ft/sec in the Rivesville boiler). "It can readily be seen" that turn-up and turn-down control will be very hard to manage. Small units would have seriously limited operating flexibility, making them "completely unsuitable" for operation by home owners.

Lovins: The next issue on Dr. Weir's list is alleged difficulties of turn-down and load following. These difficulties have already been addressed and resolved at a scale of a few MW by Fluidfire and for larger boilers by Babcock & Wilcox (UK), Kymmene-Mustad, and others. The analogy of ping-pong balls is inapplicable because of completely different aerodynamic properties (and different constraint geometry). Shallow fluidized beds can change their volume by at least 100%, and in some cases up to 300%, without slumping, partitioning, or excessive entrainment in the gas stream, and heat-transfer zones can be designed (as they are by Fluidfire) for particular sensitivity to both whole-bed and local air flow, so readily achieving turn-down of a factor of four by primary air flow alone and of a factor of seven by using also the air flow to the paniers. Westinghouse has also quoted a turn-down ratio above 4:1 for industrial boilers (C.Y. Liu et al., "Environmental Assessment/System Analysis and Program Support for Fluidized-Bed Combustion," Battelle Columbus Laboratories draft report for EPA, 22 September 1976).

Feed rates for air and coal, excess air adjustment, local air circulation patterns, and flue-gas recirculation offer further parameters whose control, given the straightforwardness of the needed instrumentation and electronics, should not be and has not proved burdensome. The substantial heat capacity of fluidized beds, even in very small sizes, makes them resistant to fluctuations caused by changes in coal quality, size, or feed rate. Dr. Weir appears to ignore all these options, and particularly to assume the airflow to be controlled only by one primary adjustment (in fact the adjustment of local airflow to paniers effectively varies their geometry and is a very simple way of combining wide-ranging adjustment with adequate sensitivity). Evidently Dr. Weir has not investigated the flexibility and control sophistication which have in fact been achieved, contrary to his expectations, in the small-scale units which I cite, most of which have automatic start-up.

Paraphrase of Weir: Another problem is limestone utilization for the removal of sulfur dioxide. As previously mentioned, a Ca/S ratio of between 3 and 4 is expected at Rivesville, a limestone utilization of 25 to 33 percent. The designers of Rivesville—Pope, Evans, and Robbins—predicted a much higher utilization: 42 percent. Extrapolating from their data, it is easy to show that in order to meet Southern California SO_2 removal standards, limestone utilization consistent with the January 1977

Rivesville estimate would be 5 percent, and for the original estimate, 15 percent. While limestone utilization is generally lower than lime, lime utilization figures exceeding 99 percent were consistently obtained in an SCE scrubber. Obviously, utilization factors of 99 percent are better, technically and economically, than factors of 5 to 15 percent.

Lovins: Dr. Weir discusses sulfur capture and limestone utilization in fluidized beds. His data seem to be based entirely on preliminary results from Rivesville, which, as already indicated, is a misconceived design that is not typical of fluidized beds now operating and even more atypical of those that can reasonably be expected to be built in the future. Dr. Weir partly acknowledges this point when he quotes an EPRI manager as saying that Rivesville has limited design flexibility and built-in operating problems such that it will not be possible to optimize the operation of the atmospheric fluidized-bed concept.

I have already pointed out that though Dr. Weir considers it "extremely doubtful that 83% SO_2 removal could be obtained in the single stage fluidized bed boiler," sulfur capture well over 90% has been obtained in many atmospheric tests—over 98% in some pressurized tests—as reported in papers cited by my *FA* and Oak Ridge references. Reagent utilization would accordingly be of the order of 40-70% (higher for dolomite in pressurized beds)—though it is not a governing factor, as neither reagent cost nor reagent availability is a significant problem, even without assuming regeneration.

Paraphrase of Weir: Rivesville's design objective is 83 percent SO_2 removal, or an "exit concentration" of about 420 ppm. California may mandate 50 ppm. A fluidized-bed boiler with scrubber might achieve this level, but one claimed advantage of the fluidized-bed concept is *dry* SO_2 removal.

Coal burned in a fluidized-bed boiler in the Los Angeles Basin would certainly have a far lower sulfur content than the 4.3 percent coal used at Rivesville. But it is very unlikely that 83 percent SO_2 removal could be achieved with a single-stage fluidized-bed boiler and low-sulfur coal, for the Rivesville designers' data indicate that other things remaining the same, removal of SO_2 decreases as the sulfur content of the coal decreases. Extrapolating from Rivesville data, a fluidized-bed boiler burning "an extremely low sulfur coal which does not exist" would achieve about 10 percent SO_2 removal. To meet Los Angeles standards, one might propose multiple fluidized-bed boilers installed in series (an extension of the carbon burn-up cell idea) or the addition of scrubbers. But it would be difficult or impossible to meet environmental standards in Southern California with fluidized-bed boilers, even if very low-sulfur coal was burned. If the required SO_2 removal could be achieved, which is doubtful, baghouses or electrostatic precipitators would be needed to remove particulates.

One is led to the conclusion that fluidized-bed technology is not an answer to SCE's needs.

Lovins: The Los Angeles Basin presents a unique problem—one of the few places in the world where one *might* consider combining a fluidized bed with either low-sulfur coal or a scrubber. It may be of interest that the post-1981 emission standard which the Enköping plant is warranted to meet—240 mg S/MJ—is approximately equivalent to burning 1% sulfur oil with no scrubber, whereas Dr. Weir's firm must now comply with a 0.25% sulfur-in-oil standard and may in the future have to drop that to the equivalent of 0.10%. On the other hand, the warranted sulfur capture in the Enköping plant is 80% to comply with the post-1981 Swedish standard (though better than 90% is expected), and presumably a fluidized bed capturing 98% of sulfur—a level that BCURA has achieved at elevated pressure with a Ca/S atom ratio of only 2—could meet the possible future Los Angeles Basin standard without a scrubber. (It is also noteworthy for California, expecting a temporary surplus of residual oil if Alaskan crude is refined in the state, that a fluidized bed can be built—as at Enköping—to burn any sort of oil or gas initially, then to switch to coal or other solid fuels once the coal infrastructure is in place.)

Dr. Weir's argument of diminishing returns is fallacious. He extrapolates out-of-date Rivesville projections to show that if one burned coal that had a lower sulfur level than any available unrefined coal, little of that sulfur would be captured. Not only are his quantitative conclusions unsupported by the limited data he cites, but he never points out that with actually achieved levels of sulfur capture in properly designed fluidized beds, the situation to which he refers would never arise because such low-sulfur coals would never have to be burned. The remedy, if sulfur capture were inadequate, would instead be to pressurize the bed and use dolomite, or, if preferred, to retain an atmospheric bed but raise the Ca/S ratio above the 2.0 that Dr. Weir assumes. (Dr. Weir also assumes the very stringent sulfur emission standard corresponding to 0.1% sulfur oil.)

Though Dr. Weir's discussion of carbon burn-up cells is obscure—he suggests that one *may* need to add a feature which he had earlier suggested was *essential*—he seems to have overlooked the tendency of many fluidized-bed designers, especially in Europe, to add a waste-heat boiler, economizer, recuperator, second-stage heat exchanger, or some similarly named device. Some designers also use a fuel-rich bed which gasifies, then complete combustion above the bed with excess air and flue-gas recirculation. None of these schemes has anything to do with carbon burn-up cells.

As for particulate emissions, I have addressed this issue earlier, agree a baghouse may be needed, and include its cost in my analysis.

The discussion of fluidized beds in my *FA* article did not contemplate their use for large-scale utility power stations, let alone try to cope with the unique requirements of the Los Angeles Basin in the detail that evidently concerns Dr. Weir. As I have just suggested, however, they may still be suitable for his requirements, and offer a much more elegant (and cheaper) approach to his problem than conventional stations with scrubbers do.

Paraphrase of Weir: EPRI says "All previous efforts, including the Rivesville plant, have limited value for the design of a practical FBC [fluidized-bed combustion] system applied to the utility industry."

Lovins: The quotation overlooks the relevance of much past and present fluidized-bed work to industrial cogeneration and district heating—proper and profitable activities for U.S. utilities today. If present fluidized beds are not prototypes for gigawatt power stations, it is as much because gigawatt power stations are unattractive at the margin as because fluidized beds are not a panacea for every conceivable design problem.

Paraphrase of Weir: According to *Electrical World* (December 15, 1976), ERDA awarded a contract to Curtiss-Wright to build a 13 MW pilot plant that will use a pressurized fluidized-bed combustor with a cyclone separator and a combined-cycle power plant. Turbine blades will be subjected to a stream of gas at high temperature that contains particulate matter and alkali-metal salts.

Lovins: The Curtiss-Wright plant is a peculiar design which will probably spend many years failing to do what Stal-Laval can do already. I have pointed out (*SEP*, n 6, pp118-9) why the Stal-Laval design philosophy of limited turbine inlet temperature and mixed closed-and-open-cycle gas feed is advisable to avoid the blade erosion problems that Curtiss-Wright rightly anticipate.

Paraphrase of Weir: Amory Lovins, British Representative of Friends of the Earth, wrote enthusiastically in *Foreign Affairs*, October 1976, of fluidized-bed combustors that "burn more cleanly than a normal power station with the best modern scrubber." Let's review the current status of the technology. The largest fluidized-bed boiler is a 30 MW unit, and it was only able to operate for 40 hours before being shut down. By way of comparison, the EPA indicates that 10,000 MW of scrubbers are operating or under construction in the U.S.

Lovins postulated that fluidized beds could "be a tiny household device—clean, simple, and flexible—that can replace an ordinary furnace or grate and can recover combustion heat with an efficiency over 80%."

He cited the work of Douglas Elliott and his firm, Fluidfire Development, Ltd. Lovins said Fluidfire "has sold many dozens of units for industrial heat treatment or heat recuperation. Field tests of domestic packaged fluidized-bed boilers are in progress in the Netherlands and planned in Montana." [Dr. Weir consistently referred to Fluidfire as "Wildfire".]

The Fluidfire system uses natural gas for the annealing of small metal parts. It does not burn coal or use limestone, does not generate steam, and does not control pollutants. It is an energy consumer, not an energy generator.

Lovins: If Dr. Weir troubles to verify the cited references, he will find that they fully support my carefully phrased contentions, and that it is he, not I, who misunderstood or did not properly review those references. I understand that other recent reviews of the state of the art (e.g. at Oak Ridge and by Rob Ketchum of the Science & Technology Committee of the U.S. House of Representatives) have tended to support my views of the potential of fluidized beds and the backwardness of the ERDA program.

On the specific points Dr. Weir raises, the statement about clean burning is consistent with the SO_x, NO_x, and other emission data that I cite; fluidized-bed boilers with capacities of at least tens of MW(t) are a normal item of commerce today; the 40-hour shutdown refers to the early history of the misbegotten Rivesville plant (which, as already indicated, I have no intention of defending), not to years of successful and routine operation of dozens of fluidized beds abroad (to say nothing of the 59-MW(t) Fives-Cail-Babcock Ignifluid boiler operating since 1968 in Morocco or the 23 smaller Ignifluids made by FCB); Fluidfire has made and sold a wide range of fluidized beds, including not only the gas-fired metallurgical furnaces to which Dr. Weir refers but also waste and fume burners, diesel waste-heat recuperators (including a 2.3-MW(t) steamraising boiler installed in a Shell tanker), and coal-burners.

Paraphrase of Weir: Lovins speaks of the town of Enköping, Sweden, which, he says, "is evaluating bids from several confident vendors for a 25 MW fluidized-bed boiler to add to its district heating system." The boiler will be used to produce hot combustion gas for steam raising in a waste-heat boiler. The system uses two-stage combustion, has no heat exchanger immersed in the bed, and according to the EPRI, is "totally irrelevant" for utility applications.

Lovins: The Enköping system to which Dr. Weir alludes does not raise steam; it raises 16-bar (190°C) hot water for the Enköping district heating grid. It could as well raise steam from its 1000°C gas temperature if that were its object, so it is hardly "totally irrelevant" even for electrical utility applications (let alone for other tasks that should concern utilities). Contrary to Dr. Weir's statement, half its heat exchange is to horizontal tubes immersed in the bed.

It is 25 MW(t) not because the vendor would not build it larger but because the customer could not afford a 100 MW(t) version at this time, and will therefore build it in stages by retrofitting existing 25- and 50-MW(t) standard oil-fired district heating boilers to fluidized-bed operation, starting probably in 1978. As I have mentioned, the Enköping 25-MW(t) fluidized-bed boiler meets stringent Swedish air quality standards while burning any sort of oil, gas, coal, peat, wood, trash, etc. without a scrubber (using only a baghouse). It is being built turnkey with a complete two-year guarantee and a warranty of all performance specifications, was the subject of vigorous international competition among five vendors anxious to supply on those terms, has a first-unit turnkey price of $100/kW(t) including building, baghouse, high-sulfur-oil tank, and all other ancillary equipment except a coal pier, and has a run-on price (considered, like the first-unit price, to be realistic, neither a loss-leader nor inflated) lower than or comparable to that of an equivalent conventional boiler capable of burning only low-sulfur oil. These remarkable features should, I feel, have elicited some comment from Dr. Weir. Instead he has preferred to rely on evidently sketchy information supplied by EPRI, which can hardly be expected to know much about it.

Paraphrase of Weir: "It is obvious that Lovins is overstating the case" when he says the fluidized beds "are now ready to be commercially applied to raising steam and operating turbines." The most enthusiastic supporters of fluidized-bed technology within the federal government and EPRI do not expect this within ten years.

Lovins: My statement about commercial availability is justified by the Enköping plant, the National Coal Board assessment and the Babcock & Wilcox (UK) Ltd announcement mentioned earlier in this response, the Stal-Laval/B&W(UK)/AEP project, the Fluidfire sales, and the activities of about a dozen other vendors here and abroad. I cannot be responsible for Dr. Weir's reliance on poor sources of information, and do not think he has justified his claim that my discussion of fluidized-bed technology is "misleading." Indeed, it is hard to reconcile Dr. Weir's account of EPRI's views with Mr. Maaghaul's paper on the EPRI fluidized-bed program (presented in May 1976 at the Fourth International Conference on Fluidized-Bed Combustion), which envisaged a 200-MW(e) demonstration plant operating on a utility grid by early 1982 and a full-scale commercial unit operating in 1987—about the commissioning date for a reactor ordered today. (EPRI apparently expects pressurized fluidized beds to be working in the utility industry at about the same time.)

Paraphrase of Weir: Finally, "it is believed" that fluidized-bed combustion is *not* a technology that SCE would choose for coal-fired generation of electricity. Home owners wishing to use coal for heating will find that other techniques would be technically superior and simpler to operate.

Lovins: Dr. Weir's concluding paragraph answers a specific question I never asked—whether fluidized beds are suitable for huge power stations in Los Angeles's airshed. Worse, it probably answers this question incorrectly, owing to a seriously defective knowledge of the state of fluidized-bed technology, especially in Europe. But even if this question were answered correctly, it would have little to do with my thesis: that fluidized beds offer a prompt and attractive method of burning a wide range of fuels, including high-sulfur coal, relatively cleanly, cheaply, and efficiently to supply heat (and, if desired, cogenerated electricity) at scales of the order of 10^3-10^8 W.

Contrary to Dr. Weir's last sentence, I am not aware of any similarly attractive small-scale method of using coal for home heating nor has he shown such an alternative for the very important industrial applications. I conclude that Dr. Weir's paper is ill-informed, technically narrow, and of very limited relevance to my article.

[Lovins's arguments gained further support with the 1978 publication of W. C. Patterson and R. Griffin's state-of-the-art survey, "Fluidized Bed Energy Technology: Coming To a Boil" (INFORM, 25 Broad St., New York, NY 10004.]

18

PERRY/STREITER vs. LOVINS

Economists Charge Gross Exaggeration, Urge Prudence

One of the best critiques of his work, in the estimation of Amory Lovins—who nevertheless disagrees with it—was written by Harry Perry and Dr. Sally Streiter of National Economic Research Associates, Inc., a private firm of consulting economists widely employed by electric utilities. The NERA paper entered and responded to in the hearing record was a draft. Mr. Perry and Dr. Streiter would allow us to use only the final version. In agreeing to do so, we are perhaps stretching a point: the common thread that ties this book together is the fact that the material in it first appeared in the record of a Congressional hearing, and this final version didn't. The point isn't stretched very far, however, since the two versions don't differ greatly. The later version elaborates some points, and makes some sizable excisions, but the two versions are identical through much of their length. We have omitted NERA's supporting references and made minor changes in Mr. Lovins's comments either for clarity or to make them more responsive to NERA's revised, final text.

Dr. Streiter and Mr. Perry would not agree to have their material "chopped up with comments," but said it could be "followed or faced by comments." To append Lovins's comments at the end of the NERA paper would not be a significant improvement over the treatment of similar material in the hearing record, but a parallel-column approach, though inferior in our opinion to the "dialogue" format, does make the argument distinctly more intelligible. The following material will be arranged in that fashion.

NERA: In a widely read article published in *Foreign Affairs* last October, Amory Lovins proposed that the high technology or "hard" path to energy supply on zhich the United States seemed to be set was the wrong path. It was wrong because it would lead to nuclear proliferation and the end of the world, because it induced inappropriate cultural values and, anyway, it was too expensive. Another path was open, he

Lovins: On the whole the NERA critique is more constructive and responsible than its predecessors. It begins, in tone and to some extent in substance, the kind of professional discourse I have been seeking. Most of the criticisms it makes are made and answered elsewhere in the record, but it raises one novel point: the discussion of price elasticity. All points are answered here, at the risk of repetition, because

NERA

proposed, and the paths were mutually exclusive. His path would involve, first, conservation, through the use of insulation, redesign of engines, furnaces and other appliances, reuse of waste heat through cogeneration, and redesign of buildings. This could reduce energy demand over the next fifty years to below the current level while maintaining standards of living. The reduced demand would then be filled by renewable energy sources, namely solar energy and biomass conversion with a little windpower thrown in. To get us over the hump (demand rises 30 percent between now and 2000 in Lovins' scenario, before falling off again in 2025), coal could be used in fluidized beds, but not to generate incremental electricity. Nor would he use solar energy to generate electricity. In fact, electricity is the chief villain, led by nuclear which he would like to see banned.

Lovins

they are stated with greater clarity than in most previous critiques and deserve more than a passing cross-reference. I am particularly glad that the authors have chosen to emphasize policy issues rather than quantitative criticisms, both because the former are most important and because, according to Dr. Streiter, NERA's checking of my numbers (though not always back through the original source) had disclosed no errors.

NERA's soft supply mix leaves out present hydro and microhydro. Electricity is not "the chief villain" but an inappropriate form at the margin, at least from centralized sources: see *Soft Energy Paths*, pp. 141-4.

o

NERA

Lovins' critics have generally gone after his numbers without too much success; indeed, in a very public discussion, Nobel laureate Hans Bethe conceded that he had been mistaken about the potential of solar heating.

Lovins

Professor Bethe's concession was with regard to seasonal storage of solar heat for houses, not "the potential for solar heating" in general. He apparently has some residual reservations about the cost and feasibility of the solar technologies I propose (other than the use of seasonal heat storage), but I have been unable to ascertain what these reservations are or how he justifies them.

o

NERA

We have reviewed in Section III of this paper the current status of some of the technologies he proposes, and conclude that, while many of them are actually or potentially feasible in a technical sense, their introduction has been thus far delayed because they have been uneconomic. Changes in the price of the fossil fuels clearly render at least some of them more economic, but not so clearly superior as to warrant a "dirigiste autocratic" approach (to borrow

Lovins

I agree that introduction of practical soft technologies has been delayed because many of them cannot compete with historically cheap and heavily subsidized fossil fuels, but that does not mean any of them are "uneconomic" compared with other, hard-technology alternatives to those cheap fossil fuels. I have not proposed a "dirigiste autocratic" approach to energy investment, but rather the use of conservative and symmetric economic criteria which, I

NERA

a phrase) to divert the economy exclusively into soft technologies. The basic supply question that remains unanswered and unanswerable today is what the supply schedule looks like. Even the prices quoted by Lovins for current delivery are very soft, for which we cannot really fault him, but the costs of mass production and the time involved for new technologies to take hold are not at all clear.

Lovins

believe, would consistently favor soft-technology over hard-technology marginal investments. The "costs of mass production" are in some cases speculative (though less so than the costs of hard technologies), but my argument does not rest on estimates of mass-production costs; rather, on empirical data from existing devices. The "time involved for new technologies to take hold" is indeed "not at all clear" (*SEP*, pp 95-8) but is also not critical; in the unlikely event that my arguments about simplicity and scale are invalid and soft-technology deployment takes longer than expected, the transitional era, which itself uses rather attractive technologies, can simply be stretched out.

o

Will solar technology be like the hand-held calculator, improving with volume and getting cheaper by the week, or will it be like the nuclear industry where the early models were cheap and bottlenecks in supply of labor and materials drove the price up? Aluminum and plate glass, the raw materials for solar collectors, are heavily dependent on electricity and natural gas in their production, and there are good reasons to suppose that incremental energy supplies to these industries will be substantially more costly than historic sources, driving up the price of solar as production increases. Biomass takes land—what will happen to the price of land if a biomass industry grows up to half the size of all current agricultural production?

The potential speed of introduction and eventual market share of the soft technologies has not yet been fully explored, but there is a wide disparity between Lovins' and other estimates of the market share potential of soft technologies. ERDA's estimates suggest 7 to 10 quads by 2000 for soft technologies, whereas Lovins goes for 32 quads.

o

Some solar collectors use aluminum, but not many; it is prone to corrosion. Few modern ones use glass (they mainly use plastics such as Tedlar®). In general, increases in the cost of materials should affect soft technologies less than hard ones, as the material intensity of the former is usually less and the labor intensity greater. The marginal land requirement of the biomass systems I assume is nil [since it uses wastes, not crops]. If "mass production . . . may make . . . [soft technologies] much more expensive" through escalation of materials prices, the same should be all the more true of hard technologies.

The reasons for discrepancies between ERDA's and my estimates of soft-technology penetration rates are given in *SEP*, pp.95-8: briefly, I assume a stabilizing rather than rapidly growing market to be penetrated, prompt removal of institutional barriers (including lack of capital-transfer schemes), end-use matching, the widespread use of relatively simple short-lead-time technologies many of which are unknown to ERDA, and a *scheduled* move toward long-run-marginal-cost pricing

Lovins

of all depletable fuels. (The schedule short-circuits one or two decades' lag in price signals to investors: knowing in 1980 roughly what fuel prices will be in 1990 is almost equivalent to having the 1990 prices in place in 1980. The conventional, *e.g.* CONAES, assumption of slow exponential increases in real fuel prices with no anticipatory schedule produces a long "dead time" early in the program, effectively delaying much soft-technology penetration by 10-20 years.) It is also noteworthy thay my soft-technology growth estimates are considerably slower than most authorities' nuclear growth estimates even though the latter technology has a current lead time of 12-13 years, not weeks or months, and is enormously more complicated to design, build, and manage.

[In an addendum to his response to NERA, Lovins observes: "I inadvertently omitted a major argument in favor of assuming rapid deployment rates for soft technologies. This is that the soft-technology contribution consists of many diverse components—dozens of kinds of technologies—whose individual constraints on deployment rate are relatively *independent of each other*. Thus, instead of having one or two monolithic technologies (syngas, nuclear power) subject to generic constraints everywhere at once, one has *e.g.* microhydro being constrained by factors different from those which constrain passive solar heating of new houses, which is constrained by factors different from those which constrain solar process heat or feedlot-manure digestion, etc. Dozens of relatively small wedges, each growing at a modest rate because of unique constraints that are not generally shared by other wedges, can thus independently add up by sheer strength of numbers to very rapid *total* growth for the diverse soft technologies. Further, technologies that diffuse into a vast consumer market, rather like CB radios and snowmobiles, are quicker to deploy than those, like power stations, that require a specialized process of technology delivery to a narrow market."]

On the demand side, we have examined the various projections for energy consumption in the year 2000, most of which are considerably higher than Lovins' 95-quad scenario. We agree with him that the lockstep GNP:energy hypothesis is faulty, and that there is considerable potential for dampening demand through conservation measures which are economically justified. However, we believe Lovins has overlooked a subtle point which is totally damaging to his estimate. We explain it further in Section II. He has assumed that doubling energy efficiency will halve demand, whereas the effect will be also of halving the price and, if there is any demand elasticity, the full savings will not be gained. We therefore conclude that his demand scenario for 2000 has underestimated increases in demand over 1976.

Before launching into the numbers, however, we have tried to tackle the area most of Lovins' critics have avoided; we have endeavored to take his social policy arguments as seriously as he evidently does and consider their meaning and merits. We turn to this now in Section I.

As noted below, I have not overlooked the point NERA thinks I have overlooked.

O

In the sphere of politics as of personal values, could many strands of observable social change be converging on a profound cultural transformation whose implications we can only vaguely sense: one in which energy policy, as an integrating principle, could be catalytic?

NERA

> The first path is convincingly familiar, but the economic and sociopolitical problems lying ahead loom large, and eventually, perhaps insuperable. The second path . . . offers many social, economic and geopolitical advantages, including virtual elimination of nuclear proliferation from the world.

[from Lovins's *FA* article, pages 95 and 65]

O

Amory Lovins is a consultant physicist but his main message is that of a social physician. Although criticisms of his work have tended to center on the strategies he proposes, their costs, benefits and feasibility, they have by and large ignored the main message. Yet it is a message which in various forms is receiving increasing acceptance in widely disparate areas of social policy. In education, in health, and now in energy, a line of criticism is emerging which runs as follows: "What you are doing does not work. You are not managing to do even what you say you want to do. You are pouring resources into the wrong things. There are other and better ways, which you are not even considering. I think they are better because they will lead to a social organization I prefer. You, however, can reject my values but not my proposals because my proposals meet *your* test which is that they are more economic."

The radicals are not talking basically about tinkering on the edges of policy. Lovins—despite all the critics who to a man would like to see his soft energy proposals receive more research funds and be incorporated into the system where appropriate—holds fast to the view that there are two mutually exclusive paths. Charles Reich's popular book, *The Greening of America,* was based on the belief that opposition to the Vietnam war and development of a popular counterculture would truly revolutionize society. Lovins' vision per-

Lovins

O

My writings are not those of a "social physician" and a soft path is not bound to "lead to a social organization I prefer." A soft path permits, but *does not entail*, types of social change I prefer, and I have carefully separated my personal preferences from my analytic assumptions. The main social consequence of a soft path is that it permits maintenance of the status quo—that it *avoids* the social change otherwise entailed by a hard path. I am thus not sure which unnamed "radicals" NERA has in mind: my proposals reflect a deep conservatism, both political and fiscal. It is not only tendentious, therefore, but absurd to state that "Lovins' vision perceives energy as the ideal catalyst to revolutionize the world." Juxtaposing my work with Reich's or Illich's does not mean that we have said the same thing.

A remarkable feature of the soft path is that whichever of Harman's three "perceptions" one is in, one can end up doing exactly the same things—for different reasons. (One can install a solar collector because it is cheaper if one is in Perception A, because it is benign if one is in Perception B, and because it is autonomous if one is in Perception C, but the result is exactly the same.) I fear that the authors have read more into Section X of my *FA* article than was there: its purpose was to raise, *independently* of my thesis, salient value questions that I thought should be borne in mind in energy policy, not to support

NERA

ceives energy as the ideal catalyst to revolutionize the world.

The father of this approach to major changes in national priorities is Ivan Illich, who has turned his prolific pen on many sacred cows. In education, his theme is to "deschool society"—he holds that pouring more teachers and more classrooms into the development of our children does harm, not good. Massive increases in education budgets have not even raised reading scores, and although Illich would not consider a reading score a useful test, his point is that the aim of the education system seems to be to produce reading scores and it cannot even do that. Thoughtful people in education report that, even if they do not share Illich's social values, they cannot deny his conclusion that budget expansion has not been the panacea they had hoped.

In health, the same sort of criticism is current. Medical care costs are growing faster than any other item in the economy, consuming 8.6 percent of the U.S. Gross National Product. This is a higher proportion than any other country, yet the U.S. health indices compare unfavorably with those of other developed countries in terms of such measures as infant mortality and even life expectancy. The critics of the health care system suggest that its goals and methods are wrong: that there is an overemphasis on the more dramatic and costly forms of intervention and too little on the basic forms of prevention and care which affect many more people and are relatively cheap.

Even the rhetoric is the same (and note, this is the Director General of the World Health Organization, not a maverick critic).

> The director general of the World Health Organization today [May 3, 1977] urged the breaking of what he termed the 'chains of dependence on unproved, oversophisticated and overcostly health technology.'

In a keynote address to the annual

Lovins

my arguments for a technical-fix soft path.

assembly of the 150-nation agency, Dr. Halfdan Mahler called for a technology that was 'more appropriate because it is technically sound, culturally acceptable and financially feasible.'

Compare this with Lovins:

> The soft energy path, besides having much lower and more stable operating costs than the hard path . . . appears generally more flexible . . . . Its technical diversity, adaptibility, and geographic dispersion make it resilient . . . [and] ideally suited for rural villages and urban poor alike . . . . Soft technologies do not carry with them inappropriate cultural patterns or values. . . .

O

When this line of argument is applied to energy delivery systems, it is tempting, though foolish, to ignore it. It is in large measure foreign to the thought of those of us who reason in equations or see the market system as arbitrating our various wants and needs in a rational manner. In our terms, the critics are arguing that there are massive externalities and grossly imperfect markets. Lovins does indeed argue for rational pricing of energy, for long-run marginal costs, and asserts that, if the paraphernalia of regulation, subsidization and controls were working properly, then that would be a first step to introducing his program which is "more economic." It is hard to argue against the idea that 52¢ natural gas made insulation relatively unattractive, or that price controls on oil have contributed to our international vulnerability.

O

However, tempting as it is to revert to familiar territory, we leave analysis of the economics of his proposal until later in this paper. For the message is also, very clearly, my way would be better even if it were not economic.

O

This is an excellent restatement of my argument.

O

Unfortunately, "What are the societal values which Lovins and his co-physicians [guilt by association?] are advocating?" runs off the rails again. The sentence quoted just above it points out that there are massive exter-

NERA

If it [a 100-percent solar system] cost more [than a nuclear and heat-pump system], there would still be good reasons to use it anyway; but its being cheaper makes a neater argument for people who value narrow economic rationality above unemployment, inflation, centrism, vulnerability, proliferation, and other concomitants of the nuclear option.

What are the societal values which Lovins and his co-physicians are advocating?

Lovins

nalities and grossly imperfect markets that far outweigh private internal costs—far from an argument that "my way would be better even it it were not economic" even after externalities had been internalized and true markets introduced.

O

NERA

First, Lovins' message is that a domestic nuclear program encourages international nuclear proliferation. "We must soon choose one [path] or the other—before failure to stop nuclear proliferation has foreclosed both." How reasonable is his stance on proliferation?

Those who oppose nuclear power domestically frequently stress the international implications and the relationship of commercial nuclear power to weapons proliferation. The linkage is real and the problem enormous, yet the solution is not as easy as Lovins would like us to believe.

The linkage comes about through two mechanisms, people and materials. Nuclear reactor technology requires a substantial cadre of trained people for its operation, and those same people once they know about nuclear power, also have sufficient knowledge to make nuclear bombs. Second, light water nuclear reactors both consume and produce fissile material—enriched uranium and plutonium, respectively. While this fissile material is presently covered by international safeguards [particularly provisions of the Treaty on the Nonproliferation of Nuclear Weapons (NPT) which is enforced by the International Atomic Energy Agency (IAEA)], the safeguards are not foolproof and not all reactors are subject to them. France, South Africa

Lovins

This discussion should also have mentioned the ease of making clandestine production reactors based on fresh LWR fuel—a far simpler operation than using natural uranium.

NERA

and Brazil, for instance, refused to ratify the NPT. Furthermore, some small research reactors are not subject to the safeguards. It was a small research reactor that India used to manufacture its bomb, and estimates of Israel's nuclear weapons capability are routinely based on the plutonium-producing capacity of its research reactor at Dimona.

O

The reactor-bomb linkage is sufficient but not necessary. Reactors are not the only route to bombs. Countries can acquire the knowledge necessary for weapons without training a cadre of commercial reactor specialists; in fact, the information is relatively easy to obtain. On the materials side, the reactor linkage is clearer: a commercial reactor program makes access to weapons-grade materials dramatically easier. While it is not a particularly attractive *direct* route toward obtaining weapons, its existence provides a country with the *opportunity* at almost any time to extract large amounts of plutonium from spent fuel, should it decide to do so.

Only highly enriched uranium and plutonium are suitable for the construction of nuclear weapons. Because the former is still a product of highly advanced and relatively expensive technology, the control of the access to highly enriched uranium has been relatively simple. Nuclear fuel itself does not provide the necessary uranium since light water reactors are fueled with low enriched uranium (about 3 percent U_{235}) rather than the highly enriched uranium (about 90 percent U_{235}) required for the production of nuclear weapons. Therefore, the construction of nuclear weapons using highly enriched uranium requires a country to make substantial economic and technical investments in enrichment technology. While the uranium route could be facilitated with the advent of laser isotope separation, at least with current technology, the direct enriched

Lovins

O

In Chapter 11 of *SEP* (which was apparently not available to NERA in June [1977]—it was only published at the end of May) I meet the argument that there are non-power-reactor routes to bombs, pointing out that all such routes can be made prohibitively difficult—both technically and politically—by a consistent program of denuclearization. I also point out why it is incorrect that power reactors are "not a particularly attractive *direct* route toward obtaining weapons," owing to their large size, negligible or nil plutonium performance penalty, favorable credit terms and training programs, and above all their nil political cost owing to their civilian "cover." Further, on *SEP* p 187 I suggest that it is not necessarily correct that uranium enrichment today is only available to a few very advanced countries: it may in fact be available to sophisticated terrorists.

NERA

uranium route to nuclear weapons remains particularly unattractive.

O

To date, the United States has exported very limited amounts of highly enriched uranium and has refused to export the enrichment process itself. Together with the Soviet Union, the U.S. now maintains a virtual monopoly on the supply of enriched uranium to reactors in the western countries today. However, by the mid-1980s, about ten countries are expected to possess enrichment facilities built in conjunction with their commercial reactor programs. This, together with the expected laser technology for uranium enrichment, will greatly increase the potential access to facilities that could conceivably be utilized to produce highly enriched uranium for weapons use.

O

At the present time, the critical factor in the proliferation problem is plutonium. Plutonium is produced as a by-product of nuclear fission in a commercial reactor and, if separated and accumulated, can be fashioned into a relatively inefficient but still deadly nuclear weapon. The separation of plutonium from spent reactor fuel requires a relatively simple and inexpensive chemical process which is within reach of most of the countries that are now building commercial reactors. As a result, a country with commercial reactors and spent fuel will generally be in a position to relatively quickly and easily extract the plutonium required for nuclear weapons. For these countries, the roadblock to nuclear weapons is not technical, but political, including the Nuclear Nonproliferation Treaty, IAEA safeguards, bilateral safeguards

Lovins

O

The past U.S. exports of highly enriched uranium can hardly be called "very limited" (*SEP*, p 184). Such claims do not seem consistent either with published export records or with published reports (*e.g.* from the Institute of Nuclear Materials Management) of various shipping mishaps with very large single shipments of highly enriched uranium.

As noted on *SEP* p 187, centrifuge enrichment is a real problem today, so low-budget enrichment is not limited to laser techniques tomorrow. Ultimately eliminating uranium as an item of commerce would therefore be a desirable precaution. Centrifuge enrichment places a different complexion on "denatured" thorium-cycle fuels. [Lovins has amplified this in an article, "Thorium Cycles and Proliferation," published in the February 1979 *Bulletin of the Atomic Scientists.*]

O

I welcome NERA's agreement on the importance of the proliferation problem and on the proposition that "although there are some disputes about the economics of reprocessing, the balance of opinion seems to be that using plutonium does not make economic sense in the U.S. at present anyway." But NERA's statement that a bomb made from power-reactor plutonium would be "relatively inefficient" is correct only for crude designs; with modest design sophistication, yield and predictability could be made as high as for normal military bombs.

[Lovins has expounded this point in a technical paper, "Nuclear Weapons and Power-Reactor Plutonium," submitted to *Nature* in January 1979.]

NERA

and the general political and security implications of a particular country's decision to build a bomb.

Indeed, the plutonium route does not even require commercial reactors. A country could produce a relatively crude plutonium-producing reactor using natural uranium and other materials over which there are few, if any, international controls and a simple reprocessing plant for extracting the plutonium.

Almost no one who has thought about or investigated the topic would deny that here is a very, very big problem. President Carter has proposed an indefinite postponement of plutonium recycling for commercial purposes and a ban on U.S. exports of plutonium reprocessing plants. U.S. supplies of uranium seem to be sufficient to permit light water reactors to function without reuse of the plutonium as a fuel for at least the next twenty years, and although there are some disputes about the economics of reprocessing, the balance of opinion seems to be that using plutonium does not make economic sense in the U.S. at present anyway.

Lovins

o

Carter has postponed the construction of the plutonium-fueled breeder reactor at Clinch River indefinitely, citing the dangers of plutonium and the potential availability of thorium as a breeder fuel, although the feasibility of thorium is not clear. Therefore, current U.S. policy can be summarized as follows:

Minimize the availability of plutonium abroad, by discouraging reprocessing: the assertion that it is not economic anyhow is made more believable by a domestic posture postponing plutonium recycling and the plutonium breeder.

Do not give countries an incentive to develop their own front-end and back-end facilities: the policy here is to guarantee a stable and low cost supply of enrich-

o

The thrust of NERA's argument is that other nuclear countries will press ahead with nuclear exports and the plutonium economy no matter what the U.S. does. This reflects a tacit political judgment contrary to mine, and NERA should say so, not pretend the matter is one of fact. I place perhaps more stress on the domestic political problems of nuclear exporting countries and the unprofitability of nuclear sales for the vendors. It is of interest here that past nuclear deals between France and West Germany on the one hand and Iran, Pakistan, and Brazil on the other are rapidly collapsing for a variety of fiscal and political reasons. Further, there is no "world market for nuclear reactors": all the vendors have lost money and

NERA

ment services and to provide storage facilities for spent fuel that is not reprocessed.

Lovins concludes that the way to solve the problem is to "phase out [the U.S.] nuclear power program and its support of others' nuclear power programs." This neglects the realities of the world market for nuclear reactors. The once near monopoly position of the U.S. in world reactor sales has been rapidly eroded by other industrialized countries such as France, West Germany and Canada. Any effort to effectively control the sale of reactors and reactor technology requires the cooperation of a fairly large number of industrial countries. In fact, these countries have been meeting regularly to develop safeguards and export controls to limit the proliferation potential of commercial reactor sales. Unilateral efforts by the U.S. to phase out its nuclear program, restrict access to low enriched uranium, or ban the sale of commercial reactors are likely to be counterproductive in the long run. By taking the proliferation problem seriously, the U.S. has been able to use the economic and political leverage that it does have to upgrade safeguard guarantees and to convince other countries to agree not to export additional enrichment and reprocessing technology to countries like Brazil and Pakistan. Germany, France and other countries have all apparently agreed on this point (although prior sales to Brazil and Pakistan will apparently go forward).

O

An interesting solution to the problems of proliferation and inadequate controls—a market-sharing cartel—was recently proposed by Senator Ribicoff. First, Ribicoff would reduce the competition to sell reactors (which he views as the crux of the nuclear problem) by allocating a *pro rata* share of the market to supplier countries, based upon their productive capacities.

Lovins

most if not all exports have been at special concessionary prices.

O

France and West Germany have already agreed not to export fuel-cycle facilities except under the existing contracts with Pakistan and Brazil respectively, and there is substantial doubt that even those will finally be executed. [In late 1978 it appears that the transfer of sensitive technologies to Brazil and Pakistan will not take place.]

NERA

All orders would be filed with the IAEA and placed on the basis of agreed-upon minimum sales for each supplier. Second, the spent fuel problem could be solved if fuel enrichment, reprocessing and fabrication were concentrated in large, multinational, IAEA-controlled plants. Under this agreement, uranium enrichment services would be provided to customer nations as a credit for the plutonium contained in the spent fuel which the members deposited. Ribicoff believes that, under such a system, economies of scale and reliability of supply could be assured. The main problem he foresees is convincing France and West Germany to agree not to export the elements of the fuel cycle. Although both France and West Germany now have uranium enrichment technologies, large-scale production will not begin for several years and they are dependent meanwhile on U.S. enriched uranium for their own domestic reactors. Ribicoff sees this as a pressure point to force agreement. Such an approach now seems unnecessary since these countries have agreed not to export additional critical fuel cycle facilities and have severely tightened bilateral safeguard requirements.

o

In the end, the proliferation problem is only partially a function of access to a complete commercial fuel cycle. Germany, Japan, Canada, etc. could have developed nuclear weapons long ago. They have not because they perceived that such development might actually reduce their security rather than increase it. Controls over the commercial fuel cycle must continue to be supplemented by other security guarantees which make it unattractive for a country to want to develop nuclear weapons in the first place. Whether countries like South Korea, Taiwan, Israel, South Africa, etc. develop nuclear weapons depends much more on

Lovins

o

NERA argues that "too much of the burden of the proliferation problem has been placed on commercial nuclear power programs. . . ." I have, for that very reason, proposed a coherent package of national policies going far beyond nuclear power and including major initiatives in non-nuclear energy policies (both domestic and foreign) and in strategic arms control. NERA appear to have completely ignored this proposal and noticed only the civil denuclearization proposal, so concluding that my solution "simply will not work." That civil denuclearization is not a *sufficient* condition for nonproliferation does not stop it from being a

whether credible defense alliances with the U.S. continue, than it does on the existence of a commercial nuclear energy industry.

The burden of nonproliferation cannot be put on the commercial nuclear energy industries completely. Export policies, economic incentives to stay away from critical fuel cycle facilities, and safeguards *can* reduce the probability that a country will suddenly enter the nuclear weapons club. But such policies are only adjuncts to overall international defense and mutual security programs. Pakistan probably wants the opportunity to develop a bomb because it does not perceive existing security guarantees as adequate in the face of Indian military capabilities; the Brazilians and the Argentinians probably feel somewhat the same. Far too much of the burden of the proliferation problem has been placed on commercial nuclear power programs and far too little on the basic economic, social and political institutions that create the incentives for nations and subnational groups to acquire nuclear weapons.

We agree that Lovins has identified an important problem. But his solution simply will not work. Improved international political institutions for minimizing the risk of proliferation are certainly called for. The U.S. has the power to take the lead in negotiating international agreements that will serve to minimize the risks that additional countries will either want to or be easily able to develop nuclear weapons. But unilateral U.S. action cannot eliminate the risk of proliferation and it might even increase it.

A source close to the SALT talks puts it as follows: there are basically two views about proliferation. The first is that it is a disaster, the world is likely to blow up, but you can't stop it. This was essentially the private view of the Nixon/Ford administration. The second is that it is a disaster, the world is likely to blow up, but you can stop it.

necessary condition. While NERA have produced a generally competent diagnosis of the problem, their discussion of proposed solutions is not, I feel, as sophisticated as NERA's resources permit, and does not really address the geopolitical issues I raise at *SEP* pp 171-218. Nor did I claim (if I am meant to be among the unnamed "superdoves") that "the answers are easy"; only that they are practical and worth doing in view of the high stakes.

NERA

This is essentially the view of Carter and Vance. Many serious observers believe that it is perhaps *the* ultimate question, but apart from the superdoves, no one thinks anwers are easy.

Lovins

o

Lovins' second message is that our present energy delivery system is all-pervasive and controlling.

> . . . [T]he kinds of social change needed for a hard path are apt to be much less pleasant, less plausible, less compatible with social diversity and personal freedom of choice, and less consistent with traditional values than are the social changes that could make the soft path work.
>
> While soft technologies can match any settlement pattern, their diversity reflecting our own pluralism, centralized energy sources encourage industrial clustering and urbanization . . . [and] allocate benefits to suburbanites and social costs to politically weaker rural agrarians. Siting big energy systems pits central authority against local autonomy in an increasingly divisive and wasteful form of centrifugal politics. . . .

o

My political descriptions cited refer to a postulated hard-path future energy system, not to the present system.

o

Lovins has reiterated several times in his writing that "low energy futures can (but need not) be normative and pluralistic, whereas high energy futures are bound to be coercive and to offer less scope for social diversity and individual freedom." This basic theme can be examined in two parts.

First, Lovins suggests a close linkage between social organization and energy supply systems. Alvin Weinberg has questioned the deterministic implications of this viewpoint.

> Indeed, the tie between energy intensity, bureaucracy, and centralization, which is so prevalent a theme among the anti-energy intellectuals, is largely a mystery to me. There are

o

I have nowhere claimed, as NERA implies, that energy systems solely and uniquely determine social organization. It is not "electricity" but marginal central electrification which, I argue, has specific and carefully described political costs, and neither NERA nor Dr. Weinberg has tried to refute my argument that those costs are real and large. Indeed, NERA draws the electricity/central electricity distinction more carefully below, but mars it with the argument (NERA's, not mine) that electricity implies central electricity because of "inevitable" scale economies (whose net value, I claim, is actually negative). My argument referred specifically to central electrification as an important hard-path

NERA

> many factors in our modern society that encourage bureaucratization, centralization, and vulnerability, and I would think that energy and how we generate it are hardly the most important. For example, the technologies of mass communication are probably far more significant potential sources of malign centralization than are the technologies of energy generation: Hitler was much more the product of radio than he was of central electricity.

It is hard to improve on Weinberg's comment, for it is really difficult to see how electricity dictates social organization.

Others such as E. J. Mishan have argued that economic growth and/or the automobile are responsible for the decline in neighborliness and civility, which Lovins abhors, while Illich would probably cast the blame on industrial civilization. Indeed, the main villain of Illich's diatribe on energy is the automobile and the highway system.

> The typical American male devotes more than 1,600 hours a year to his car. He sits in it while it goes and while it stands idling. He parks it and searches for it. He earns the money to put down on it and to meet the monthly installments. He works to pay for petrol, tolls, insurance, taxes and tickets. He spends four of his sixteen waking hours on the road or gathering his resources for it. And this figure does not take into account the time consumed by other activities dictated by transport: time spent in hospitals, traffic courts and garages; time spent watching automobile commercials or attending consumer education meetings to improve the quality of the next buy. The model American puts in 1,600 hours to get 7,500 miles: less than five miles per hour. In countries deprived of a transportation industry, people manage to do the same, walk-

Lovins

trend, not to the use of electricity *per se*.

NERA

ing wherever they want to go. . . .

O

Illich would not argue that people did not freely choose to love their automobiles, only that they were heavily influenced in doing so by free roads and by advertising and that the externalities in terms of loss of neighborliness and isolation were not readily perceived.

It is, however, difficult to make the same connection between electricity and social organization, unless perhaps one has in mind television and its effects on family life. Centralized energy sources are bad, we are told, because they "encourage industrial clustering and urbanization." But the chief complaint against the car was that it encourages suburbanization and isolation! The connecting argument seems to be that scale economies in electricity make large-scale energy supply systems inevitable, and require highly specialized (and therefore "remote") people to run them, whereas soft energy systems are more directly attached to the people who use them.

At one level, this appears to be obviously true, although not obviously undesirable. At another level, it is not quite as true as it seems. The combustion of coal in a fluidized-bed system takes place in the consumer's own home: the combustion of coal to produce electricity takes place far from him. Yet the coal mine is still required and the transportation network for coal delivery would surely rival the complexities of the electric distribution system. Similarly, a solar collector is placed on a building and is therefore "local." But the aluminum and plate glass industries standing behind the collector are as far away and remote and technically complex as the generating station. One might also add that they are highly centralized industries. In 1972, for example, the concentration ratio for the top four companies in the plate and float glass industry was 91 percent and for

Lovins

O

The examples given are unpersuasive. Coal can be stockpiled, and primary materials industries are irrelevant to the user once the solar collector is in place, whereas central electrification implies a continuous and indefinite dependence on a vulnerable supply delivered by remote technocrats. I do not press for "bureaucratic and centralized" solar regulation, and think consumer fraud protection should be provided at local and state level in all small-hardware areas, including solar. I also have not tried to extend soft-technology definitions beyond the energy system into the broader sphere of industrial and governmental structure.

NERA

the top four primary aluminum producers, 80 percent.

The advocates of solar energy might also be well advised to remember the fate of early heat pumps, which were killed by lack of local knowledge and standardization, and press hard for "bureaucratic and centralized" decisions to eliminate fly-by-night local contractors and the indignity of $12,000 solar roof installations that leak and crack. Solar energy will give each individual more control of his own destiny, presumably until the friendly local solar engineers become less available to fix the leaking collector than the friendly local utility is to fix the distribution outage.

Lovins

o

Again, biomass conversion, which is Lovins' chief answer to liquid fuel supply, would take, he tells us, ten times the current U.S. brewery capacity. It is not clear that this is likely to show fewer economies of scale and less concentration than refineries, especially given the transportation problems associated with the light and bulky initial product.

NERA appears to be under the misapprehension that I proposed expanding the beer industry to make biomass fuels. The transportation problems "associated with the light and bulky initial product" imply less centralization, not more. I envisage a combination of scales: some analogous to a pulp mill; some to a milk packaging plant (surrounded by the crop-residue territory analogous to a "milkshed"); some mobile plants on a truck able to go to blowdowns, sites where abundant crop residues are temporarily available, etc.

o

It is possible to argue on many grounds that pluralism, dispersion of power and decentralization of economic decision-making are desirable things; indeed, many economists would argue that democracy is enhanced by consumers making their own economic choices among alternative products, rather than a single four-year choice between alternative politicians who then make choices for them. At some points, Lovins seems almost to argue that a *highly centralized* decision be made to progress along the soft path. There is no real need for this. The very character of the soft path is that it need not be

I do not argue that "a *highly centralized* decision [needs to] be made to progress along the soft path"; such decisions are not made that way. A soft path probably cannot even tolerate central management. I envisage rather a sum of many smaller decisions, at all levels from individual to federal, that collectively amount to a national consensus. Thus the disagreement cited is illusory, and I am proposing exactly the sort of "gradual adoption" that NERA advocate. How the "decision" (a surrogate term for a myriad of decisions) gets made is also unrelated to the question of exclusivity, which I address below.

NERA

adopted in its entirety. If the soft technologies prove economic, then they can be phased in, their very virtues of flexibility permitting their gradual adoption. Nor is the hard path so inflexible that it could not be phased out if it turns out to be desirable to do so. The effort to resolve the false dichotomy raised by Lovins seems to be more antithetical to pluralism than a less radical policy which, by pricing sensibly, encourages new technologies and removes market barriers.

Lovins

o

Perhaps what Lovins is really getting at is the nuclear argument again. He argues that the safety risks of nuclear are unacceptable to present and future generations.

> Governments should suspend their nuclear programs, until enough infallible people can be found to operate them for the next few hundred thousand years and until all those affected by such programs have been consulted (which may present technical difficulties).

In fact, Lovins' hostility to nuclear power is less obvious in his *Foreign Affairs* article than in his other writings. His efforts on behalf of soft energy futures are, to a very considerable extent, efforts to first politicize and then outlaw nuclear.

o

NERA's imputation of motives is gratuitous and mistaken. Being safe and nonproliferative is only one of many important criteria for a sound energy policy—albeit one that deeply concerns me. NERA seem to be trying to read in a tacit nuclear argument as a substitute for facing my explicit exclusivity arguments.

o

Lovins suggests society cannot handle the nuclear genie. This will, of course, remain moot regardless of any intellectual arguments that could be raised, but serious efforts have been made to evaluate the social effects of nuclear power. The Ford Foundation's Energy Policy Study Group was set up in 1975 in the belief that the nuclear debate was poorly structured and undisciplined: the ground rules were that the participants be not only highly technically qualified but also open-minded. An unusually distinguished group of

o

Both the Ford/MITRE report [not mentioned in the draft to which Lovins originally responded] and NERA's views on these issues are tendentious, often judgmental, and at variance with a substantial part of the professional literature. They do not deal convincingly with uncertainties or opposing arguments. The Ford group's discussion of health effects misstates the state of knowledge about actinide pathways and toxicity, and omits the main term—long-term dose commitment from ^{222}Rn produced by ^{230}Th (78,000-year half-twenty-one people grappled with most of the problems raised by the critics of nuclear power, including the soft arguments raised by Lovins and others, and recently published its 400-page report.

On the issue of health effects of nuclear power, the group concluded that "the adverse health effects of nuclear power are less than or within the range of health effects from coal" and that

> Nuclear wastes and plutonium can be disposed of permanently in a safe manner. If properly buried deep underground in geologically stable formations, there is little chance that these materials will reenter the environment in dangerous quantities. Even if material were somehow to escape eventually in larger quantities than seems possible, it would not constitute a major catastrophe, or even a major health risk, for future civilizations.

This would not be sufficient to satisfy Lovins, who contends further that our civil liberties would be jeopardized by the requirement to guard the wastes against the threat of terrorism.

> . . .[I]f nuclear power were clean, safe, economic, assured of ample fuel, and socially benign per se, it would still be unattractive because of the political implications. . . . For example, discouraging nuclear violence and coercion requires some abrogation of civil liberties; guarding long-lived wastes against geological or social contingencies implies some form of hierarchical social rigidity or homogeneity to insulate the technological priesthood from social turbulence . . . .

The Ford report concludes that the possibility of terrorism must be taken seriously, but that, as with airline hijacking, modest improvements in security could buy a substantial amount of protection. They recommend strengthening of physical security arrangements for nuclear facilities, but conclude that "improved security measures can be in-

life)—whose truncation after 30 years reduces the total nuclear hazard calculated by several orders of magnitude.

The Ford/MITRE report did not address, and its chairman declined to receive, many views which continue to animate the nuclear debate, including soft-path arguments. (For example, the report *assumes* growing electrical demand and nowhere addresses the key argument of end-use structure. Its consideration of renewable sources is derisory and riddled with technical and conceptual flaws.) As an exercise dedicated to molding the debate into a narrow framework, ignoring its rich and diverse structure, the report was from the start fated to be largely irrelevant—an in part competent and in part surprisingly sloppy answer to ten-year-old and largely superseded questions.

NERA

troduced without endangering civil liberties," and note that while the NRC may soon require security clearance for 30,000 employees in the nuclear industry, this compares to 5 million persons already subject to security clearance.

The Ford report is not the final word on these issues of values, which will continue to be debated for many years, but it does provide a substantial counterbalance to the gloomier views of Lovins.

O

A second strand of Lovins' domestic arguments is his championing of participatory democracy and local control.

> In an electrical world, your lifeline comes not from an understandable neighborhood technology run by people you know who are at your own social level, but rather from an alien, remote, and perhaps humiliatingly uncontrollable technology run by a faraway, bureaucratized, technical elite who have probably never heard of you. Decisions about who shall have how much energy at what price also become centralized—a politically dangerous trend because it divides those who use energy from those who supply and regulate it.

Lovins' energy democracy would appear to be characterized by high levels of informed public participation in energy decisions or the use of local elected representatives as a surrogate. Past political experience does not lend much enthusiasm for the level or effectiveness of public participation. For example, Nie and Verba report that widespread political activity "is found largely for those acts that are relatively easy" and "only a small minority of the citizenry is active beyond the act of voting." Mitchell found that voter participation

> [i]n the general purpose units is much higher than in the special districts because citizens believe themselves to be more vitally affected by the traditional general governments

Lovins

O

The energy systems I suggest would not require much political management once in place. Further, the political response reported ignores the success of local groups of all kinds, from Women's Institutes and Leagues of Women Voters to fraternal lodges, farmers' unions, Chambers of Commerce, and Scouts. The apathy often observed stems largely, I feel, from perceived impotence to influence remote political machinery.

NERA

than by the special districts providing mundane, if essential, services that private firms do not sell in the market—services such as fire protection, sewage disposal, park districts, etc.

Lovins

o

Is Lovins proposing new special energy districts to add to the present 80,000 units of governmental responsibility in the United States, which include some 21,000 special districts? Or, is he proposing to reorganize all political districts around the "central" energy theme?

Such local autonomy might also have its negative side. Water pollution control (sewage disposal) and air pollution control used to be local enterprises, but because of the effects of local decisions on the environments of neighboring jurisdictions, federal legislation has largely preempted local control. Education is also an essentially local enterprise, but the courts in various states have held that local funding discriminates against children in poor areas and have required a broader funding base. Lovins does not tell us whether he expects local control of energy to be funded by taxes or whether he simply wants local solar and insulation distributors to be smaller scale than local utilities; in fact, it is not absolutely clear what he does have in mind.

I am not proposing new energy jurisdictions. Present ones suffice. Only in a few cases, such as district-heating systems, would new local institutions be necessary, and they would seldom if ever raise jurisdictional problems. Nor is the education example fair (see my response to Dr. Pickering). "Local control of energy" does not need to be "funded" by taxes or otherwise; it is a function of the technical and political structure of the energy system. (It is, however, true that district heating in Europe is commonly done by municipal corporations in just the same way as water and sewage.) The capital-transfer schemes I propose should deal satisfactorily with the equity problem NERA raise.

o

One could delve further into trying to understand what Lovins is trying to tell us about what he apparently believes to be the strong linkage between energy delivery systems and social organization. On the foreign side, we felt he had the right problem and a nonsolution; on the domestic side, we are tempted to conclude that he has a nonproblem.

I regret that NERA have concluded this section without joining issue with my sociopolitical arguments, which many senior politicians seem to agree with me are a key issue and which, articulated in my *FA* article, seem to have struck chords in many readers who were subliminally worried by just these issues.

NERA

o

Our efforts to examine the socio-political motivation of the Illich/Lovins/Schumacher school of energy critics leave us in a position where, although we may disagree, perhaps violently, with some of the premises, an alternative premise has been offered which forces us to examine the conclusions. The soft path is not only more "benign" and humanistic, it is cheaper.

The "soft" path envisaged by Lovins is best seen in the alternative pictures of our energy future presented in his article. The "hard" way substitutes coal and nuclear power, via electrification, for most of the oil and gas, with a gross energy consumption of over 150 quads by 2000, 230 quads by 2025. The "soft" path reduces demand via conservation (which we examine below) to 95 quads in 2000 and a mere 60 quads in 2025, and fills it by 2000 with 32 quads of soft technology, the rest with oil, gas and coal; by 2025 the entire demand is met by soft technologies, with no nuclear, coal, oil, gas, or, indeed, one gathers from the text, any electricity at all except existing hydroelectric resources.

What are these miraculous soft technologies? Their chief characteristic is that they "rely on renewable energy flows." (Lovins also waxes eloquent about them, asserting that they are "flexible, resilient, sustainable and benign." Well, but any utility president would say the same of electricity. Benignity is in the eye of the beholder.) Soft technologies are also characterized as matching end-use needs in scale, quality and geographic dispersion, a concept which he develops at some length.

But before the soft technologies fill the demand (which prospects we examine in Section III), the total demand has to be reduced by conservation. In this section, we examine other people's projections for energy demand in 2000; the relation between GNP and energy use; the concepts of gross and

Lovins

o

The reference to "the Illich/Lovins/Schumacher school of energy critics" and the earlier use of this tactic of hashing together disparate views and arguments are weaknesses in an analysis of particular and distinct arguments (as opposed to "motivations").

The primary energy for the soft path in 2025 should be 65q, not 60; the soft supply should be 35q in 2000 (not 32) and 63q in 2025 (not 60). But note that these and the hard path figures given are all illustrative, not definitive.

My article mentions wind-electric as well as existing hydroelectric sources, and would, had space permitted, have mentioned new microhydroelectric capacity too, rather than only the first of these as NERA suppose. The "chief characteristic" of soft technologies—renewability—is a misconception; all their properties are equally important, including the relative simplicity (from the user's point of view) and the diversity that NERA fail to mention.

net demand; and the implications of conservation through efficiency on gross energy demand. We have not examined in detail the potential for conservation, but we note in passing that a recent study by Arthur D. Little makes a persuasive case for the existence of virtually cost-free conservation through insulation in new buildings: savings of up to 60 percent in energy use are obtained using a design in which the extra initial costs of insulation are (in most cases) entirely offset by the reduced size and reduced initial cost of the heating, ventilating and air-conditioning systems required. The existence of this type of conservation potential is the foundation for the discussion of the effects of conservation on gross energy demand.

o

How plausible is Lovins' 95-quad scenario in total quantity and in mix? What do other serious students of the energy scene foresee?

Projections used to be done with a ruler. These projections, based on the 1960s, would have told us that overall energy use would grow 3.6 percent annually, for a year-2000 total of 177 quads. But rising energy prices have put an end to simple forecasts, so now we do complicated ones with computers instead of rulers. *Projections* make assumptions about prices, technology, population and GNP, and estimate consumption. *Scenarios* make assumptions about consumption and estimate what would be required in the way of prices, technology, population and GNP. It is desperately difficult to do either well. Even the Bureau of the Census offers three projections for the population growth rates: by the year 2000 the population may have increased 35 percent, or 16 percent, or something in between, say, 24 percent, over the 1974 level. Small wonder then that the energy forecasters, who were practically invented in the last decade, can do no

o

The citations in Table 1 have some curious omissions; excluded are one of the Ford Foundation Energy Policy Project (EPP) scenarios, all energy-industry (and EPRI and EEI) projections, and several other well-known ones such as Chase.

NERA

better than the demographers who have been at it for years.

There have been at least seven major studies of energy demand in the last three years. Their results, in terms of energy use in 1985, 2000 and various other years, are given in Table I. The official projections of 1974 and 1975 do, as Lovins suggests, range mainly between 130–170 quads for 2000, although more recent studies come in with lower estimates. ERDA's combination case, which it uses as a projection, is 137 quads, reflecting a 2.5 percent annual growth, the same as the projected labor force growth.

o

Lovins accuses the official projections of assuming a close causal link between GNP and energy consumption and, in ERDA's case, this is true. The FEA and the Department of the Interior see little change if any in the assumed energy:GNP relation, which they put at 56 or 57 thousand Btu/GNP$ (1971 dollars). (See Table II.)

The main evidence that this linkage is not an inevitable lockstep comes from international data. Schipper's study of Sweden noted that

> . . . in 1971 the United States had a GNP per capita 10 percent higher than Sweden's at the then current exchange rates. However, for each dollar of GNP, Sweden required only 68 percent as much energy as the United States. Correcting for the energy embodied in foreign trade . . . reduces the 1971 Swedish figure to 61 percent.

While the authors conclude that cultural, institutional, economic and technical factors all combine to reduce energy consumption, they stress the importance of prices. "The most important variable affecting energy use and energy efficiency is the relative price of energy with respect to other resources." For example, gasoline has been nearly

Lovins

o

The summary of international results is fine as far as it goes (it omits much useful fine structure) and in agreement with my statements. The cited Swedish-American comparison by Schipper and Lichtenberg deals not with economic but with technical coefficients, and therefore factors out all differences due to climate, settlement, and transportation patterns, hydroelectric fraction, composition of GNP, etc.

twice as expensive in Sweden as in the United States for many years, and Schipper shows per capita consumption of less than half the U.S. level in Sweden. On the other hand, the mean price of electricity was similar in both countries and consumption levels were almost equal.

More recently, a study by Darmstadter, Dunkerley and Alterman for Resources for the Future compared energy consumption patterns for nine industrialized countries and found that, with the exception of Canada, the energy:GNP ratio was lower, ranging from 54 percent of the U.S. level in France to 86 percent of the U.S. level in the Netherlands. This is at least partly due to differences in geography, existing social infrastructure and taste as well as prices, and the authors warn against using the study to argue uncritically for emulation of the energy styles of the Europeans.

> [T]here are practical limits to the opportunities for instituting foreign-inspired energy husbandry in the United States. The United States is not about to acquire the population density of Japan, the geographic compactness of Germany, nor—very likely—the rail network of France, so that in the best of circumstances, some intercountry energy/GDP variability is bound to endure because of historical or deeply rooted reasons. . . .

Clearly then it is impossible to argue a lockstep GNP:energy relationship in absolute terms. Countries can and do vary in the energy content of their GNP.

O

Within a single country, however, given the geography and social infrastructure, one might perhaps expect the energy:GNP ratio to be more stable. An analysis done by Shew for NERA showed an unmistakable decline over the period 1947–73 in energy intensity

O

The results cited from Shew's unpublished work may reflect the tenuous relationships between GNP and welfare, or the shift in composition of GNP toward services (and transaction costs), or both. It is gratifying that NERA "can probably reject, with Lovins, the

NERA

of GNP in the United States associated with rising per capita income, although he observes that in general, international comparisons show an overall association between energy use and per capita income. He suggests that this may be due to disparities in the range of data.

> [T]he income range of the post-war U.S. data is less than a quarter of the range of international per capita income observed. If energy intensiveness of output first rises with per capita GNP, as a consequence of increasing industrialization, and then declines, as service industries assume a larger role in output, this apparent discrepancy might be explained.

We can probably reject, with Lovins, the *lockstep* energy:GNP argument and, in common sense terms, it would be surprising if it were otherwise. Energy as a factor of production will presumably be substituted for labor while it is cheap, and less so when it is expensive. If energy has been cheap relative to labor costs in the U.S., one would expect to find more energy-intensive production methods. Energy as a consumption item will be substitutable for every other type of consumption, the tradeoff depending on the price. If Swedes choose to buy other things than energy, it should hardly surprise us, particularly since gasoline, for instance, is priced at $1.35 per gallon. It would not surprise us if the Swedes found cheese more to their liking than beef, especially if beef were $5 per pound.

Lovins

lockstep energy:GNP argument, and in common sense terms, it would be surprising if it were otherwise." (The net energy:GNP ratio improvement projected below by Resources for the Future, Inc., for 1975–2000 is 38%.)

The Swedish example fails to note that many items of personal expenditure for Americans are supplied free by the Swedish State to Swedes, so costly beef and gasoline can be paid for out of a possibly larger discretionary income. I agree, however, with NERA's following recognition of the importance of analyzing end-use.

o

NERA

It does not make much sense to expect that *gross* energy use should be related to anything very much. What interests us, or what should interest us, is the work we get out of energy. People want warm houses and miles travelled, not gallons of distillate or gasoline. If technologies change, we may well be able to extract more work (or net energy) from a gallon of gasoline, and

Lovins

The end-use efficiency of devices competing with electrical ones is of course taken into account in my computations: see *e.g.* *SEP*, p 135.

rising prices would be one obvious agent for technological change. With this in mind, the Resources for the Future model projects a reduction in the gross energy:GNP ratio from 68 Btu/dollar in 1975 to 47 by the year 2000; the Institute for Energy Analysis offers 42 for its low case and 50 for its high case, while DRI-Brookhaven suggests 44. (See Table II.)

Table I shows "gross" and "net" energy figures, a distinction which is now standard but somewhat misleading. Electricity, it is commonly asserted, is a wasteful energy source because only a third of the energy contained in the original fuel is turned into electricity. The net/gross distinction is based on this supposed waste. Gross energy is all the energy used either directly by consumers or in electric generating plants. It corresponds to the Btu requirements for supply. Net energy is the energy that goes to consumers plus the energy that comes out of electric generating plants. Hence, the net/gross ratio depends on the proportion of electricity being used in the projection.

This distinction points up the supposed inefficiency of electricity while ignoring the very substantial end-use inefficiencies of other types of fuel use. The assumed 3:1 disadvantage of electricity is misleading because it does not take into account the inefficiencies at all stages in the sequence of producing, transporting and using fuels directly. (Measured efficiency of home furnaces is close to 50 percent, although theoretically a clean efficient furnace could work at nearer 80 percent.) A more useful definition of net fuel use might be the units of work (heat, miles, motor power) required or supplied for end-use. That sort of measure, while not currently available, would have two types of use: first, it would enable us to examine energy:GNP linkages in a more useful hypothesis, namely, is the end-use quantity of work related to GNP. Second, it would clarify some of

NERA

the confusion surrounding the concept of conservation.

Conservation, in the broadest sense, means "using less." But it actually encompasses several different phenomena which we can review in terms of the traditional supply and demand curves.

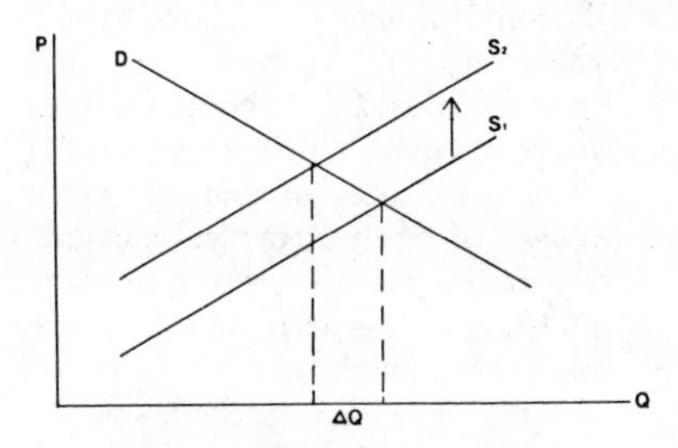

(1) Upward shift (price increase) in supply function. No change in taste or technology. The obvious example of this is the oil price increase. People use less at higher prices: either they drive fewer miles or have colder homes or, in the long run, they switch to fuel-saving technologies (smaller cars, insulation).

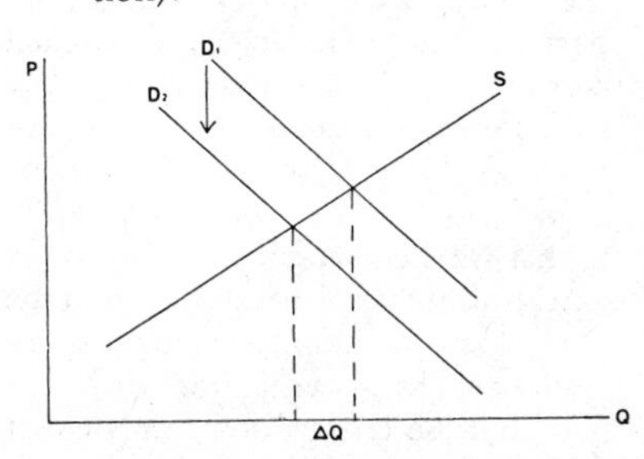

(2) Downward shift in demand function. No change in supply schedule or technology. These are demand changes which may be caused by the availability of substitute commodities or which may reflect changes in taste and lifestyle rather than economics. Many of the social physicians hope to see bicycles replace automobiles because of their lifestyle implications. Exhortations that 65° is healthier than 72° for an indoor temperature are also attempts to shift the demand curve.

(3) Shifts in end-use conversion efficiencies through technical change.

To review this, we have to specify our axes more clearly: supply and demand for what? With fixed technical efficiency options, the question does not arise. But suppose that a simple and costless invention increased the end-use efficiency. What would be the effect? We must assume that demand for energy is not demand for gross Btus but for units of work (heat, miles, motor power), and that gross energy demand is a derived demand. Then a change in technical possibilities would shift a gross/net schedule in the following way:

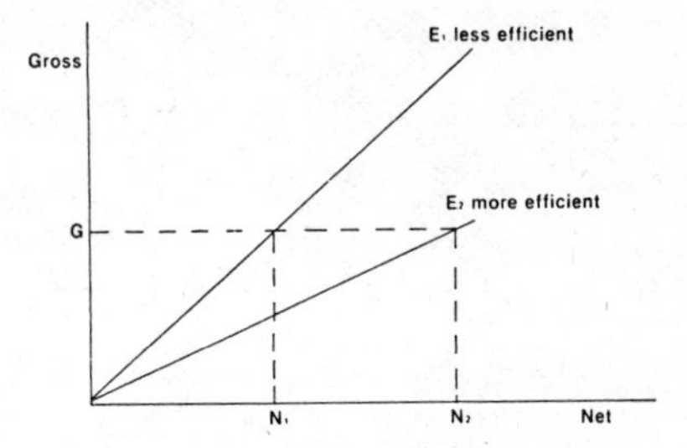

(E_2 yields more units of net energy per unit of gross energy.) Assuming that demand is for net energy, this has the effect of shifting the supply curve of net energy downward, while leaving the demand schedule unmoved. The same amount of gross energy at the same price (on our assumption of a costless invention) can now provide more net energy at the same price.

If there is any elasticity of demand, ΔQ net will be positive, although so long as the elasticity of demand is not greater than unity, changes in *gross* energy consumption will be negative. In more mundane terms, more miles per gallon may induce people to drive more because it is cheaper per mile than before: in any event, this effect will limit

NERA

the impact of technical efficiency in reducing gross Btu consumption.

Expanding the above diagram, we can see that one might have hoped the efficiency change would reduce gross consumption from G_1 to G_3. It will, however, only reduce it to G_2.

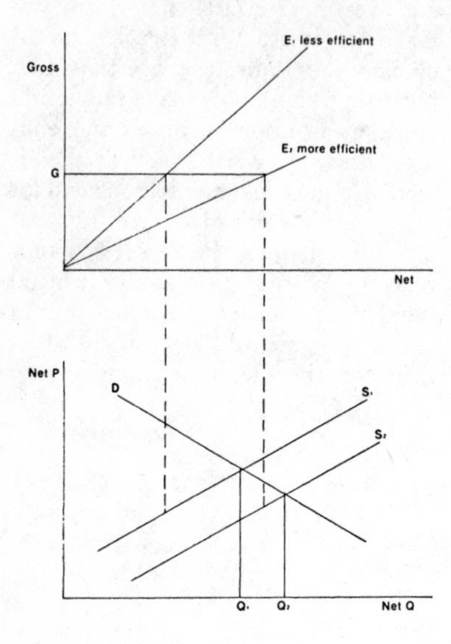

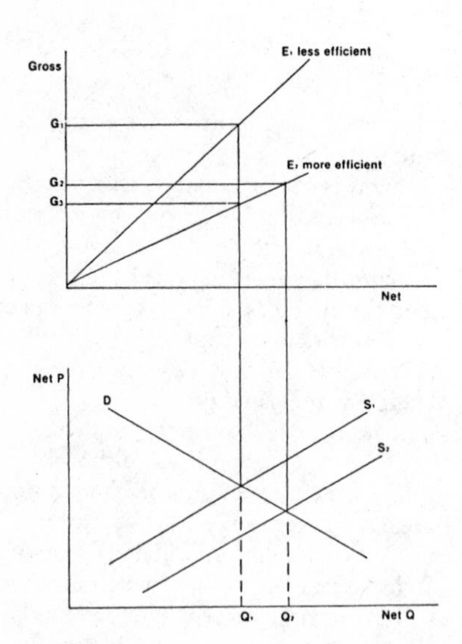

This is an extremely important result. Lovins' assertion that a 95-quad scenario for 2000 is feasible depends crucially on his assumption that conservation through technical efficiency changes alone can reduce gross energy demand. He asserts:

> Theoretical analysis suggests that in the long term, technical fixes *alone* in the United States could probably improve energy efficiency by a factor of at least three or four. A recent review of specific practical measures cogently argues that with only those technical fixes that could be implemented by about the turn of the century, we could nearly double the efficiency with which we use energy.

Among the technical fixes he proposes are thermal insulation, heat pumps, more efficient furnaces and car engines, less overlighting and overventilation in commercial buildings and recuperators for waste heat in industrial processes. While all of these measures could "conserve" in the broad sense of "using less energy," all the fixes he suggests, with the possible exception of overlighting, have the effect of reducing the marginal cost of net energy or work units. They shift the supply curve downward. Lovins says:

> My own view of the evidence is, first, that we are adaptable enough to use technical fixes *alone* to double, in the next few decades, the amount of social benefit we wring from each unit of end-use energy; and second, that value changes which could either replace or supplement those technical changes are also occurring rapidly. If either of these views is right, or if both are partly right, we should be able to double end-use efficiency by the turn of the century or shortly thereafter, with minor or no changes in lifestyles or values save increasing comfort for modestly increasing numbers.

In other words, the reduction from 155 quads to 95 quads is to be accomplished

NERA

by increasing, in fact doubling, end-use efficiencies; what Lovins has not realized is that if there is any demand elasticity at all he will not be able to keep all the efficiency gain: the reduction in the marginal cost of work units or net energy will increase the net energy consumed and cut into his gross savings.

A doubling of end-use efficiency is equivalent to halving the price. This is not a small point. If the elasticity of demand of end-use energy is, say, −0.5, *net* demand will *increase* by a little more than 40 percent; gross demand will therefore not be halved, but will be reduced to 70 percent of the previous level, or 40 percent above Lovins' projection. What this means for Lovins' implied scenario is that a relatively costless technical fix, such as more efficient furnaces and automobile engines, or the heat pump, or even ASHRAE standards on new homes, will not save nearly as much energy as he hopes. Comparing his 95-quad scenario with the 155-quad projections he imputes to others, and assuming that costless technical fixes applied to 120 of the 155 quads account for the 60-quad difference, we can estimate that, on his assumption of doubling efficiency and our assumption of −0.5 demand elasticity, the 120 will go to 84 (70 percent of 120), not 60 (50 percent of 120), and the total demand would be 119 not 95 quads. (If technical fixes are not costless, the shift in the supply curve will be smaller, the net increase in demand will be smaller, and the gross savings greater from any given technical improvement.)

A revision of the 95-quad scenario to 119 would bring it much more in line with recent estimates for the year 2000, although still comfortably below the older "official estimates." Nevertheless, the growth implied over the next twenty-five years by a 95-quad scenario is only 20 quads, whereas a more realis-

Lovins

O

This discussion fails to note that my analysis does not assume "downward shift in demand function" types of conservation. It also seems to confuse (i.e. to lump together) what I have distinguished as "energy system efficiencies" up to the point of end use with the efficiency of the end-use devices themselves. Most of the difference in primary energy use between hard and soft paths is due to improvements in energy system efficiencies (see *SEP*, Fig. 2–3, p 47). This assumes that NERA's definition of "net energy" here is their second one—delivered functions (goods and services)—rather than their first (energy delivered *to* end-use devices).

The effect NERA identifies is well known to energy conservation analysts: the classic example is that if people insulate their homes better, they are likely, other things being equal, to keep the rooms warmer—assuming that they started off in a country (such as England) where the rooms are likely to have been below a luxurious temperature to start with. I am well aware of this effect and have taken account of it in my calculations—as NERA could have ascertained by asking before jumping to conclusions. I did not explain this rather subtle point in the *FA* article both because it is so well known to specialists and because neither the *FA* audience nor the available space permitted it.

NERA's argument that my supposed omission of this point is "crucial" and invalidates my estimates of savings rests on several fallacies. First, other things are not equal: price is increasing fairly rapidly on an anticipatory schedule toward long-run marginal cost at the same time that end-use efficiency is increasing, both encouraging that efficiency improvement and discouraging (through price elasticity of demand) a compensatory increase in consumption of the end-use service. (It would indeed

NERA

tic estimate would suggest planning for growth of at least 40 quads.

To be fair, Lovins does not ignore the other two types of conservation discussed above: he favors lifestyle changes (demand curve downward shifts) for philosophical reasons and he favors marginal cost pricing (supply curve upward shifts), a position with which we cannot possibly quarrel. Therefore, it is possible that not all of his 60-quad savings come from shifts of technology, but rather from a mixture of all three phenomena. However, if we make the crude assumption that energy is underpriced 20 percent, an increase to marginal cost, with a −0.5 elasticity would result in an end-use reduction of 9 percent, bringing us from 155 to 141 quads, and leaving 46 quads to be dealt with by technology shifts before we got to the 95-quad scenario.

The numbers outlined above are only indicative of possible problems: Lovins' thesis is, however, that technical fixes can reduce demand and soft technologies supply the reduced requirement. The latter assertion is examined in Section III, but as we have seen, the reduced demand scenario is probably substantially overdrawn.

Lovins

be hard in practice to identify the latter effect separately; it would appear as part of price elasticity of gross energy demand in most cases, though it might be detected by careful *engineering* analyses of, say, the temperatures maintained in houses.)

Second, the end-use services in question may be subject to saturations: it does not automatically follow that if one buys a car that gets twice as many miles per gallon one will buy the same amount of gasoline but drive twice as far, since there is a limit to how long one cares to sit in a car (now nearly an hour per day).

Third, one important motive for increasing end-use efficiency is to save money because one does not have a surplus of discretionary expenditure available: money saved by insulation may be urgently needed to pay the tax bill, piano lessons, or food bill, and may not be available for putting up the thermostat. Of course, it is possible that the money saved by conservation investments may be spent on something more energy-intensive than it was before, say a snowmobile, rather than on something less energy-intensive such as a concert ticket: this is the well-known "Hannon boomerang effect" (see *SEP*, n 33, p 21). But this is not necessary *a priori*, and it is indeed reasonable to suppose that many householders sensible enough to insulate their houses will be energy-conscious enough not to increase their energy-intensity (which is not easy to do in any case: few if any products are more energy-intensive than the heating oil and other direct fuels on which end-use efficiency improvements save money). For all these reasons, the effect to which NERA attach such importance is in fact quite small, and even had I omitted it, would not significantly have affected my conclusions.

NERA's quantitative assessment is defective in several further respects. It assumes a 155-q base case from an unstated source (my Figure 1 datum is 160 q) and ignores the heavy subsidies used to inflate demand above the level that would otherwise, in a hard path, be associated with long-run-marginal-cost pricing and with the heavy cash-flow demands of financing electrification. The NERA argument about end-use efficiency gets muddled with the hard/soft-path comparison because much of the saving on the latter path is due not to end-use efficiency improvements but to energy-system efficiency improvements—reduction of conversion and distribution losses by a different pattern of energy supply investments. NERA assume without support a demand elasticity of minus 0.5, to which the stated results are quite sensitive. And NERA assumes that technical fixes applied to 120 q of hard-path primary energy (not delivered energy, which in the hard path in 2000 might only *be* 90 q!) are costless, whereas in nearly all these cases there would be a marginal investment requirement and, as NERA states, the calculated discrepancy would thus be smaller: the marginal investment for the efficiency improvement would absorb part of the monetary savings from the energy saving. NERA acknowledges the possibility of concomitant price rises, but greatly understates them—depletable fuels in world markets today are probably underpriced as against long-run marginal costs from hard technologies by a factor of order 3, not by just 20%—and estimates a very low price elasticity of demand (−0.5, compared to long-run estimates ranging from −1 to −2 or more in the professional literature for electricity, and −0.5 to −1.5 or so for fuels).

NERA provides no sensitivity analysis. I suggest, however, that for all the above reasons, this objection to my estimates is wholly without substance, and the conclusion that "the reduced demand scenario is probably substan-

NERA

The basic assumptions underlying Lovins' discussion of the "hard" and "soft" energy supply paths are that (1) the two paths are mutually exclusive, (2) both paths represent difficult but very different problems, and (3) many genuine soft technologies are now available and are now economic.

A number of Lovins' critics have been puzzled as to why the two paths are "mutually exclusive." Lovins offers no explanation in his original *Foreign Affairs* article or even in the rebuttals that he has written to his critics' comments. The only rationale advanced for the mutually exclusive argument is that "commitments to the first *may* foreclose the second, [and] we must soon choose one or the other. . . . " [emphasis added]. Not only is this statement conditional, but there are no reasons given for why the assertion should be accepted as fact. On the contrary, many transitions in the sources from which energy is harnessed have taken place in the past in an orderly and compatible way. Between the 1850s and the late 1970s transitions that were not mutually exclusive have occurred as society moved in a series of steps from wood to coal, to oil, to gas and most recently to the use of nuclear energy.

Lovins

tially overdrawn" is correct only in the sense that I have *under*estimated the scope for technical fixes (*Sep*, p 37).

NERA's statement that "Lovins gives no reasons in his original *Foreign Affairs* article or even in the rebuttals that he has written to his critics' comments" for arguing that the soft and hard paths are exclusive is incorrect. See *e.g.*, in the article, p 86, last paragraph, and pp 95–6, section IX; and the review cited in the letter to Dr. Seamans, to say nothing of my Senate testimony (published well before NERA's critique). The many arguments in other responses in this record were not requested by nor cited by NERA. NERA's statement that "The only rationale advanced for the mutually exclusive argument is that 'commitments to the first *may* foreclose the second,' and . . . there are no reasons given for why the assertion should be accepted as fact" inexplicably overlooks the above passages. Further, NERA state that "On the contrary, many transitions in the sources from which energy is harnessed have taken place in the past in an orderly and compatible way"—failing to note that such a smooth, evolutionary transition is just what I suggested and is no way incompatible with my argument about exclusivity. The historic transitions NERA refer to from the 1850s onward, too, were sequential, and while of course "not mutually exclusive" to *each other*, did exclude other transitions that might have taken place instead.

For the avoidance of doubt, I shall now summarize the grounds of my exclusivity argument yet again, and hope this lays the matter to rest once and for all.

First, soft *paths* are not identical with soft *technologies*. The use of hard technologies entails political problems that define a hard path, but the use of soft technologies only permits, and does not define, a soft path: the soft technologies must be deployed noncoercively within the framework of the *political* conditions that define a soft path and together with the other *technical* elements of a soft path, namely, greatly increased end-use efficiency and transitional fossil-fuel technologies.

Second, I have never stated that hard and soft *technologies*, as hardware, are incompatibile or exclusive, and in particular, that they are technically incompatible. The proposed soft path shows this clearly: starting with a hard-technology mix, it shifts over 50-odd years to a nearly all-soft mix, with transitional technologies coming in and going out in between. Thus throughout the transitional period, hard and soft *technologies* would coexist and their proportions change.

Third, my exclusivity statement concerned *paths*, not *technologies*, and thus dealt with two broad patterns of internally consistent evolution for the whole energy system *in its full social and political context*—the context in which the above-mentioned shift from hard- to soft-technology dominance occurs. I argued that these two paths were mutually exclusive in three senses:

1. They are *culturally* incompatible. Each entails an evolution of perceptions that makes the other kind of world harder to imagine. People who most vehemently reject the concept of exclusivity are often the same people who best illustrate it by the difficulty they have in even imagining the existence of approaches to the energy problem different from what they have been doing for the past few decades.

2. The two paths are *institutionally* antagonistic. Each entails organizations and policy actions that inhibit the other. Again, where we are now is a good example: the rigidity of some of our institutions, notably the utility sector, is a result of past commitments to a nascent hard path and is manifestly inhibiting proper consideration and implementation of a soft path. Each path

NERA

Lovins

accumulates institutional barriers similar to those which today are locking us into more of the cheap oil era, not something different.

3. The two paths *compete for resources*—money, time, fuel, materials, skills, work, political attention—so that one can reduce the other to "garbled and incoherent fragments" that no longer make sense as an internally consistent program. Yet again, this is precisely what our nuclear commitments since the 1940s have done—until, perhaps, the past year or two—to soft technologies. Indeed, were this not a real risk, defenders of nuclear power would hardly be arguing that adoption of a soft path would foreclose *their* allegedly valuable option. They cannot have it both ways.

o

While one might characterize all of these changes as merely being a replacement of one "hard" energy path by another, this is more a matter of semantics than of substance. In the 1850s the decreasing use of wood would have been viewed, using the Lovins terminology, as a replacement of a soft energy form (biomass) with a hard one (coal). However, the majority of these changes among fuel forms took place side-by-side with the availability of other proven soft technologies—windpower, hydropower and even solar energy—each of which could have become the dominant fuel source. Why they did not, if they were more economic or better in some other way, remains a mystery.

Other critics of Lovins have reviewed the analysis of the "hard" and "soft" supply paths and have frequently arrived at different interpretations of the state of the various technologies and their costs. It requires a wide background in the state of the art of various soft technologies which Lovins offers in place of the more conventional ones to

o

Wood was replaced largely because it was being overused for applications that nowadays would be more appropriately met by other means, such as solar space heating, and was therefore becoming inconveniently scarce. Coal also appeared to be cheaper than it really was. The reverse effect is now visible in, say, Vermont, where about half the houses now use wood stoves and 8% are heated entirely with wood (*SEP*, p 96) owing to the high prices of fuel oil. At various times in U.S. history, solar and wind technologies—though far inferior to today's—have played an important role: the U.S. has used a cumulative total of more than 6 million wind machines, for example, and 30% of the houses in Pasadena in 1897 used solar water heaters. These renewable sources were temporarily suppressed by cheap (and subsidized) fuels and power.

[In an addendum to his response to NERA, Lovins says: "I stated that 'Wood was replaced largely because it was being overused. . . . ' Charles Berg, former Chief Engineer of the Federal Power Commission, tells me that

NERA

evaluate how accurate his assertions are.

Lovins

in general (with a few exceptions) this was not the main cause of the shift away from wood. Rather, he says, technical innovations in a number of widely used and fuel-intensive industrial processes happened simultaneously to encourage shifts to superior processes that were designed to use coal and that did better with coal than with wood. Thus, regardless of the relative availability and price of coal and wood, the historic accident that these processes led to a technical preference for coal over wood in major applications was the driving force for the transition—rather as if the introduction of new processes for, say, cement- or glass-making that depended on natural gas accelerated the shift toward that fuel. I am glad to defer to Dr. Berg's superior knowledge of industrial process history and to withdraw my earlier explanation of the wood-to-coal shift."]

o

Claims that new technologies are both economic and ready to be applied widely on a commercial scale are common in the scientific and engineering community—even when there is little or no basis for such claims. It is only necessary to cite a few examples, chosen deliberately from the energy industries, to illustrate how carefully claims of the availability and economic competitiveness of a new technology must be evaluated. The evaluation must include not only an examination of the technical and economic status but, just as importantly, the nature of the environmental, political, institutional, legal, regulatory and financial barriers that must be overcome before the technology can be deployed widely enough to make an important contribution to energy supply. Making such evaluations has been just as difficult for new "hard" technologies as it has been for "soft" ones.

o

I agree that the criteria identified must be taken into account, and I have tried to do so.

NERA

o

Even if the evaluation is favorable, there are very long lead times and large costs involved in commercialization. For example, the first fission chain reaction to prove the basic scientific principle needed to assure that atomic power could be produced from uranium was demonstrated in 1942. Despite the enormous commitments in funds and other resources that were made by the United States government and others to convert this knowledge to practical use, the first small commercial reactor did not begin operation until 1959, nearly seventeen years later. In 1955, at which time small experimental reactors were already operational, the National Planning Association predicted that by 1980, nearly forty years after the basic scientific principle had been proven, the United States would be supplying only 10 percent of its energy from nuclear fuels. In fact, in 1976 it supplied less than 3 percent of total energy demand and is projected to supply only 8 percent by 1985.

The slow commercial development of nuclear fuels is not just the result of the special problems associated with nuclear energy but typical of the difficulties in commercializing any new energy resource. The development of oil shale, coal liquefaction and coal gasification technologies have been even less successful than those for nuclear fuels. Oil from shale was produced in a number of countries in significant quantities in the 19th century but had largely disappeared by the turn of the century with the discovery of large oil reservoirs. For example, in 1860 a total of 53 U.S. companies were producing oil from coal and shale, but the discovery of liquid petroleum in 1859 soon ended the U.S. shale oil industry. Interest in oil shale was revived in the 1920s due to the relatively high price of petroleum and several pilot plants were erected, including two by the Bureau of Mines. However, oil shale activity was

Lovins

o

The comparison of obstacles between soft technologies on the one hand and fission and coal gasification on the other is misconceived, owing to the extremely difficult and (in the former case) completely novel technical problems of both and their large scale, high capital intensity (requiring slow serial development), and long technical lead times. Both are also subject to generic rate constraints and sell to a highly specialized market. As I point out in *SEP* (Chapter 5 and elsewhere) soft and, in general, transitional technologies do not have these properties—which is why, for example, the past year's [1976–7] solar heating developments outside the ERDA program have done what ERDA expected its own program to do in five years. It is unfair to tar soft technologies with the brush of the hard-technology defects they inherently avoid.

again halted because large discoveries of petroleum caused a decrease in its price. From time to time since then, notably after 1945, as oil reserves appeared to be becoming limited, studies were made that showed that oil from shale was competitive, or nearly competitive, with petroleum. Despite these repeated claims in reports made by both industry and government, no commercial oil shale plants are currently operating. Moreover, their future looks almost as bleak today as it did fifty years ago.

Coal gasification has had a history similar to that of oil shale. As with oil shale, claims have been made repeatedly by government research reports and by large industrial firms that coal could be gasified at *competitive prices* to make a high Btu gas that could be substituted for natural gas. In spite of the clear evidence that coal gasification is feasible, it has simply not been economic. Prior to 1940, low Btu coal gasifiers were used in the United States and during the 1940s and 1950s several attempts to develop new high Btu gasification processes were made. By 1973, at least twelve new processes were ready for commercialization, but costs were estimated to be three to four times the cost of natural gas. Construction has not yet been started on the first commercial coal gasification plant and there is no indication that commercial plants will be constructed soon.

Coal liquefaction to produce substitutes for crude petroleum and its products has had a similar history. In 1962, the National Coal Association announced that "[c]onversion of coal to fluid fuels is a technical reality," and in 1964 it reported that

> . . . Consolidation Coal Co. is designing a pilot plant to advance its technically proven process for producing economical, high-test gasoline from coal. . . . Engineering and economic scrutiny indicates that a plant using 10,000 tons of coal per day could produce gasoline in com-

petition with the petroleum industry. And four years later, it announced that "[g]asoline from coal . . . is off the bench, and the product is in the barrel. . . . "

In all of these cases, the technology is clearly feasible and has been for some years, yet the economics have never enabled the technology to reach the commercial phase.

O

The most important transitional technology advocated by Lovins—fluidized-bed combustion of coal—has had an early history no different from those of the other technologies described above, and it is difficult to determine or understand why there exists optimism in some quarters regarding its future. R&D has been conducted on fluidized-bed combustion for about fifteen years. A few small plants using a modified fluidized-bed principle have been operating for a number of years but without the pollution control measures that would be required now on new plants. How the application of such measures would affect operability and costs is unknown.

Even the most aggressive proponents of fluidized-bed combustion believe that its initial applications will be in relatively small plants—something that Lovins favors. But the use of coal in dispersed small installations introduces a whole host of complex factors that will affect both the utility and costs of fluidized-bed coal combustion. What type of industrial infrastructure will have to be developed to construct plants using this new technology? How rapidly could it be developed? What type of commercial infrastructure will have to be created to purchase and distribute coal in the small quantities that will be needed at many individual plants? How will the environmental problems associated with the storage and later distribution by the coal broker be solved? What methods will be used

O

This discussion of the state of the art of fluidized-bed coal burners reveals a very limited knowledge of recent developments, particularly in Europe. (See my response to Dr. Weir and the references cited in my *FA* article, the Oak Ridge companion-piece, and *SEP*.) The conclusions reflect extraordinary parochialism—carried to the extent of regarding the obsolete Rivesville project as an important test-case—and are wholly without merit. My contentions about the technology are based not on speculation or laboratory studies but on commercial devices that I have seen. The authors' pessimism about the future development of clean coal-using technologies may well be a self-fulfilling prophecy if the quality of US research management in this field is not greatly improved, but bears little relation to the remarkable achievements which I cite from current European experience.

to dispose of the mixture of ash and spent limestone that will remain at these multitudes of small plants?

These and other problems associated with the use of coal in small plants (using fluidized-bed combustion or any other coal combustion technology) are soluble. However, the costs may be very high and thus reduce the potential use of coal in fluidized beds to a low level. Moreover, overcoming these problems can only reduce the rate of penetration of the new technology into the market.

One must therefore conclude that the status of fluidized-bed systems for burning coal in an environmentally acceptable way is still uncertain. A number of technical problems must be resolved before it is possible to "know" whether this type of process will work successfully with coal or even to estimate whether the costs will be competitive with alternative coal combustion processes that were once used extensively in small installations. Hopefully, the required information will be developed at an ERDA-sponsored experiment now being conducted in West Virginia on a 30-megawatt unit or in some other fluidized-bed combustion experiment. Until then, those who claim that the technology and the economics are proven must be viewed with the same caution one reserves for the proponents of oil shale, coal gasification and coal liquefaction.

Finally, with respect to coal technology in general, Lovins concludes "neglected for so many years, coal technology is now experiencing a virtual revolution." There has been no revolution in coal technology to date—super-critical gas extraction, flash hydrogeneration and flash hydrolysis have been the subject of experiments for twenty to fifty years. The only revolution that has occurred is that very much larger funds are being spent on R&D—mainly by the government—than were ever spent before. Whether this will result in the development of any radically new technology that will enable coal to be used in large amounts remains to be seen—probably in fifteen to twenty-five years.

o

The history of solar energy development is very similar to that of many other promising technologies. The use of solar energy for water heating and for space heating and cooling has also received a great deal of research attention for a number of years. In 1964 Farrington Daniels said:

> There is no gamble in solar energy use; it is sure to work. It has been demonstrated that solar energy will heat, cool, convert salt water into fresh water, and generate power and electricity. The problem is to do these things cheaply enough to compete with the present methods based on fuel, electricity, animal power or manpower in any given locality.

He was able to be this certain of the technical feasibility because the sun had been used successfully but intermittently and always on a small scale for nearly a hundred years for these very purposes. Important research on home heating was conducted in the 1930s and 1940s at MIT. Sporadic research continued until the early 1970s when a massive infusion of federal funds increased the size of the R&D effort by orders of magnitude.

Solar energy and other renewable resources may now be able to compete with the nonrenewable resources in many applications where their use was uneconomic over the past fifty years, since the prices for nonrenewable resources have increased sharply over the past four years—by as much as a factor of four—while those for most renewable resources have increased more modestly, if at all. It is, however, quite misleading of Lovins to assert on the basis of current cost comparisons that:

In the United States (with fairly high

o

The NERA critique states that aluminum "is probably the most highly energy-intensive industry in existence." This is not correct. Though aluminum is very energy-intensive, some other industries are much more so: candidates include some other light metals such as titanium, nuclear-grade zirconium, etc.; semiconductor silicon; and some surprising entries such as maple syrup, which some analysts believe may be the most energy-intensive commercial product in the U.S. economy.

NERA | **Lovins**

> average sunlight levels), they [solar heating, and, imminently, cooling] are cheaper than present electric heating virtually anywhere, cheaper than oil heat in many parts, and cheaper than gas and coal in some.

The *current* cost of solar panels is virtually irrelevant to the discussion. The question is not what one or two panels would cost, but what one or two million panels would cost. Solar collectors use aluminum and plate glass. Aluminum is probably the most highly energy-intensive industry in existence, and is currently located almost exclusively in the Northwest where electricity prices average 4 mills per kilowatt-hour. The marginal cost of new electricity to serve an expanded aluminum industry is nearer 2 cents. Plate glass prices include a substantial component for natural gas, where, again, the marginal cost is clearly above the current contract average. The costs of expanded supply of the soft technologies cannot be estimated simply by assuming that mass production will make them cheaper—it may make them much more expensive. Estimates of the speed with which any new energy technology can be introduced and its eventual share of the market, depend, or should depend, on calculations like these. We doubt whether anyone has done very sophisticated calculations in this area, but it seems quite possible that increased production of solar collectors would tend to drive the price *up*, not down.

o

Additionally, *capital* cost comparisons of alternative sources which Lovins uses freely, make no sense as a matter of economics. A $20 investment which lasts for five years is not necessarily cheaper than a $40 investment that lasts for fifteen years: the capital costs alone are only part of the equation. The relevant number for comparison purposes would include the expected life of the investment and the running costs

o

It is reasonable to suppose that solar penetration will be slow if the institutional barriers in ERDA are not promptly recycled and if the other measures I suggest are not implemented. The passage quoted [reflecting Lovins's optimism concerning solar technology] is fully justified by recent literature (see *e.g.* the ERDA/MITRE study discussed in my response to Dr. Forbes, the OTA and Waterloo studies cited in my 16

NERA

(labor, fuel, materials) expected over that life. Lovins may be right in asserting solar is cheaper, but his methodology is not the right support for the assertion.

Lovins is particularly optimistic about the status of solar technology.

> Ingenious ways of backfitting existing urban and rural buildings (even large commercial ones) or their neighborhoods with efficient and exceedingly reliable solar collectors are being rapidly developed in both the private and public sectors. . . . Good solar hardware, often modular, is going into pilot or full-scale production over the next few years. . . .

The actual state of development is not this favorable. There are over 120 companies manufacturing or in the advanced stage of developing solar collectors, but it is still not certain which general methods will be most successful. Other manufacturers are in the process of developing new or modifying old designs of some of the needed ancillary components, but whether these components will be compatible with the collectors is yet to be determined.

In a new industry it might appear that having many companies starting to manufacture equipment would be an advantage in getting the technology deployed quickly. On the other hand, with no standards for manufacturers to meet, it is very likely that many solar energy systems that are marketed will be substandard. Even if they were not, the existence of a hundred different types of units, designed without minimum industry specifications, can only result in many custom-type installations that will make maintenance difficult to obtain and costly. If these problems actually do occur in early installations, it could have a strong negative effect on development of a solar industry.

Collectors can now be purchased commercially almost anywhere in the

Lovins

September 1977 letter to Science, and the series of *Science* "Research News" articles). The latest [1979] estimates of the U.S. population of solar buildings, incidentally, is about 10,000—twice as many as there were solar water heaters in Florida in 1950. [Over half of the 100,000 are passive, and over half of *those* are retrofits (mainly with attached greenhouses).]

It is correct that "it is still not certain which [collector types] . . . will be most successful," but my statements are based on existing commercial systems that are already successful. NERA's argument that mass production may *raise* unit costs, where it applies to soft technologies at all, applies with even greater force to hard technologies (e.g. aluminum, copper), and is indeed essentially the argument NERA used to explain higher nuclear costs.

Industry "standards" such as NERA proposes are premature and, as presently conceived, would be likely to stifle many innovative active and passive designs; the remedy against charlatans among the equipment vendors is the existing consumer-fraud machinery (perhaps with some more expertise added), not rigid standards that discriminate against simple or novel designs. The pessimism about retrofits is not shared by most people who have thought hard about and achieved excellent results with simple designs, passive retrofit, and neighborhood-scale retrofit. [Aerial surveys in 1978 showed ample roof area and orientation for single-building heating and photovoltaic retrofits in Denver and in the San Fernando Valley (Los Angeles) respectively.] Lifetimes are speculative in some cases (though less so than reactor lifetimes) but are well known in other cases and, with decent design, have been shown to exceed 30 years in harsh climates with essentially zero maintenance. I do not use capital cost as a *substitute* for life-cycle cost, but as an

NERA

United States, but the number of operating systems is still small—about a thousand buildings throughout the country. Until some long-term experience is gained with these early installations, actual lifecycle costs will remain speculative. The difficulty in retrofitting many existing installations is very great and most experts are not optimistic about developing economic ways to use solar energy for space heating and cooling in already constructed buildings. If this view is correct, the rate of market penetration of solar energy for these uses can only be relatively slow. More success can be expected in retrofitting hot water heaters to use solar energy, but the total amount of energy used for this purpose is small.

Under the impetus of higher prices for conventional energy systems, incentives for using solar energy, large R&D expenditures to improve solar energy systems and massive sums spent on demonstration projects to establish the best methods for commercializing solar energy, wider and more rapid use of solar energy can be expected. However, there is at this time no technical justification for the great optimism expressed by Lovins concerning the technical feasibility, economics and the projected high rate of market penetration of solar energy systems.

Lovins

adjunct to it and a first-order signal for it (*SEP* pp. 68–9, 135–6).

NERA's conclusion that "there is at this time no technical justification for the great optimism expressed by Lovins concerning the technical feasibility, economics and the projected high rate of market penetration of solar energy systems" stems, I suspect, from familiarity with ERDA's programs, which do indeed move at a glacial pace, rather than with the exciting things now happening in the field with even greater rapidity than I thought possible when I wrote the *FA* article. I would even contend that the most innovative and promising solar work in the country is for the most part being done by people who would hang up if ERDA telephoned them, and who fervently hope that ERDA will do no more to get in their way.

o

NERA

Because it will be necessary to have liquid fuels for transport in the Lovins scenario of the future, a soft technology is needed to produce these liquids. Conversion of biomass to methanol by enzymatic routes was proposed. According to Lovins

> . . . exciting developments in the conversion of agricultural, forestry and urban wastes to methanol and other liquid and gaseous fuels *now offer practical, economically* interesting technologies sufficient to run an efficient U.S. transport sector [emphasis added].

Lovins

The biomass section shows yet again a lamentable unfamiliarity with the literature I cite. The correct cost comparison is not with methanol, say, made from historically cheap natural gas, nor with Texas oil at pre-embargo prices, but with synthetic gas, Alaskan oil, year-2000 OPEC oil, nuclear-electric cars, etc. On this measure, as I state, biomass technologies are already economically attractive, since the price of running a car on wood- or crop-waste-derived alcohols is comparable to the price of running it on taxed gasoline. (See *e.g.* the many detailed

NERA

Alcohols and other potential nonpetroleum types of fuels were produced commercially from wood and other biomass feedstocks for many years. These operations ceased quickly, however, when catalytic methods were developed for producing these chemicals using coal, oil or gas as the raw material feedstock. In 1944, as part of the Synthetic Liquids Fuels Act, Congress appropriated funds

> . . . to determine the manufacturing steps and costs of a small industrial plant scale of a process developed by the Bureau of Agricultural and Industrial Chemistry for the production of alcohol and other liquid fuels from agricultural residues.

After several years of experimentation, the R&D was halted because it had become obvious that other ways of making synthetic fuels would be much more economic than processes based on agricultural residues.

It is a gross exaggeration to imply that technologies exist or are even on the horizon for the *economic* production of alcohols or any other fuels from agricultural or forestry wastes. There is no doubt that alcohols can be produced by either enzymatic or more conventional processing of such wastes, but the costs are currently so much greater than alternative routes using other feedstocks that massive technical breakthroughs would be needed that have not yet been seriously studied even in a preliminary way. Moreover, the overall environmental impacts on the land and the soil are unknown as are the types of adverse environmental impacts that the conversion plants would create. Even in the unlikely case that the costs appear attractive, there is not sufficient information to compare the relative environmental impacts of using conventional technology for producing alcohols from conventional fuels with the yet-to-be-studied, let alone commercially developed, systems suggested by Lovins.

Lovins

assessments by Dr. Reed of the Solar Energy Research Institute.) Indeed, contrary to NERA's assertion, some biomass fuel technologies, mainly anaerobic digestion of feedlot manures, are not only conceivably economic but in commercial operation. The residue source terms that NERA quotes from a single source are only slightly smaller than those I assume, based on a far more detailed set of diverse literature citations, and the calculated net yield (nearly 6 q—see my response to Dr. Lapp) does suffice to run a US transport sector at the best European efficiencies of today, which NERA is probably not assuming.

While NERA's plantation cost estimates are inconsistent with many careful calculations in the recent literature—*e.g.* the California kelp-culture estimates—that is irrelevant, since I assume no such technologies.

NERA | **Lovins**

In addition, it is doubtful that in practice the amount of biomass could be made available to produce the liquid fuels needed to operate the transport sector. As far as biomass potential from residues and municipal wastes is concerned, the Stanford Research Institute has concluded:

> About 1 quad of forest and agricultural residues and municipal waste are now used to produce energy and up to 10 quads are available. However, collection costs are too high for most of this resource to be used to produce energy unless competing fuels rise in price or raw biomass resource costs fall 50 percent below current forecasts.

The potential for biomass based on plantation growing is even less likely to be cost competitive than that from currently produced biomass. Even if it were attractive economically, it is very doubtful that there would be enough suitable extensive land areas not already being used for agricultural purposes to develop large biomass plantations. Using a number of smaller plots would increase the cost of delivering the biomass to a central plant. If many small plants were used to reduce transport costs then, except under unusual circumstances, costs of production would be higher.

O

There is no evidence that the hard and soft energy paths are mutually exclusive—a basic premise of the Lovins argument. Enough evidence does exist to conclude that hard and soft technologies can exist side by side, both being used to the extent that they provide the lowest-cost (including all external costs) source of energy.

O

These two sentences make it clear that NERA has construed the exclusivity of paths to mean the technical incompatibility of hard and soft technologies—an absurd thesis contrary to my transitional diagram.

O

Until now, however, the soft supply technologies have been unable in most instances to compete with the hard

O

Many reasons are given in the FA article for a prompt choice, including depletion of premium fuels, commit-

NERA

ones. In fact, a transition from mostly soft technologies—biomass (wood), windpower, solar energy to a more limited extent and hydroelectric—took place worldwide in the past hundred years. This occurred not because the two types of energy supplies are mutually exclusive—in fact, they continue to coexist almost everywhere—but because under the existing conditions and under the constraints in operation, at least up until now, the hard technologies were generally less costly.

Major transitions have also occurred among the hard technology energy forms as the world moved from wood to coal, to oil, to gas and then to nuclear energy. These transitions were orderly; the use of wood did not disappear suddenly and, despite the massive inroads of oil and gas as a source of energy in the United States, coal production still remains at the level of thirty years ago. All the evidence, in fact, supports the view that there is no mutual exclusiveness to the use of many diverse supply forms, both hard and soft.

But let us assume for the moment that we must make a choice as to which of the two supply paths to select. How certain are we that the soft technologies can be successfully developed for widespread commercial use? Based on past experience with the development of new energy technologies, or even with new technologies in most other areas, one can only conclude that we are in no position to place great reliance on these still-to-be-demonstrated energy technologies. The fact that they are soft and as Lovins says "flexible, resilient, sustainable and benign" is no assurance that they will be any easier to commercialize than hard technologies. We must be as conservative in appraising the new soft technologies as experience has taught us to be for new hard technologies.

The more information one could develop about the competing systems before selecting one, the more confident

Lovins

ment of resources, and nuclear proliferation. I do not advocate deciding on clearly inadequate data, but suggest that we already have enough evidence for a *prima facie* case to be made, and assume not "still-to-be-demonstrated energy technologies" but only the best present art already in or entering commercial service—in contrast to the big coal-gas plants and breeders which, we are told, are essential to our energy future but have not been demonstrated anywhere.

NERA

society could be that it had selected the "better" path. At the moment, the data base on which to make a decision is weak. Since there is no urgency in making a selection, it would be wise to delay a commitment to a single course (should that ultimately appear necessary for some reason) and to develop more reliable technical and other information on which to make the choice.

o

The prudent course seems clear. Since there is no reason why two paths should not both be used at the same time—each where it is most suitable—a policy should be adopted of developing and using as much of the renewable resources as can be justified. If commercial experience with technologies in their current state and if R&D on these resources result in improved technology, the soft path will become increasingly important as a source of energy supply. Under these circumstances, hard technologies will be phased out gradually and in an orderly fashion. If, on the other hand, those soft technologies that are now being advocated are not successful, no unnecessary risks will have been taken in providing an adequate energy supply that is so necessary for economic well-being. The hard technologies will then still be available in some form to provide the energy needed.

Lovins

o

Like Forbes, NERA call for a "prudent" course of "two paths . . . at the same time"—a category mistake arising from a confusion of technologies with paths, as the political conditions defining the two paths cannot coexist within the framework of Aristotelean logic. "Improved technology" is desirable and will occur but is not necessary to my thesis. I agree that "a policy should be adopted of developing and using as much of the renewable resources as can be justified"—by the conservative criteria of long-run-marginal-cost economics, preferably internalizing externalities. That plus clearing away institutional barriers is precisely what I suggested as means of implementing the soft path, so that "hard technologies will be phased out gradually and in an orderly fashion." If soft technologies take longer than expected, transitional ones are still available to ensure adequate energy supplies for as long as necessary, at lower cost (economic, environmental, and social) than would have been the case with continued construction of hard technologies.

TABLE 1
COMPARISON OF ENERGY CONSUMPTION PROJECTIONS
FOR THE UNITED STATES
Total Energy Consumption

Study/Case	**1975**	**1980**	**1985**	**2000**	**2010**	**2025**
			(Quads)			
	(1)	**(2)**	**(3)**	**(4)**	**(5)**	**(6)**
(1) RFF Base Case						
Gross	74.9	86.6	93.0	114.2	138.9	171.4
Net	60.2	68.2	70.5	85.2	105.5	128.4
(2) DRI-Brookhaven						
Base Case-Gross			100.0	156.2		
Energy Tax Case-Gross			93.5	117.9		
(3) Energy Policy Project						
Historical Growth Case-Gross			116.1	186.7		
Technical Fix Case-Gross			91.3	124.0		
(4) FEA National Energy Outlook						
$13 Oil Case-Gross			98.9			
(5) Department of Interior						
Gross		87.1	103.5	163.4		
Net		68.9	77.5	110.2		
(6) Institute for Energy Analysis						
Low Case-Gross			82.1	101.4	118.3	
High Case-Gross			88.0	125.9	158.8	
(7) ERDA-48-Gross						
Historical Base Case			107.3	165.5		
Improved End-Use Efficiency Case			97.0	122.5		
Coal and Shale Synthetics Case			107.3	165.4		
Intensive Electrification Case			106.8	161.2		
Limited Nuclear Case			107.0	158.0		
Combination Case			98.1	137.0		

Source:

(1) Resources for the Future, draft report to be submitted to the National Institutes of Health, March 1977.

(2) Data Resources, Inc. and Brookhaven National Laboratory, *The Relationship of Energy Growth to Economic Growth Under Alternative Energy Policies*, prepared for U.S. ERDA, March 1976.

(3) Energy Policy Project of the Ford Foundation, *A Time to Choose: America's Energy Future*, (Cambridge, Mass.: Ballinger, 1974).

(4) Federal Energy Administration, *National Energy Outlook*, (Washington, D.C., U.S. Government Printing Office, February 1976).

(5) Walter G. Duprees, Jr., and John J. Corsentino, *U.S. Energy Through the Year 2000 (revised)*, U.S. Department of the Interior (December 1975).

(6) Institute for Energy Analysis, *Economic and Environmental Implications of a U. S. Nuclear Moratorium, 1985–2010* (Oak Ridge, Tenn.: September 1976).

(7) Energy Research and Development Administration, A *National Plan for Energy Research, Development and Demonstration: Creating Energy Choices for the Future*, Vol. 1, June 28, 1975.

TABLE 2
COMPARISON OF ENERGY CONSUMPTION PROJECTIONS
FOR THE UNITED STATES
Energy:GNP Ratio

Study/Case	**1975**	**1980**	**1985**	**2000**	**2010**	**2025**
			(10^3Btu/$71)			
	(1)	**(2)**	**(3)**	**(4)**	**(5)**	**(6)**
(1) RFF Base Case						
Gross	68	62	59	47	44	40
Net	55	49	45	36	34	31
(2) DRI-Brookhaven						
Base Case-Gross			50	44		
Energy Tax Case-Gross			49	39		
(3) Energy Policy Project						
Historical Growth Case-Gross			56	55		
Technical Fix Case-Gross			47	36		
(4) FEA *National Energy Outlook*						
$13 Oil Case-Gross			57			
(5) Department of Interior						
Gross		56	57	55		
Net		45	42	37		
(6) Institute for Energy Analysis						
Low Case-Gross			51	42	38	
High Case-Gross			54	50	47	

Source:
(1) Resources for the Future, draft report to be submitted to the National Institutes of Health, March 1977.
(2) Data Resources, Inc. and Brookhaven National Laboratory, *The Relationship of Energy Growth to Economic Growth Under Alternative Energy Policies*, prepared for U.S. ERDA, March 1976.
(3) Energy Policy Project of the Ford Foundation, *A Time to Choose: America's Energy Future*, (Cambridge, Mass.: Ballinger, 1974).
(4) Federal Energy Administration, *National Energy Outlook*, (Washington, D.C., U.S. Government Printing Office, February 1976).
(5) Walter G. Dupress, Jr., and John J. Corsentino, *U.S. Energy Through the Year 2000* (revised), U.S. Department of the Interior (December 1975).
(6) Institute for Energy Analysis, *Economic and Environmental Implications of a U.S. Nuclear Moratorium, 1985-2010* (Oak Ridge, Tenn.: September 1976).

TABLE 3
COMPARISON OF ENERGY CONSUMPTION PROJECTIONS
FOR THE UNITED STATES
Percentage of Energy Consumed as Electricity

Study/Case	**1985**	**2000**	**2025**
		(Percent)	
	(1)	**(2)**	**(3)**
(1) RFF Base Case	38.5	43.0	43.7
(2) FEA Project Independence			
Business as Usual Case			
($7 oil, no conservation)	37.5		
Accelerated Case			
($11 oil with conservation)	37.0		
(3) FEA *National Energy Outlook*			
Reference Case	34.1		
(4) Energy Policy Project			
Historical Growth Case	32.3	39.1	
Technical Fix Case	25.4	24.8	
Zero Growth Case	26.0	31.1	
(5) Department of Interior			
December 1972 Case	34.6	41.9	
December 1975 (revised) Case	37.8	48.1	
(6) National Petroleum Council			
Intermediate Case	35.5		

Source:

(1) Resources for the Future, draft report to be submitted to the National Institutes of Health, March 1977.

(2) Federal Energy Administration, *Project Independence Report*, (Washington, D.C.: U.S. Government Printing Office, November 1974).

(3) Federal Energy Administration, *National Energy Outlook*, (Washington, D.C.: U.S. Government Printing Office, February 1976).

(4) Energy Policy Project of the Ford Foundation, *A Time To Choose: America's Energy Future*, (Cambridge, Mass.: Ballinger, 1974).

(5) Walter G. Dupree, Jr., and John S. Corsentino, *U.S. Energy Through the Year 2000 (revised)*, U.S. Department of the Interior (December 1975).

(6) National Petroleum Council, *U.S. Energy Outlook, A Summary Report of the National Petroleum Council*, U.S. Department of the Interior (December 1972).

EDITOR'S NOTE: The authors of the NERA critique would permit us to use it only on condition that we also publish a memo concerning Lovins's response to their critique. For NERA to seek the last word in a book published by the friends and co-workers of the man they denigrate seems a bit eccentric, but there it is: in order to bring you the foregoing, we must also, willy-nilly, bring you the forthcoming.

NERA's parting shot has not left Lovins at a loss for words. He has furnished an extensive comment and will be disappointed, I know, by my decision not to use it. But for the purposes of this book, the argument has to end someplace. And to answer the NERA memo as comprehensively as Lovins has done—perhaps dignifying it beyond its deserts—would smother most readers in economic esoterica. I propose, instead, to offer a little editorial comment of my own—not challenging every questionable point, but commenting only where the failure to do so would seem seriously unfair to Lovins and his thesis. His own more technical response, fully meeting the points raised below by NERA, can be obtained by sending a stamped self-addressed letter envelope to FOE Books, 124 Spear, San Francisco, CA 94105.

TO: Energy Research Group
FROM: Sally H. Streiter and Mik Chwalek
RE: Lovins' Response to NERA's Critique
DATE: February 13, 1978

The following are comments on Lovins' comments on our review of his *Foreign Affairs* article (circulated to you on November 4, 1977). After several readings of his response, it still strikes us as unduly self-congratulatory and more semantic than substantive. But there are some points which need a rejoinder:

Efficiency Improvements and Demand Elasticity. In his introductory note Lovins praises the clarity of the critique and notes that it raises "one novel point" which he intends to answer, along with all other points, later in his notes. We assume, although we are not told, that the reference is to the effects of efficiency improvements coupled with demand elasticity on total energy demand. On page 6, he does "answer" this point by telling us, in essence, that it is so well known that there seemed little reason to be pedantic. And, besides, space did not permit him to explain this "rather subtle" point to the *Foreign Affairs* audience. At any rate, Lovins assures us that this point is taken into consideration in his calculations, "which NERA could have ascertained *by asking* before jumping to conclusions . . ." (emphasis added).

There are a few problems with this comment. First, how much space should it take to explain a rather subtle, yet well-known effect to a *Foreign Affairs*-calibre audience? Second, assuming that such an explanation was too space-consuming, Lovins has by now written three books or probably a total of 1,000 pages on the subject of elegant energy use. It seems a bit odd that he has not yet been able to find sufficient space to mention this point (not even in a footnote!), particularly given the importance of improved efficiency and resulting demand reductions (i.e., conservation) to his general argument. Third, we suspect Lovins' last comment concerning this effect—"even had I omitted it, [it] would not significantly have affected my conclusions"—is a dead giveaway that he has *not* considered it previously.

Editor: Lovins did not say that the effects of demand elasticity were well known. He said in one instance that they were "well known to energy conservation analysts," and in another, "well known to specialists." For NERA to say that Lovins alluded to demand elasticity effects simply as "well known," without any qualification, is either careless or deliberately misleading.

The space it might take to explain demand elasticity to a *Foreign Affairs*-calibre audience would be space wasted if, as Lovins maintains, "the effect to which NERA attaches such importance is in fact quite small."

The "dead giveaway" that NERA purports to see in Lovins's language is not visible to me. For NERA to "suspect" that Lovins had not previously considered demand elasticity effects, in the face of his assurances that he had, is a gratuitous affront without any evident justification.

NERA: However, on page 6 of his response to NERA, we do have a review of the price elasticity problem. Lovins argues (a) energy prices are increasing anyway, so the effect of simultaneous technical efficiencies in reducing the net price will be offset to some extent. (This is true but irrelevant, because absent the technical efficiencies the price would rise even more and demand would accordingly be further reduced.); (b) end uses can become saturated: no one wants to drive all day.[1] (This translates as "elasticity is low."); (c) people will spend money saved by more efficient use of energy for piano lessons, not for putting up the thermostat. (This translates as "elasticity is low.") So the effect would be small.

In the next paragraph he makes an amusing, fatal error. NERA's analysis used a −0.5 elasticity figure, based on Lou Guth's studies, as a basis for a rough, calculated example. According to Lovins, who has just been arguing that elasticity is low, our figure of −0.5 is *much too low*: he thinks the literature supports −1.0 to −2.0 for electricity and −0.5 to −1.5 for fuels. Without entering into a debate about which numbers are correct, we can only note that if elasticity were greater than −0.5, our reworking of his 95-quad estimate to 119 quads would have to be revised substantially upward. Moreover, if elasticity were greater than unity, increasing efficiency would boomerang into *increased gross energy consumption* as technical efficiency reduced net energy prices. If Lovins really believes that energy use has such undesirable externalities, *and* that the elasticity is greater than −1.0, he could logically argue for higher prices but he should resist efficiency improvements to the death, and most especially those that are costless.

(As a parenthetical note Lovins now insists that nearly all his technical fixes *would* require a marginal investment, thus mitigating the perverse effects we cited. Funny; in *Soft Energy Paths* he is still touting the technical fix which requires "practically no capital investment"[2] or where "the additional capital cost is often negative."[3])

[1]The issue is, rather, whether an average driver will do, say, 11,000 miles a year rather than 10,000 miles.
[2]A. B. Lovins, *Soft Energy Paths; Toward a Durable Peace* (Cambridge: Ballinger, 1977), p. 33.
[3]*Ibid.*, p. 34.

Editor: The economics discussed here are over my head, but it seems clear to me that NERA has failed to make its case in terms the woman in the street can understand—demonstrating, in the process, why space spent on this point in *Foreign Affairs* would have been space ill-spent.

Readers may well wonder, as I do, why an allegedly fatal flaw in Lovins's economic thinking escaped the attention of all other critics, including one as economically sophisticated as Dr. Arnold Safer, Vice President of Irving Trust and a former professor of economics.

NERA: On the same topic, Lovins accuses NERA of having confused the demand-dampening impact of end-use efficiency with energy-system efficiency improvements. It actually does not matter one bit *where* the improvement takes place along the energy chain so long as the end-use price reflects the efficiency; there will still be the perverse effects of elasticity to contend with. He has not adequately contended with them.

Editor: NERA is presumably right in saying, "It actually does not matter one bit. . . ." Right in purely economic terms, that is. But economics isn't everything, and it may matter a great deal in non-economic terms. Improvements in end-use efficiency wouldn't thrill me very much if we continued to follow the route of over-centralization, over-electrification, and over-consumption.

NERA: *Lifestyle Changes.* Lovins has repeatedly said that the soft energy path would permit but not require lifestyle changes, but at two places in his response, it becomes clear that this is not so. He castigates us for asserting that his 95-quad scenario is out of line with most other projections, saying NERA has overlooked several year-2000 energy projections which are, in fact, considerably lower than his 95-quad estimate. Specifically, he cites "several CONAES scenarios" which suggest 54 to 75 quads in 2010. It is highly misleading of Lovins to imply that the CONAES panel has given its official "imprimatur" to these scenarios. By checking with a panel member, NERA learned that the 54- to 75-quad possibility does not represent the consensus of the CONAES panel. Nor will the final CONAES demand-conservation report include this scenario within its range of possibilities (78 to 195 quads in 2010[4]). Rather, the scenario Lovins cites represents the conclusions of a small group headed by Laura Nader that, *given complete lifestyle changes and decentralized living*, it would be possible to limit *end-use* requirements to 54 to 75 quads in 2010. Gross energy requirements for the same scenario are assumed to be at least 20 percent higher, for a minimum demand of 65 to 90 quads. However, it is not Lovins' comparison of an end-use scenario with a primary-use scenario that is most disturbing here. More importantly, it is his attempt to support his allegation that improved technical efficiency *alone* can substantially reduce gross energy requirements with a reference to a CONAES scenario which is based on "radical changes in living patterns."

[4]The lower estimate is based upon a quadrupling of real energy prices and a 2 percent GNP growth rate; the higher estimate on no real energy price change and on a 3 percent GNP growth rate.

Editor: The "small group headed by Laura Nader" was, in fact, the charmingly named CONAES-CLOP Panel on Consumption, Location, and Occupational Patterns. In its final report of May 1977, the so-called Lifestyles Panel described a 53-quad and a 72-quad scenario (quads of primary energy, Lovins tells me, not of end-use energy as NERA states here). As Lovins correctly reported, CONAES-CLOP had earlier examined scenarios with even lower energy consumption.

Whatever "complete lifestyle changes" may be, Lovins assures me that the CLOP results assume not the revolutionary social transformations that

NERA implies, but "only a reasonable projection of observable value trends."

The argument is complicated by the fact that the report of the Steering Committee of CONAES remained unpublished as of late 1978, more than a year after its work was completed. But the blue-ribbon Demand and Conservation Panel has published (*Science*, 14 April 1978) "pure technical fix" projections of 77 quads and 96 quads primary energy in 2010—and another projection of 63 quads that assumes modest (already occurring) lifestyle changes. Lovins believes, and says some Panel members agree, that on less conservative technical assumptions, the 63-quad scenario could also be a "pure technical fix" scenario.

This is part of the answer, but let's see, what was the question? The question was, is Lovins wildly out of line to propose for consideration 95 quads of primary energy use in the U.S. in the year 2000? If so, Secretary Schlesinger of the Deparment of Energy is out of line too; two years after Lovins had done the same thing, the Secretary projected year-2000 primary energy consumption of 95 quads (at a real oil price of $32/bbl).

Finally, it seems remarkable to me, as a layman, that NERA considers it a mere allegation, requiring support, that improved technical efficiency alone can substantially reduce gross energy requirements. Other things being equal, what else can improved efficiency do?

NERA: Similarly, it is misleading of Lovins to announce that biomass is capable of running an "efficient U. S. Transport sector" without explaining that his "efficiency" estimate is dependent on quite a bit more than improved EPA mileage standards. The transportation sector now consumes nearly 19 quads annually; according to Lovins' comments, only 6 quads are really required. A *threefold* efficiency improvement through technical fixes could presumably achieve this. A 6-quad transportation budget would also bring U.S. per capita transportation consumption in line with that of West Germany (or, "the best European efficiencies of today," as Lovins says)—where per capita miles traveled are half that of the U.S. level, and where buses and trains are used far more extensively than they are here. But a change in the direction of West Germany, which Lovins calls for, hardly constitutes a technical efficiency improvement. In the case of biomass-fueled transport, a transition to total reliance on a soft energy technology is dependent upon a concomitant transition in lifestyles.

Editor: Lovins argues that the U.S. can achieve 50 years from now a level of technical efficiency comparable to the best European practice of today. If he is wrong about that, there is something terribly wrong with us. We ought to be able to do enough better than that to achieve the efficiency specified without flocking to buses and trains. Which is not to say that West Germans who ride on buses and trains are objects of pity.

Many of us would regard it as a definite improvement in our lifestyles if pedestrian and bicycle pathways made walking and pedaling real transportation alternatives in urban and suburban areas, and if America's railway system were upgraded to the level of Lower Slobbovia's. Speaking for myself, it would be worth more to me, cash out of pocket, than our auto-obsessed system is. But in all fairness, NERA should note that Lovins assumes no radical changes in transportation modes: he assumes you and I will continue to drive cars a lot, but lots more efficient cars.

NERA: *Social Physician v. Technical Realist.* On page 2 Lovins goes to great length to emphasize the fact that he is not a social physician, and reminds us that the soft path, unlike its opposite, permits limitless social options. He characterizes our comment, that energy is seen by him as the ideal catalyst to revolutionize the world, as both "tendentious" and "absurd." The soft path permits us, we are told, to be in any of "Harman's three perceptions." We are not told that Harman is a futures scientist whose book, *An Incomplete Guide to the Future* (cited in Lovins' *FA* article), is devoted to the premise that we are quickly approaching the "transindustrial era." It seems strange that Lovins should introduce a(nother) social physician to support the position that he himself is *not* a social physician.

Why should Lovins so resent being characterized as a social physician or a radical? A radical is one who wants to get to the roots of the problem, which is surely what he is trying to do; in the introduction to his new book, *Soft Energy Paths*, he restates:

> . . . many who work both on energy policy and in other fields have come to believe that, in this time of change, energy—pervasive, symbolic, strategically central to our way of life—offers perhaps the best integrating principle for the wider shifts of policy and perception that we are groping toward. If we get our energy policy right, many other kinds of policy will tend to fall into place too.[5]

[5]*Soft Energy Paths*, p. 6.

Editor: If Lovins went to considerable lengths to deny that he is a social physician, perhaps it's because NERA made it so plain that it used "social physician" in a pejorative sense. NERAns might take pains to deny being angels if Lucifer were the angel they were accused of being.

There is nothing contradictory in citing the work of a "social physician," if there be such an animal, in arguing that you aren't one. If I were to deny being an economist, would it be strange for me to cite an economist as my authority for that?

Score a (minor) debator's point for NERA, if you wish, concerning whether Lovins is or is not a radical. Does NERA, however, disavow being radical itself in the same sense it attributes radicalism to Lovins? In common usage, radical means something other than mere determination to get at the roots of a problem, and "radical" is not a word that all of us would apply to ourselves.

That Lovins sees agreeable social side-effects being made possible (*not* inevitable) by a soft energy path doesn't by any means prove that it was the side-effects that Lovins was mainly interested in. If I marry a woman who is a good cook and eat better as a result, it doesn't prove that I married her in order to eat better. Lovins says his energy strategy was worked out on its merits as an energy strategy, and unless NERA has information on this point that it has not shared with Lovins, I believe it should refrain from idle speculation.

NERA: *Nuclear Proliferation, Health and Safety Risks.* For the most part Lovins' comments on the nuclear section are in agreement with that section. The only difference which we can discern concerns how much risk is worth taking ("NERA's conclusions about nuclear risk are judgmental and tendentious.") The draft to which Lovins was privy did not contain the section on the Ford Foundation's conclusions that health and safety risks of nuclear power are within acceptable limits; it would be interesting to know his reaction to this Ford Foundation study. (One of the best recent pieces on nuclear risks is probably Samuel McCracken's *Commentary* piece, "The War Against the Atom," which was distributed to ERG members. Again, it would be interesting to note Lovins' response to that article.)

Editor: Lovins's opinion of the Ford Foundation/MITRE study of nuclear risks is contained in his response to the NERA critique. He has an even lower opinion, he tells me, of the McCracken article ("which many thoughtful critics of nuclear power consider a prime example of slick misrepresentation").

NERA, like most of us, quotes scripture when that suits its purpose and refrains from so doing when scripture perversely says the wrong thing. The Ford/MITRE report that NERA quotes with approval respecting nuclear risks also published one of the first authoritative studies indicating that a healthy economy could be sustained with zero (or negative) growth in energy consumption—a viewpoint with which I feel fairly sure NERA is out of sympathy. Is the report infallible only when it belittles nuclear risks?

NERA: *Mutual Exclusivity.* The exception that Lovins takes to our conclusion—i.e., that he gives no reasons (either in his *FA* article or in his rebuttals to critiques of the same[6]) for why the two paths should be mutually exclusive, save the possibility that "commitments to the first may foreclose the second"—is not at all well founded here (p. 7). Lovins bids us to review pages 86 and 95-6 of the original article; curiously, a review of these passages seems to *support* NERA's original interpretation (possible foreclosure of commitments) of Lovins' exclusivity reasoning:

> The innovations required [for a soft path], both technical and social, compete directly and immediately with the incremental actions that constitute a hard energy path. . . . These two directions of development are mutually exclusive: the pattern of commitments of resources and time required for the hard energy path and the pervasive infrastructure which it accretes gradually make the soft path less and less attainable.[7]
>
> Enterprises like nuclear power are not only unnecessary but a positive encumbrance for they prevent us, through logistical competition and cultural incompatibility, from pursuing the tasks of a soft path at a high enough priority to make them work together properly. A hard path can make the attainment of a soft path prohibitively difficult, both by starving its components into garbled and incoherent fragments and by changing social structures and values in a way that makes the innovations of a soft path painful to envisage and to achieve.[8]

Perhaps we are misreading these passages, but we fail to see how NERA's interpretation "inexplicably overlooks the above passages." Lovins' final attempt to clear the controversy surrounding his exclusivity argument—soft paths are not the same as soft technologies; the former are mutually exclusive, the latter are not—is also less than convincing. He tells us that whatever puzzlement exists does so due to the inability of his critics to "read as carefully as he wrote." Oddly enough, at no point during his chastising explanation does he bid us to refer back to the original text, where, at page 77, we learn that "[e]nergy paths dependent on soft technologies, illustrated in Figure 2, will be called 'soft' energy paths, as the 'hard' technologies sketched in Section II constitute a 'hard' path. . . ." It is difficult to understand how one path could be exclusive to another, while a technology could not, when we learn that a path is defined as nothing more than the composite of certain, similar technologies.

[6]It appears that we overlooked the correct responses: "The many arguments in other responses in this record were not requested by nor cited by NERA," p. 7.
[7]A. B. Lovins, "Energy Strategy: The Road Not Taken?," *Foreign Affairs*, October 1976, p. 86.
[8]*Ibid.*, p. 96.

Editor: That hard and soft technologies can co-exist seems obvious on its face, and is certainly part of Lovins's thinking. Otherwise, how could he envision a 50-year transition period during which hard and transitional technologies would gradually be phased out? The question is, can we follow a hard and a soft *path* at the same time? Reasons for thinking we cannot will not be repeated yet again, but the argument that clinches it for me is cultural antagonism. Values implicit in a society dedicated to a hard path are antithetical to cultural values implicit in dedication to a soft path, and vice versa. We must adopt one path and reject the other, or else succumb to a sort of societal schizophrenia.

To *identify* a soft path as one in which soft technologies are used is not necessarily to *define* a soft path as one in which soft technologies are used. I may be adequately identified as "that guy over there with a receding hairline," but I hope this falls short as a definition. In the passage NERA quotes, it seems clear to me that Lovins was identifying paths, not defining them. And immediately following the passage NERA quotes, Lovins wrote: "The distinction between hard and soft energy paths rests not on how much energy is used, but on the technical and sociopolitical *structure* of the energy system. . . ." The importance of sociopolitical factors is emphasized sufficiently to invalidate NERA's claim that "a path is defined as nothing more than the composite of certain, similar technologies."

NERA: *Cost Comparisons.* Surely one of the more basic of Lovins' contentions is his repeated pronouncement that soft technologies are *now* economically competitive with hard technologies. At page 9, that contention begins to seem less forceful. The biomass section is "lamentable" because NERA has failed to realize that the correct basis of cost comparison for biomass is not historically cheap natural gas but (say) year-2000 OPEC oil or nuclear-electric cars. On this measure "biomass technologies are already economically attractive." So, we learn, that when Lovins says his soft technologies are now economically attractive, "now" does not mean "now." Rather, they are "now" competitive with the long-run marginal replacement costs for currently *in situ* fossil fuels, which, themselves, are *now* underpriced by at least a "factor of order 3" (p. 7). It is really important to note this, because Lovins has not actually said this before. He favors "long-run incremental cost pricing" but he apparently believes that this would mean tripling energy prices for conventional sources now. That is not NERA's view of long-run incremental cost pricing, we may add, but it must be something of a shock to the many consumer groups and congressmen who have been fascinated by the soft path arguments, to learn that the soft path requires a *tripling* of current energy prices to make the alternatives economic. Perhaps Lovins could be persuaded to say that up-front.

Editor: NERA once again reads and quotes Lovins with peculiar selectivity. The sentence NERA quotes does not end "biomass technologies are already economically attractive" but continues, after a comma, "since the price of running a car on wood- or crop-waste-derived alcohols is comparable to the price of running it on taxed gasoline." Two sentences later, Lovins adds that " some biomass fuel technologies, mainly anaerobic digestion of feedlot manures, are not only conceivably economic but in commercial operation." Obviously, Lovins is speaking here of the present, not the future. Other examples of competitive biomass technologies could be adduced, including oil obtained by pyrolysis from woody residues, alcohol from foresty wastes, and "gasohol" programs in 26 states. The cleverness of NERA's quip that Lovins's "now" doesn't mean "now" thus rests on an insecure foundation.

It seems elementary that soft technologies should be compared not with historically cheap oil and gas that are dwindling, but with real and realistically-priced alternatives. Elementary also that if reasonable price policies would result in a tripling of energy prices, economic conservatism requires that we move in the direction of tripled energy prices; otherwise, the artificial cheapness of energy will continue to distort economic decisions, forcing people to make fundamentally uneconomic choices on pseudo-economic grounds.

Lovins does "say up-front" that energy prices will necessarily rise under a soft (or any other) energy strategy. He also insists that since soft technologies (and conservation) are cheaper than hard technologies, a soft path means lower energy prices than would otherwise prevail. NERA is more than slightly disingenuous in implying that Lovins's policies would treble energy prices while his critics' policies would hold energy prices roughly constant.

EDITOR'S AFTERWORD: NERA's memo and critique, like other critiques of Lovins's work, leave me with this impression: if intelligent people who obviously want badly to discredit Lovins's thesis can't do better than they have done, then that thesis must be basically sound. Friends of the Earth would be interested to hear from readers whether they came away with the same impression or a different one, and why.

Appendices follow, containing material that doesn't quite fit within the framework of the main text since it is not related to the hearings on Alternative Long-Range Energy Strategies of the Senate Committees on Small Business and on Interior and Insular Affairs. It is by and large the most up-to-date material in the book, however, and anyone who has read this far will find it rewarding.

—Hugh Nash

Appendix 1

A Review of *SEP*

The March 1978 issue of *Energy Policy* contained a review of Amory Lovins's *Soft Energy Paths: Toward A Durable Peace* by Alvin M. Weinberg, who heads the Institute for Energy Analysis, Oak Ridge Associated Universities. We reprint that review together with an exchange of letters between Weinberg and Lovins that appeared in the June 1978 *Energy Policy*.

Amory Lovins is surely the most articulate writer on energy in the whole world today. In his most recent book, *Soft Energy Paths: Toward A Durable Peace*, he displays a capacity for synthesis, a felicity of phrase, and a command of detail that is a true wonder. So persuasive is his mellifluous prose that I found myself carried away and all but convinced by his arguments.

Lovins is a revolutionary in the tradition, though not with the doctrinal cast, of Karl Marx. Unlike Marx he is a neo-Jeffersonian who sees in decentralized energy systems the true way to Utopia. *Soft Energy Paths*, together with his other books, both written and promised, will undoubtedly constitute Lovins' Das Kapital. Already Lovins has had much influence in the great energy debates: his views were evident in President Carter's introduction to the recent meeting of the International Nuclear Fuel Cycle Evaluation group. Energy policy analysts cannot ignore Lovins: they must examine both in detail and in the large his position. He argues his case so articulately that his influence for either the right or wrong path is bound to be enormous. If he speaks more clearly than he thinks, the energy community must know this.

Lovins states both his purpose and conclusion at the outset: a comparison of the hard and soft energy paths, paths that are "distinguished ultimately by their antithetical social implications. To people with a traditional reverence for economics, it might appear that basing energy choices on social criteria is what Kenneth Boulding calls a "heroic decision"—that is, doing something the more expensive way because it is desirable on other and more important grounds than internal economic cost. But surprisingly, a heroic decision does not seem necessary in this case, because the energy system that seems socially more attractive is also cheaper and easier."

Thus in reviewing *Soft Energy Paths* one must ask whether what he regards as the socially more attractive soft path is indeed cheaper and easier; and whether his Utopia has much bearing on our real world—in short, whether Lovins, for all his coruscating brilliance really knows what he is talking about.

What he claims for the soft path—decentralized, non-electric, non-nuclear, largely solar—reads like a patent medicine nostrum: "A soft path simultaneously offers jobs for the unemployed, capital for businesspeople, environmental protection for conservationists, enhanced national security for the military, opportunities for small business to innovate and for big business to recycle itself, exciting technologies for the secular, a rebirth of spiritual values for the religious, traditional virtues for the old, radical reforms for the young, world order and equity for globalists, energy independence for isolationists, civil rights for liberals, states' rights for conservatives." If only one-tenth of all this were so, we would be bereft of our senses not to embrace Lovinsism—to become energy revolutionaries just as he is.

As readers of *Energy Policy* know by now, Lovins' analysis begins with an examination of end use of energy. Since most of our end uses require energy at low availability, it is thermodynamically inelegant to base the energy system on electricity: of the energy finally used by the society not more than 10 percent requires electricity—electrochemical processes, industrial drive, illumination and a few others. Thus essential to Lovins' position is an almost religious rejection of electricity.

In assuming this position Lovins ignores the advantages of electricity. His Utopia is largely non-electrical; yet there is a mirror-image Utopia based on electricity since every end use that can be accomplished without electricity can also be accomplished electrically. The question is whether the advantages of electricity—its convenience, its cleanliness (as compared with decentralized fossil fuel systems) and its cost are worth sacrificing for its admittedly poorer thermodynamic match with its end use and its possible social consequences if it is generated by fission.

In a way one of the fundamental issues is the role of conservation. Why, after all, do we aspire toward conservation? Certainly not because conservation is a transcendent purpose of our society. Rather, it is because raw fuel is in short supply, or because the environmental or even social impacts of energy production are undesirable. Since electricity can be produced at but 30 percent first law efficiency, to go directly to end use rather than through electricity would conserve fuel. But in the long run, which is where Lovins' Utopia is aimed, conventional fuels will have been depleted and the choice is between the sun (plus some hydro and geothermal) on the one hand, and uranium (or possibly fusion) on the other, neither of which places much demand on fuel. Thus the imperative for conservation in Lovins' world must be sought not on grounds of shortage of fuel but on environment, burden of capital costs and other resources, and social desirability.

As for the environment, despite Lovins' claims a *properly* operating nuclear system is almost pollution-free. That there have been malfunctions of the system is undeniable; yet taken overall, I would insist that the nuclear system has done adequately well. Lovins simply denies the possibility of an acceptable nuclear future—it is this belief which fuels his zealotry. I believe an acceptable nuclear future—i.e., acceptably safe and acceptable environmentally—is possible, though I cannot prove our society will embrace such a future. But what is acceptable—i.e., how much risk one is willing to take—must depend on the relative costs. At some cost increment, solar is surely better than nuclear. Can that increment be decided *now* and our whole future pre-empted on the basis of Lovins' projections of the cost of solar energy?

For Lovins is totally exclusionary—he does not admit that Utopia of the future can be a mixture of nuclear and solar, of centralized and decentralized, of electrical and non-electrical. For him the paths are divergent and mutually exclusive. If we allow the hard, nuclear path to get well established, we forever pre-empt the development of the soft path.

Why is Lovins so intent on destroying nuclear energy rather than allowing it to compete in the various marketplaces with the solar alternative? His stated reason is that nuclear energy is unacceptable and can never be adequately fixed. But could it be that his claim that . . . "the energy system that seems socially more attractive is also cheaper and easier . . .", or (page 136) "the economic advantage of the solar over the nuclear system, even under Danish conditions, seems so robust as to survive any uncertainties in the cost calculations" in fact is nonsense; and that *only* by pre-empting nuclear on grounds other than economics can solar make serious inroads on our energy system?

This certainly seems to be the case. Let us take the comparison of capital cost per daily barrel of oil given on page 134 by Lovins and quoted almost verbatim by President Carter in his talk before the International Nuclear Fuel Cycle Evaluation meeting. The figures for coal, nuclear, and solar given by Lovins are:

Energy System	$/Bbl/Day	Form Supplied
Coal	170,000	Electricity
Nuclear-electric (LWR), mid-1980's	200–300,000	Electricity
Solar space heat, mid-1980's, no backup, assuming flat plate collectors and seasonal storage	50–70,000	Heat

I shall not quarrel with either his coal or nuclear figures, although I would argue that the load factor he assumes in arriving at his nuclear figure is too low (especially when he ascribes to the *distribution* system the same low load factor he ascribes to nuclear plants themselves), and I am incensed that Lovins ignores the difference between electricity and heat. On the other hand, his figure for solar heat is quite incomprehensible. The latest MITRE report, MTR-7485 (June 1977), gives the cost of an installed residential solar heating system, after all possible rationalization, in a small dwelling to be about $28/ft^2$. This is to be compared with Lovins' estimate of $10/ft^2$. Lovins claims correctly that a solar collector in a sunny area such as Albuquerque, New Mexico, receives the equivalent of 25 watt/ft^2 (270 watt/m^2), 24 hours a day, throughout the year. Using his assumed efficiency of 0.42, the collector would deliver 10.5 watt/ft^2, which corresponds not to $50-$70,000/Bbl/day but to $196,000/Bbl/day! This generously assumes that $28/ft^2 includes the cost of seasonal storage large enough to utilize every unit of delivered energy throughout the year (which MITRE does not include). For more representative regions of the United States the figure would be about $300,000/Bbl/day. And without the advantage of seasonal storage the cost is higher yet because average collector yields are lower in winter.

I dwell on this not insignificant point because it illustrates the great weakness in Lovins' economic arguments. He consistently underestimates costs of systems which he favors—by selecting projections (not data) that favor his case—and overestimates the costs of competing systems that he does not favor. Moreover, he takes far too seriously his back-of-the-envelope estimates of costs, especially of systems that have hardly been tested.

This is not a detail; this is almost the essence of the matter. For the ultimate question, the matter around which the great energy debate swirls is precisely, Can the sun replace uranium? Or to paraphrase and expand the question, What is lost, what is gained, if uranium is outlawed? Clearly, if our future energy path is based on an estimate of the cost of solar energy that may be four to six times too low, we embark on the soft path at the gravest peril.

But Lovins' objection to the hard path is far deeper than economic. It is grounded in his conviction that the whole social structure is determined by the character of the energy system, and that a decentralized, low energy society is far better than the one we now have. It is therefore extraordinary that Franklin Roosevelt in 1936 in his speech before the World Power Conference saw in *centralized* generation of electricity with transmission to outlying areas the key to the decentralized society: "Sheer inertia has caused us to neglect formulating a public policy that would promote opportunity to take advantage of the flexibility of electricity; that would send it out wherever and whenever wanted at the lowest possible cost. We are continuing the forms of overcentralization of industry caused by the characteristics of the steam engine, long after we have had technically available a form of energy which should promote decentralization of industry . . ." Has the world changed so much that white is now black? Are we in an Orwellian never-never land in which Roosevelt's vision has been discredited and must be replaced by that of the energy revolutionaries? Is the TVA area where I live "dirigiste and authoritarian" because our electricity is generated centrally though we distribute it through more than a hundred municipal utilities and rural electric cooperatives?

And, does the bomb make a world without nuclear power the peaceful Utopia Lovins claims in the last part of his book? Can he deny that the problem of nuclear weapons is not far from the problem of war itself: and that a total rejection of nuclear energy, as he suggests, far from making war—and in extremis therefore, nuclear war—unlikely, might exacerbate the tensions between nations without providing a countervailing stabilizing deterrent? Does Lovins really believe that renunciation of nuclear power will lead to Durable Peace?

Lovins, the revolutionary, must remember that revolution is usually strewn with disasters entirely unforeseen by its prophets. His insistence that Utopia is to be achieved only by rejecting nuclear energy or indeed centralized electricity, that Utopia somehow cannot be forged out of a combination of centralized and decentralized, nuclear and non-nuclear, electrical and non-electrical—this is far too strong medicine to be taken at face value. Unfortunately, the social forces that Lovins unleashes by his brilliant back-of-the-envelopism neither he nor his opponents will be able to control. Let us hope that common sense as well as social inertia will conspire to maintain our society on some middle road that embraces all ways to salvation from energy catastrophe.

Lovins: It is yet again my pleasant duty to celebrate human diversity by responding to a review by my friend Alvin Weinberg. There is a certain pleasing symmetry in the way each of us ascribes to the other an Orwellian inversion of reality. Dr. Weinberg, for example, describes me as an 'energy revolutionary', whereas my colleagues and I think of ourselves—and are increasingly viewed in the financial and political-science communities—as conservatives, not only in values but in methodology. My arguments rest not on values, important though I think them, but on classical criteria of economics, engineering, and political economy—precisely the criteria that electronuclear advocates must perforce reject if their technology is to survive.

Continuing the inversion, Dr. Weinberg states that I advocate a 'decentralized' and 'non-electrical' future. Quite the contrary: for the orthodox reason of minimizing costs (including social and environmental costs), I seek a degree of centralization and electrification appropriate to our spectrum of end-use needs. It is as silly to run a smelter with little wind machines as to heat houses with a fast breeder—both are a mismatch of scale (and the latter of energy quality too). Since industrial societies are now far more central-electrified than can be justified on even the narrowest criteria of private internal cost, I seek to redress the balance, but not to abolish electrification or large scale. Both have an important, limited place which they have long since saturated, and we can take advantage of the big electric systems we have without multiplying them further. I explicitly deny[1] Dr. Weinberg's 'almost religious rejection of electricity.' What I decry is the *inappropriate* use of electricity; and since electricity-specific end-uses are saturated several times over, more central generators are inappropriate. With truly cheap photovoltaics (which I do not assume), electricity might well play a larger role[2] without incurring autarchy, technocracy, vulnerability, etc.

Advantage costs

Nor do I 'ignore the advantages of electricity',[3] such as convenience for the user (whose ability to switch on and off is inconvenient for the utility) and cleanliness for the user (not for the utility's neighbors). These advantages are real, but must be paid for in production cost—which Dr. Weinberg inexplicably calls an 'advantage'—and perhaps in a quality rent. The resulting marginal price (typically about $0.063/kWh delivered, or over $100/bbl enthalpic equivalent, in the USA) seems excessive for most end-uses when virtually the same benefits can be had more cheaply by other means. These costs, and associated externalities, must be weighed along with often-low first and second law efficiencies—which are not, of course, evil *per se*[4] but only measure enthalpic and negentropic depletion.

Our previous exchange in these columns (when Dr. Weinberg reviewed my book *Non-Nuclear Futures*) showed that whether an 'acceptable nuclear future' is possible rests on many trans-scientific value questions. Dr. Weinberg—who invented the term 'trans-scientific'—would doubtless agree that these are not questions of fact. He thinks a properly operating nuclear system is benign because he is confident that its persistently toxic and explosive materials can be managed indefinitely with the requisite diligence (and without unwelcome social side-effects) despite human fallibility and malice—that is, that proper operation is achievable in practice. I do not share this confidence. Accordingly, saying that the nuclear system can be 'pollution-free' or 'acceptable' is the result of value-laden restrictive definition, rather like saying that cyanide is a more healthful food than candy because it does not promote tooth decay. Such value conflicts can only be resolved by the political process, not by a cost-risk-benefit calculus performed by some elite.[5] Agreeing, Dr. Weinberg says people must decide whether solar energy's admittedly greater benignity is worth what he views as its higher costs.

Here, then is the crux of our disagreement: what are the relative private internal costs and social external costs (many probably unquantifiable) of hard *v* soft technologies in their full sociopolitical context; and, more importantly, by what process is this comparison to be translated into social choices?

While I think Dr. Weinberg errs on both points, the first we can argue about, while the second he has simply misread. He states that I wish to 'destroy . . . nuclear energy rather than allowing it to compete . . . with the solar alternative'. Precisely the contrary. Nuclear energy, as I think Dr. Weinberg would agree, would never have arisen in a competitive market—and, I contend, cannot survive in a semblance of one now. As nuclear power has had to wean itself from military subventions, and as its true costs and problems have become clearer, nuclear orders have dried up around the world—fastest in the USA, where official projections of nuclear capacity in 2000 are falling so fast they should hit zero next year.[6] This collapse[7]—mirrored in the UK, France, Germany, Japan, and elsewhere—is not a US political eccentricity (though Dr. Weinberg would presumably accept any political verdict). Its fundamentally economic character is shown by the identical trend in Canada, which has had none of the US regulatory problems. The Invisible Fist strikes again!

Market test

This shrivelling before chill market winds has led many nuclear advocates, sometime defenders of economic rationality, to fight for government bailouts. Senator Long, for example, wants to raise the US taxpayers' share of nuclear power investments from the present ~20% to ~90%—a position Barry Commoner calls 'free enterprise for the consumer and socialism for the energy industry'. I contend rather that soft technologies have substantially lower private internal cost than nuclear and other hard technologies, and would hence displace them in the market if allowed to—for example, if freed from asymmetric institutional barriers and if compared directly in price with hard technologies rather than with the subsidized historic fuels that both are meant to replace; and that soft technologies also have far lower social external costs—which are already arguably dominating major political decisions about energy, and some of which, notably nuclear proliferation, I consider crucial. I do not shun but rather beg for the test of the market, and suggest measures, such as long-run marginal-cost pricing, to make the test fair and symmetrical, since today we have not an energy market but a fuel bazaar.

As for comparative private internal costs (ignoring externalities), I would be the last to suggest my estimates are uniquely right or should be taken on faith by anyone, and have published them so they may be refined by people who consider such costs decisive. But I take issue with Dr. Weinberg's challenges, and shall concentrate here on the solar heating number (not, alas, among those President Carter quoted), though the same case can be made in other end-use categories. Specifically:

- I do not 'ignore the difference between electricity and heat' but calculate it case by case, eg in comparing solar heating with a nuclear-electric heat pump.[8]
- Dr. Weinberg finds my solar heating costs 'incomprehensible' only because he assumes an installed price of $\$301/(m^2 + \lesssim 0.1\ m^3)$, presumably now, where I assume $\$150/(m^2 + m^3)$ in the late 1970s and $\$100/(m^2 + m^3)$ in the mid-1980s (1976 \$), corresponding roughly to $\$136/m^2$ and $\$86/m^2$ respectively. (The latter figure is used also by Blegaa *et al* in the June 1977 issue of *Energy Policy*.)[9] Dr. Weinberg also apparently assumes a barrel of oil is worth 6.37 GJ rather than my assumed 5.80, inflating his cost by 10%. Otherwise our calculations agree. Mine are based on national average insolation—125 W/m^2 in Denmark, 180 W/m^2 in the USA—not on the New Mexico desert's 270W/m^2.
- Sufficient storage to eliminate backup—demonstrated in the USA, Canada, and Denmark—accounts for only 10^{-1} of system cost.[10]
- Dr. Weinberg believes that active solar space-heating systems of good quality cannot be installed at prices as low as I state. Contrary to MITRE's calculations of $\$301/m^2$, empirical data—the basis of my calculations—show that today, contractor-installed prices of about $\$150/m^2$ are readily available, and prices of around $\$100/m^2$ are available with careful shopping, in relatively mature solar markets such as California[11] (I correct for different climatic needs elsewhere). The ERDA 1980 target is a routine $\$108/m^2$. Several designs[12] could be commercially installed in 1976 at $\$112\text{-}124/m^2$ (eg Reynolds' 'Torex 14'), and some (eg Calmac 'Sunmat', a rollable design which I do not assume) at $\$71\text{-}80/m^2$, both including all pumps, controls, etc. Designs integrated with the roof can install now at $\$100/m^2$ (1975 \$ + 12%).[13] Even in Canada, where the solar market is in its infancy, installed system prices now run[14] at about $\$194/m^2$. Of course, vastly dearer systems are available, but one need not buy them. More mature solar markets with direct franchising (fewer middlemen) should also bring real prices down.
- Contrary to Dr. Weinberg's charge that I underestimate soft-technology costs and overestimate hard-technology costs, I have included conservatisms consistently biased in the opposite sense least favorable to my argument.[15] For example, on the nuclear side I use rather low base costs, omit the marginal capital cost of reserve margin and of all past and future services, and neglect a debit of enrichment electricity against output and all real escalation after 1976 or-

dering (now running at about \$141 kWe^{-1} y^{-1} in the USA). On the solar side, I omit potential savings from mass production and simplified designs (even those already available), assume single-building rather than neighborhood systems (though the latter are several times cheaper[10,12,14]) and assume active heating even though passive heating often has a marginal capital cost of about zero and would be used in all new construction [and many retrofits].

- Even without these conservatisms, 100% active solar space heating at the mid-1980s price of \$100/(m^2 + m^3) (1976 \$) will cost about \$2.9-4.1/GJ delivered in the USA (perhaps \$5.8 in the UK) for a single building, or \$1.2-2.5/GJ for the dominant neighborhood retrofit systems in the USA (\$3.5 in the UK), all in the price range of present fuels.[10] But without the solar systems, we shall presumably be stuck with marginal prices at around \$5.0-6.4/GJ for heat from syngas, \$7.4/GJ from a PWR-powered heat-pump (COP = 2·5), or \$17.5/GJ for a PWR-powered resistive heater.[10] Doubling my solar prices to levels even ERDA and Hans Bethe consider realistic today (\$215/m^2) would still lead to heat prices indistinguishable from the nuclear-heat-pump prices—not 'four to six times' higher. This robustness appears to justify my broad economic conclusions. The documented details, published for peer review in autumn 1976, remain unrebutted.[16]
- I suspect that if we use the dearest kinds of soft technologies, as I have assumed, their long-run costs will overlap those of the cheaper kinds of hard technologies within an uncertainty band roughly half an order of magnitude wide. Such small, fuzzy marginal-cost differences are not 'almost the essence of the matter' but a dreadful basis for public policy. Sensitivities and external costs (such as nuclear proliferation) should dominate such decisions—especially since in the long run even rather large differences in energy price are probably macroeconomically unimportant.[17] Dr. Weinberg, who recently noted that his own nuclear cost estimates have proved ten times too low, should appreciate this—unless he feels that 'we embark on the . . . [nuclear] path at the gravest peril'. If my cost estimates—both private and social—are as far wrong as he thinks, the market and political process will say so, and I shall be content; but so far the question has not been fairly put.

I think that social structure is *partly* determined by the structure of the energy system, and that a less centralized energy system (my analysis does not assume a 'decentralized . . . society'), using resources more efficiently, *can* be 'far better than the one we have now'—*if* we get there, as I think we can, by pluralism rather than dirigisme. Yet Roosevelt's remark is not simply a Leninist call for 'collectives plus electrification'; it came at a time when many rural areas lacked their *first* kWe–h, whose utility was vastly greater then that of the *marginal* kWe–h we now debate. Today, with better technologies, Roosevelt's idea could be far more easily and quickly achieved with soft technologies—as David Freeman wishes his new TVA to do. And if Dr. Weinberg does not think TVA potentially authoritarian, he should reread the TVA statutes or try disputing his electric bill.

Finally, Chapter 11 makes abundantly clear that abandoning nuclear power—or equably accepting its demise in the market—will not *by itself* stop nuclear proliferation or bring peace. Rather, if we simultaneously shift nuclear programmes' public resources and political commitment to soft energy paths at home, encourage them abroad, and link these efforts with our work for strategic arms reduction, the combined package offers an excellent technical and political chance of stopping proliferation and of reducing the broader basis for conflict.

The past two years have seen a curious role reversal. People like me, who sought consideration of soft-path ingredients in case there were options there worthy of serious effort, now find ourselves on the cutting edge of policy, whereas prophets of nuclear salvation now find themselves pleading for heroic measures to keep their ailing option alive on the off-chance that someone might need it someday. Dr. Weinberg skillfully diverts attention from the conditions that led Jack O'Leary to remark on 28 November 1977 that 'the nuclear option has essentially disappeared' in the USA—and, arguably, elsewhere. But if nuclear power is indeed dead at the margin,[7] it is high time we concentrated on what we should have been doing instead. I look forward to recycling Dr. Weinberg into that exciting task.

Amory B. Lovins
Friends of the Earth
San Francisco, USA

---

[1]Amory B. Lovins, *Soft Energy Paths: Toward A Durable Peace,* Ballinger, Cambridge, Mass, 1977, and Penguin Books Ltd, Harmondsworth, Middlesex, 1977, pp 141–114.
[2]*Ibid,* p 143.
[3]*Ibid*, pp 138–140.
[4]*Ibid*, pp 137–138.
[5]A.B. Lovins, 'Cost-risk-benefit assessments in energy policy', *Geo Wash L Rev,* Vol 45, August 1977, pp 911–943.
[6]C.F. Zimmerman and R.O. Pohl, 'The potential contribution of nuclear energy to US energy requirements', *Energy,* Vol 2, 1977, pp 465–71. As Burke and Lovins note (see Ref 7 below) 1977 data continue the trend.
[7]T. Burke and A.B. Lovins, open letter to Prime Minister Callaghan, FOE Ltd, 9 Poland St, London W1V 3DG, 21 December 1977. See also A.B. Lovins, testimony to Hearings on the Costs of Nuclear Power, Subcommittee on Environ-

Alvin M. Weinberg replies:

I am delighted to learn, directly from Amory Lovins himself, that he concedes a place for centralized generation of electricity in his soft Utopia. Having made this concession, the differences between him and me come down to matters of degree—how much hard, centralized electricity is appropriate in an ultimate energy system. Amory Lovins implies that he now knows the proper ultimate mix. I urge caution, even humility, on this. For example, should a successful electric car be developed, the mix will be quite different than if transport always depends on fluid energy vectors.

I cannot take seriously Mr. Lovins' flippant rejection of the MITRE study. This was an honest, responsible effort that dotted 'i's and crossed 't's. Mr. Lovins' claim that solar heating systems are available at $100/m^2 in 'relatively mature solar markets in California' simply does not agree with what I have been able to discover after spending a month in San Diego trying to locate cheap solar systems. In particular, I find no evidence that the Calmac 'Sunmat' will last long enough before requiring extensive replacement to confirm his estimate of $71–80/m^2.

I agree with Amory Lovins—let the marketplace decide. If nuclear is dead, there is no need for him (and Friends of the Earth) to use the cost

---

ment, Energy, and Natural Resources Committee on Government Operations, US House of Representatives, Washington, 21 September 1977.

[8]See Lovins, *op cit*, Ref 1, pp 70, 114, 135–136.

[9]S. Blegaa *et al*, 'Alternative Danish energy planning', *Energy Policy*, Vol 5, No 2, June 1977, pp 87–94.

[10]A.B. Lovins, 'Soft Energy Technologies', *Ann Rev En*, Vol 3, in press, 1978.

[11]This can be seen from analysis of the costs in B.A. Greene, *Residential Solar Hot-Water Heating and Space Conditioning Systems in Northern California: A Brief Survey*, LBL-5229, Lawrence Berkeley Laboratory, August 1976. Bruce Anderson (Total Environmental Action, Harrisville, NH) routinely and profitably builder-installs active air and water systems at $108–129/m^2 (personal communication, 22 December 1977). Many other existence proofs are available.

[12]US Office of Technology Assessment, *Application of Solar Technology to Today's Energy Needs*, 1977.

[13]J.D. Balcomb *et al*, *Research on Integrated Solar Collector Roof Structures*, LA-UR-75-1335, Los Alamos Scientific Laboratory, 1975.

[14]K.G.T. Hollands and J.F. Orgill, *Potential for Solar Heating in Canada*, 77–01, report to National Research Council of Canada, University of Waterloo Research Institute, February 1977.

[15]Lovins, *op cit*, Ref 1 pp 67, 110–112, 128–130, 135–136.

[16]All 30-odd critiques published to November 1977 are collected, with full responses and supplementary material, in US Senate, Small Business and Interior Committees, *Alternative Long-Range Energy Strategies*, 2 Vols, US Government Printing Office, May and December 1977.

[17]Lovins, *op cit*, Ref 1, pp 71–72.

estimates that spring from their hopes as much as from the marketplace, let alone adroit appeals to politics, to twice kill the nuclear cat. But if his intent is to achieve the best mixture of hard and soft, centralized and decentralized, solar and non-solar, and if he is willing to concede *some* role for nuclear energy in the future, then rather than recycling myself into an elderly solar utopian, I invite Amory Lovins to join with me in helping devise an acceptable nuclear future.

Amory Lovins adds a final postscript:

That *some* centralized electricity generation is and will remain appropriate has never been in dispute and is hardly a 'concession'. That this implies a need to build more big power stations, however, is absurd in view of the manifest saturation of present or even asymptotic electrical needs in all the industrialized countries studied so far (over a dozen). That it implies a need to replace in kind, after their retirement, the thermal electric stations we now have is very doubtful. There may be an industrialized country in which this is the case, but I have not found one. There are some fundamental arguments[1] that appropriately electrical needs are not likely to increase significantly, and may well decrease, if we use efficiently the electricity we have. Of course, new uses for electricity may be developed, but I have not found major ones that even in principle look attractive. Successful electric cars, if developed, look (except perhaps in very special cases) markedly inferior to fuelled competitors (which might use motor-generator hybrids or fuel cells, but would not be externally supplied with electricity), and the biomass-residue fuel base looks sufficient for transport in the cases studied so far. If this expectation proves wrong, then the logical response will be to replace retiring central stations with dispersed renewables such as photovoltaics, not with more of the same. The applications, where *central* electric capacity makes sense are extremely limited (chiefly smelters), and are saturated by present hydroelectric capacity.

I did not flippantly reject the MITRE solar study—I did, however, question its relevance. It may be quite accurate on the costs of the collectors it considers, but there are other kinds. For example, Bruce Anderson's $108–129/m^2 is achieved by assembling the collector in the field, thus avoiding four markups (factory wholesale, retail, installation) on a packaged collector. Dr. Weinberg's reply crossed in the post with a list I sent him (to settle a bet) of a couple of dozen North American installers of good flat-plate systems at under $214/m^2, with many around and a few below Anderson's figure. As for the Calmac 'Sunmat', I did not assume it, the cost estimate is OTA's, not mine, it is made of thylene propylene terpolymer, 'superior to Neoprene for environmental, sunlight and heat resistance' (OTA), and glazed with the widely used glassfibre-reinforced plastic 'Kalwall', and while it has not been on the market long enough to prove a long lifetime, neither is there evidence to the contrary (c.f. reactors).

It is perhaps tedious, but apparently necessary, to reiterate that my solar costs are empirical and my nuclear costs are based on Bechtel's (lower than empirical). An important conservatism omitted from my previous response is that I assumed that a heat pump with a COP 2.5 would save a factor 2.5 in electrical supply capacity. This is extremely unlikely, since in a winter peaking grid such as in the UK, the coldest days reduce the the COP of most heat pumps to 1.0 or so, and resistive heaters are usually added. The average fuel saving remains 2.5 for the year, but the capacity saving is close to zero. (The capacity saved would probably be gas turbines, but the reason they are at the bad end of the merit order is that they send out dearer electricity than reactors, so that using a nuclear baseload price is also a conservatism.)

I am not willing to concede *some* role for nuclear energy in the future, largely because (as Chapter 11 of *Soft Energy Paths* argues) it proliferates nuclear weapons permanently, profusely, and unavoidably. This has a higher place in my own priority of values than, say, energy price, but if Dr. Weinberg's priorities are the reverse, I think he will be led to the same conclusion, and we need not argue about how we got there, which is entirely *de gustibus*.

While I should rather recycle a nuclear utopian into a realist than into any other kind of utopian, I should like to remind my friend Dr. Weinberg—in the spirit of Abraham Maslow, who said that if the only tool you have is a hammer, it's remarkable how everything starts to look like a nail—that a hand is a very versatile instrument, and can hold a great many things besides hammers.

---

[1]A. B. Lovins, 'Re-examining the nature of the ECE energy problem', ECE(XXXIII)/2/1G, UN Economic Commission for Europe, Geneva, February 1978.

Appendix 2

Letters to Science

Science for April 28, 1978, contained a letter by Jay James, Jr., disputing Lovins's cost figures for nuclear power and Alaskan oil. We reprint that letter here, followed by Lovins's reply in *Science* of September 22, 1978.

Energy Costs: Nuclear Versus Oil

President Carter quoted some interesting figures at the opening of the International Nuclear Fuel Cycle Evaluation (INFCE) meeting about the capital requirements of various energy resources, per "barrel of oil per day [bpd], or its equivalent derived at the ultimate site of use." The figures were: zero to $3500 for conservation, $10,000 for North Sea oil, $20,000 for Alaskan oil, and $200,000 to $300,000 for nuclear power. The President concluded that "there is a tremendous cost" for the use of nuclear power.

This argument derives directly from Amory Lovins' recent writings *(1)*. Aside from the inappropriateness of judging the cost of an energy source by its capital requirements alone, the numbers themselves are three times too low for Alaskan oil and three to five times too high for nuclear power. Apparently Lovins equates a kilowatt of oil thermal energy with a kilowatt of nuclear electric energy, ignoring thermodynamic losses and the cost of an oil-fired power plant. The interim replacement of short-lived oil field investments is also ignored. On the nuclear side, Lovins used very high costs for nuclear power plants, and electric grids, an extremely low capacity factor, and an additional 43 percent "miscellaneous" category, to reach a total of $5000/kW. A more realistic calculation (2) gives $1650/kW in 1976 dollars, including fuel cycle facilities and electric grids. Transferring to Lovins' oil base at 0.0324 bpd/kW (the oil requirements of a new, combined-cycle, oil-fired plant), the capital requirements are $56,700/bpd for nuclear power and $66,100/bpd for oil.

Even more important than these costs is that to produce oil one must own, or capture, the land that has the oil under it. Many of the countries participating in INFCE do not have the option of an Alaskan oil investment, because their territories do not contain oil deposits. Presumably they might acquire oil-producing territory by aggression, but such a strategy would surely be costly.

The energy experts attending the INFCE meeting must surely have noticed the use of Lovins' soft numbers by President Carter. It would seem desirable for the President to ask for a less extreme viewpoint from his speech-writers if the United States is to have a useful impact on world energy policies.

Jay James, Jr.

614 Canon Drive
Kensington, California 94708

Lovins on Energy Costs

Jay James, Jr. (Letters, 28 Apr., p. 381), of the Electric Power Research Institute, claims that capital costs of energy systems quoted by President Carter, presumably from my writings, are miscalculated. But James's first citation *(1)* shows that his objections are both wrong and irrelevant. Specifically:

1) James says I consider only capital costs and equate thermal with electrical energy. In fact (*1,2*), I add capital to fuel and other operating costs to obtain delivered energy costs. For consistent accounting, I compute capital costs in terms of a standard rate of delivering enthalpy (not free energy), such as 1 barrel of oil per day (bpd), which is ~ 67 kilowatts thermal (kWt). I then allow later for the First Law efficiency of end-use devices in each thermodynamic category of need, such as furnaces and heat pumps for low-grade heat.

2) James suggests I should not compare nuclear with oil investments, because oil is short term. In fact, I mention oil-system capital intensity (*1*) only as an *historic* baseline two orders of magnitude below the capital intensity of marginal electric systems. The energy systems I compare with each other in delivered price (*1, 2*) are all *long-run* marginal sources, all meant to *replace* oil urgently. Other systems are of little strategic interest.

3) James states, without citation, that President Carter's estimate of $20,000/bpd for an Alaskan oil system is "three times too low." In my book (*1*), I state no capital cost specifically for delivered Alaskan oil, though I give a range of ~ $10,000 to $25,000/bpd—derived from the Bechtel data base (3)—for 1980's U.S. frontier oil and gas. The exact value for Alaskan oil, which is strongly site- and date-dependent, has been authoritatively estimated to be about $19,900/bpd (4). This is remarkably

References

1. A. B. Lovins, *Soft Energy Paths* (Friends of the Earth-Ballinger, Cambridge, Mass., 1977); *Foreign Affairs* 55, 65 (October 1976).
2. Analysis by the writer, based on data from EPRI Technical Assessment Group, *Technical Assessment Guide* (Electric Power Research Institute, Palo Alto, Calif., 1977); I. A. Forbes and J. C. Turnage, *Exclusive Paths and Difficult Choices: An Analysis of Hard, Soft, and Moderate Energy Paths* (Energy Research Group, Framingham, Mass., 1977).

close to the President's estimate, but not to James's ~ \$60,000/bpd (which would imply, at my 12 percent per year fixed charge rate, an implausible capital charge of \$19.7 per barrel). Even if the long-run alternatives I consider did include oil, the short physical lifetime of oil field investments would not greatly alter the economics as James implies (5).

4) James states that the capital costs I assume for a marginal nuclear station and its associated transmission and distribution (T & D) capacity are "very high." In fact, they are Bechtel's data (3), converted from 1974 to 1976 dollars and ordering with appropriate indices (6) and assuming that the real escalation rate after 1976 is zero (7, 8). The 55 percent capacity factor I assume (*1*), considered "extremely low" by James, is broadly consistent with empirical data (9)—though I include conservatisms ample to allow for $\gtrsim$ 80 percent in case the vendors' hopes of greatly improved performance are realized.

5) James implies that my analysis uses a whole-system nuclear capital cost of \$5000 per delivered kilowatt electric (kWe), including "an additional 43 percent 'miscellaneous' category." The value I use (*1*) is in fact \$3495/kWe delivered, enthalpically equivalent to \$235,000/bpd and hence within President Carter's range of \$200,000 to \$300,000/bpd. The \$3495/kWe is calculated (*1, 10*) from section 4 above [assuming marginal T & D losses (*11*) of 10.7 percent] and is conservative. To show this, I also estimate (*1*)—outside the comparative analysis—that a realistic value might be about \$5000/kWe delivered (~ \$336,000/bpd). The extra ~\$1500/kWe arises not from "miscellaneous" but from estimates for specific terms (*12*) explicitly omitted from the \$3495/kWe (*1*).

6) James's "more realistic calculation" yielding \$1650/kWe is unstated and undocumented. He cites an estimate by Forbes and Turnage, to whom I have responded elsewhere (*13*), but it states a value of \$1975/kWe, is judgmental, cites no sources, and states no grounds for preferring its lower costs and higher capacity factors (65 percent for generation and transmission, 100 percent for distribution) to those of my references. James thus fails to explain why my, or the President's, estimate of nuclear system cost is "three to five times too high"—presumably meaning it should be about \$700 to \$1500/kWe.

7) James's nuclear cost of \$1650/kWe converts, at my 5.8 gigajoules (GJ) per barrel, to \$111,000/bpd of delivered enthalpy, not to \$56,700 as he states. The origin of his "\$66,100/bpd for oil" is equally obscure. His implication that the nuclear system is less capital-intensive than an oil-electric system is absurd (*14*). Accordingly, while it would be foolish to extract oil in order to burn it under a boiler—even in an efficient combined-cycle plant such as James assumes—a recent comparison shows that such a plant would send out 24 percent cheaper marginal electricity than a pressurized water reactor (PWR) (*15*).

8) James presumes that if a nuclear power station is not built, a fossil-fueled one must be built instead. But more electricity, from any source, is not a sensible answer to our current energy supply problem: heat (now 58 percent of U.S. delivered enthalpic needs) and portable liquid fuels (34 percent) (*1*). The premium end uses that are electricity-specific—now 8 percent of total end-use energy needs in the United States (*1*), ~ 7 percent in Western Europe (*2*)—are already saturated (*16*). Electricity is too costly and slow to make further saturation worthwhile. While James seems to think the energy problem is how to expand domestic supplies to meet extrapolated homogeneous demands, I think it is how to meet *heterogeneous* end-use needs with a minimum of energy (and other resources) supplied *in the most effective way for each task.* Accordingly, debating which kind of power station to build is like debating which is the best champagne when all one wants is a drink of water. That is surely the point of President Carter's economic comparison (*17*) of nuclear with non-electric investments—and of mine (*1, 2*) with both soft technologies and the far cheaper improvements in end-use efficiency.

9) The key question for petroleum-dependent countries is, What investment can relieve that dependence fastest per dollar invested (subject to other constraints)? Nuclear power—with its complexity, inherently long lead times, and narrow markets—fails that test, while a soft energy path—relatively simple, fast, accessible, and diverse—passes it (*1, 2*). This rate advantage of the soft path is *independent* of countries' access to transitional fossil fuels. Moreover, while soft technologies can substitute in every end-use category, nuclear power can readily displace only baseload electricity. Hence replacing *every* oil-fired power station (thermal and gas turbine) in the countries of the Organization for Economic Cooperation and Development (OECD) with nuclear power overnight would reduce OECD oil consumption by only 12 percent and reduce the imported fraction of that consumption from about 65 percent to 60 percent (*18*)—at the cost of increased dependence on imported uranium and capital. It is thus all the more irrational to suggest that, without nuclear power, nations must war over oil.

As several thousand pages of critiques and responses on the soft-energy-path thesis show (*13, 19*), this is not the first time someone has decried "soft numbers" before verifying references. May I renew my earlier plea (Letters, 24 June 1977, p. 1384) that analysts get on with substantive refinement, extension, and application of soft-path concepts?

Amory B. Lovins

Energy and Resources Program,
University of California,
Berkeley 94720

References and Notes

1. A. B. Lovins, *Soft Energy Paths: Toward A Durable Peace* (Friends of the Earth and Ballinger, Cambridge, Mass., 1977), especially chaps. 3, 6, and 8.
2. ——, "Re-examining the nature of the ECE energy problem" [ECE(XXXIII)/2/I.G., U.N. Economic Commission for Europe, Geneva, 1978]; "Soft Energy Technologies," *Annu. Rev. Energy* 3, 477 (1978).
3. M. Carasso *et al.*, *The Energy Supply Planning Model* (Report to the National Science Foundation, Bechtel Corp., San Francisco, 1975), two volumes, acquisition Nos. PB-245 382 and PB-245 383 available from the National Technical Information Service, Springfield, Va.; facility data sheets and updates from M. Carasso and J. M. Gallagher, personal communications, 1976.
4. D. Sternlight (Chief Economist, Atlantic Richfield Corp., Los Angeles), personal communication (1978) of detailed Arco estimate that the total present-valued investment (exploration, field cost, pipeline, Valdez terminal, and new tankers delivering to existing West Coast terminals), including interest, is about $17,500/bpd (1978 dollars) for the Sadlerochit field (lifetime ~30 years). I have deflated this estimate at 7 percent per year to 1976 dollars and added $4621/bpd for storage, refining, and distribution, all treated as marginal costs [Bechtel data (*1, 3*) converted to 1976 dollars with the Marshall and Stevens index (6)].
5. For illustration, if 30 percent of system investment were at the wellhead and 70 percent downstream, with respective lifetimes of 13 and 30 years, the present value of the stream of original plus replacement investments over 40 years, at a 5 percent per year real discount rate, would be only 30 percent above the total investment if all lifetimes were 40 years.
6. For a 1.1-GWe PWR, $585 per net installed kWe (1974 dollars) is converted to $929/kWe (1976 dollars) using the 1.26 per year index from a 35-plant multiple regression (r^2= 0.71) by I. C. Bupp and R. Treitel, "The economics of nuclear power: De omnibus dubitandum" (Harvard Business School, Cambridge, Mass., 1976). For T & D—respectively $69 and $420 (1974 dollars) per net kWe of installed marginal generating capacity, taking into account supply diversity—the conversion to 1976 dollars is made with the 1.25 Marshall and Stevens Equipment Cost Index.
7. This is highly conservative. For example, a regression (r^2 = 0.76) on the 39 U.S. light water reactors (LWR's) completed through May 1977 reveals that, with each successive year of construction permit issuance (1967–1971), controlling for all other significant variables, real plant cost rose $141/kWe. If this kept up, a 1.1-GWe PWR ordered in 1976 (25th unit built by the architect-engineer, outside the northeast region, with a cooling tower) would cost ~$1474/kWe. Both these figures have been converted to 1976 construction dollars with the Handy-Whitman steam-plant construction cost deflator, and would be higher if they were in 1976 GNP dollars [W. E. Mooz, "Cost analysis of light water reactor power plants" (Report R-2304-DOE, RAND Corp., Santa Monica, Calif., 1978].
8. The assumed LWR cost is also probably too low. For example, the California Energy Commission's draft report to the state legislature on Assembly Bill 1852 [R. Knecht *et al.*, *Comparative Generation Costs* (California Energy Commission, January 1978), appendix 19] estimates $1027/kWe for the Sundesert plant, ordered in January 1976 (neglecting dedicated transmission and deflating to 1976 dollars at 6.5 percent per year). The August 1978 final draft (in press) estimates $1185/kWe.

9. The empirical average for all U.S. LWR's through 1977 was 60 percent (58 percent if weighted by unit size), 53 percent for units over 0.8 GWe. Exhaustive regressions on the entire U.S. data base lead to a predicted average, levelized over the first 10 years of operation of a new 1.1-GWe PWR, of 60 percent, taking account of a new vintage correlation that emerged during 1977 (55 percent without it). See C. Komanoff, *Nuclear Plant Performance Update 2* (Komanoff Energy Associates, New York, 1978). Komanoff and V. Taylor have also prepared an improved analysis of the Mooz data (7).
10. Also assumed are Bechtel's (*3*) \$61/kWe for marginal fuel-cycle facilities—probably ~3 to 5 times too low—updated to 1976 dollars and ordering with the Marshall and Stevens index (6), and \$100/kWe, calculated (*1*) in 1976 dollars, for the initial core.
11. Bechtel (*3*) assumes 16.4 percent at the margin.
12. The omitted terms are: real escalation after 1976 ordering; marginal investment in reserve margin, all future services such as waste management and decommissioning, and past or present services such as federal R & D, regulation, and security services; and the ~6½ to 8 percent of electric output currently needed to run the fuel cycle. Terms omitted from both the \$3495/kWe and the ~\$5000/kWe totals include costs of end-use devices, externalities, dynamic net-energy considerations, and any "miscellaneous" items.
13. U.S. House of Representatives, Committee on Government Operations, Subcommittee on Environment, Energy and Natural Resources, *Nuclear Power Costs* (Government Printing Office, Washington, D.C., 1978), part 2, pp. 1103–1115.
14. For example, if one uses Bechtel data (3), a 0.8-GWe coal-electric system with a scrubber but no fuel cycle would cost \$2200/kWe delivered (1976 dollars) at 0.62 capacity factor (*1*, *9*). If James's oil-fired plant cost the same, an oil-system cost of \$30,000/bpd (capital charge at 0.12 per year = \$9.9/b, too much to clear the market), divided by 0.46 First Law plant efficiency, would imply a system cost of \$3172/kWe—generously assigning the whole oil-system cost to the residual rather than the light fractions (*15*).
15. J. Harding, in (*13*), part 2, pp. 1778–1802. The rationale for considering such a combined-cycle plant at the margin is that California is to have an embarrassing glut of residual oil from refining Alaskan crude oil extracted for its light fractions. Saving residual oil (nearly all of the 15 percent of California oil now burned in power stations) would probably not save crude oil; the residual oil is a by-product, not a motive. California is also considering gasifying residual oil.
16. The ratio of present electricity supply to electricity-specific needs approaches 2 in the United States today, and may exceed 3 after long-run end-use efficiency improvements.
17. J. Carter, remarks to Opening Conference, International Nuclear Fuel Cycle Evaluation, Washington, D.C., 19 October 1977.
18. V. Taylor (Pan Heuristics, Los Angeles), personal communication, May 1978. The proportional import reduction would be greatest in the United States, not in Europe or Japan.
19. U.S. Senate, Select Committee on Small Business and Committee on Interior and Insular Affairs, *Alternative Long-Range Energy Strategies* (Government Printing Office, Washington, D.C., 1977), two volumes [this contains all published critiques and responses except the exchange with Forbes in (*13*)].

[This argument was continued in the letters column of *Science* magazine, though too late for inclusion here. J. Michael Gallagher of the Bechtel Corporation raised similar questions in the 22 December 1978 issue, and Lovins's response to Gallagher's letter was published in April 1979.]

Appendix 3

Regarding Costs

Lovins's most persistent critic on the issue of costs is Dr. Ian Forbes, who claims that Lovins's soft-path cost estimates are far too low and his hard-path estimates, far too high. It was hard for Lovins to come to grips with this criticism because Dr. Forbes's data and calculations were largely undocumented. Lovins's difficulties in this connection were outlined in a letter to the Subcommittee on Environment, Energy, and Natural Resources of the U.S. House of Representatives Committee on Government Operations, before which both men had testified on the costs of nuclear power.

Lovins: During my testimony of 21 September [1977] and in a note of 7 October I agreed to respond for the record to material by Ian Forbes concerning my work on soft energy paths. Perhaps I should first review the history of my exchanges with Dr. Forbes.

In March 1977 Dr. Forbes distributed a paper, "Energy Strategy: Not What But How," updating a February draft of comments on my paper in the October 1976 *Foreign Affairs*. I responded to the June revision of it in Volume 2 of the U.S. Senate Small Business and Interior Committees' record *Alternative Long-Range Energy Strategies* (December 1977). Both these papers by Dr. Forbes were notably lacking in quantitative detail. In our debate on 14 June 1977 before the annual meeting of the American Public Power Association in Toronto, however, Dr. Forbes produced overhead transparencies of various numbers and graphs purporting to show that a soft path would cost much more than a hard path. From the numbers on the screen it appeared that he had inflated my documented soft-technology costs by a large factor, probably averaging three or more. I asked for a written copy of the numbers and the documented supporting calculations which he said he had prepared. He promised to send them to me promptly—probably within a week. I never received them.

On 21 September 1977 we met again before your Committee. Dr. Forbes's written testimony contained, so far as I could tell, the numbers and graphs he had shown at Toronto. Since I had not been sent an advance copy of his testimony (though he had mine), I was again put in the position of having to respond to his quantitative conclusions at sight and without seeing supporting calculations. His written testimony and our colloquy before the Committee implied to me—though he denied it—that his cost estimates were based on pathological prototype costs or were otherwise unrealistically high. Dr. Forbes told the Committee that, on the contrary, he had prepared fully documented calculations showing that his cost estimates were realistic and reasonable, but that despite his repeated requests, a sponsor of his work, Oak Ridge National Laboratory, had refused to let him release the calculations to me. Chairman Ryan then asked the staff to ask the ERDA Administrator to tell Oak Ridge to let the numbers go, and I undertook, if they reached me in time, to respond to them.

On 17 October 1977 Dr. Forbes came up to me after a debate between Professor Rose and myself at MIT. He said that though he still did not have clearance from Oak Ridge, he was giving me the promised backup material anyhow. It is the enclosed report "Exclusive Paths and Difficult Choices: An Analysis of Hard, Soft and Moderate Energy Paths," by Dr. Forbes and his colleague Joe Turnage, and is dated 30 June 1977 (Second Draft). [They later issued a much revised version, undated, with the same title.]

I am sorry to report that this supplementary material does not live up to the expectation that I, at least, had formed on the basis of Dr. Forbes's comments before your Committee. In particular, it does not seem to me that this new paper contains the backup Dr. Forbes said it did, nor that it justifies his assertions about the reasonableness of his cost estimates. . . .

[Editor's note: The Forbes/Turnage paper, "Exclusive Paths and Difficult Choices" (hereafter, EPDC), is in most respects much like the Forbes paper reprinted in the main body of this book. It differs in containing more detail on costs. We reproduce here parts of EPDC dealing with costs, followed by some of Lovins's comments (as they appear in his letter to the House subcommittee, of which we have already reprinted the opening paragraphs).]

EPDC: Lovins overestimates the cost of "hard" technologies, particularly nuclear power. For example, a reasonable estimate of the cost in 1976 dollars to deliver a kilowatt from a nuclear plant in the mid-eighties would be about

[750 (plant) + 75 (fuel cycle) + 60 (core) + 75 (transmission)]
÷ 0.65 (capacity factor) + 300 (distribution) = $1777/kw-av.

Factoring transmission and distribution losses, this figure is increased to about $1975 per average kilowatt or about $140,000 per barrel per day—substantially below Lovins' estimate of $200,000 to $300,000/bbl.-day, but significantly higher than non-electric energy source investment. Recent estimates by EPRI and Commonwealth Edison, and practical experience with the McFarland (coal) and Davis-Besse (nuclear) units in Ohio, suggest that the capital cost for coal plants with scrubbers is about the same or only marginally less than for nuclear.

The use of the "barrel per day" unit is frequently inappropriate since it neglects the lifetime of the plant (or capital replacement rate) and does not distinguish between differing efficiencies of end use. For example, if we take the equivalent full-capacity life of a frontier oil or gas field to be 12 years and assume a total capital cost of $17,500/bbl.day (including refining, delivery costs and losses), the capital investment over a 36-year period (about the life of a power plant) would be

$$3 \times 17{,}500 = \$52{,}500/\text{bbl.-day},$$

if costs remain constant. If costs were to escalate at 5% in constant dollars, the 36-year cost would be

$$(1 + 1.80 + 3.22) \times 17{,}500 = \$105{,}400/\text{bbl.-day}.$$

In these terms, the marginal capital cost for oil and gas development is closer to one-half that for electricity.

(It is noteworthy that Lovins, while meticulous about infrastructure costs and T&D losses for electricity generation, neglects refining, transportation and other costs and losses for 'direct' fuel use.)

The "barrel of oil delivered to end use" neglects the efficiency of end use. Only about one-quarter of the electricity produced in the U.S. goes to uses that can be considered inappropriate or inefficient compared to direct use—resistance space and water heating, cooking, clothes drying and some industrial uses (remediable in many instances with the use of heat pumps, microwave ovens, etc.)—while the bulk goes to uses that are necessarily electrical or that are comparable in overall efficiency to direct fuel use. Similarly, direct fuel use is never at the full energy potential of the fuel (except in raw feed uses). The most notorious example, of course, is the automobile at about 10% overall first-law efficiency.

Lovins generally underestimates the cost of "soft" technologies. Given the current state of development, precise estimates cannot be made, but sufficient information exists to show that Lovins' "soft" path is very capital intensive.

The National Research Council of Canada's 120 ft. by 80 ft. diameter, 200 kilowatt, vertical-axis wind turbine for the Magdalen Islands has been constructed by Dominion Aluminum Fabricating, Ltd., at a price of $230,000 exclusive of foundation and grid connection. Industrial estimates of the run-on production cost are about $120,000. Assuming a price of $180,000 installed complete, a capacity factor of 30%, an associated transmission and distribution cost of only $150/kwe, and losses of 10%, the cost per delivered kilowatt is $3889, or about $275,000/bbl.-day. Such systems appear to make most sense for the applications considered in Canada—in isolated areas where wind speeds are high (average wind speed at Cap aux Meules as about 19 miles per hour) in combination with existing diesel generators.

However, their deployment in large numbers without storage is doubtful, particularly for the "soft path." Storage could add $50,000 to

$100,000/bbl.-day to the cost. Because this sytem has a rated wind speed of about 30 miles per hour (and noting the V^3 variation of output) the number of locations at which capacity factors of 30% could be achieved in the U.S. would be limited. Even at good sites, output would be limited by the requirement for spacing windmills up to 30 diameters apart to avoid sheltering effects. The carefully-selected site on Grandpa's Knob, Vermont, for the Smith-Putnam machine had an average wind speed of 17 mph (a reduction in wind speed from 19 mph to 17 mph would reduce a 30% capacity factor to around 21.5%). At 25% capacity factor and $200/kwe for storage, the cost of the NRC wind turbine would increase to $390,000/bbl.-day; at 20% capacity factor, to $490,000/bbl.-day (other parameters as above).

The cost of the Smith-Putnam machine has been estimated at $5000 per average kilowatt in 1971 dollars; this translates to about $548,000/bbl.-day in 1976 dollars (10% losses). ERDA/NASA horizontal-axis wind turbine costs are about $5000 per rated kilowatt (20 m.p.h. winds). If we *assume* that these machines could be produced *with* storage at just 40% of the current cost without storage and achieve 35% capacity factor, the capital cost per delivered kilowatt would be $6825 ($150/kwe T&D, 10% losses), or $483,000/bbl.-day.

Flat plate collector systems for solar space and water heating are currently selling at prices averaging around $20 to $25 per square foot uninstalled, varying from under $10/ft^2 uninstalled to over $40/ft^2 installed. Efficiency, reliability and durability vary considerably. Tests of 100 solar water heating systems in New England have yielded generally disappointing performance (with a few notable exceptions); several Canadian installations have showed signs of significant deterioration within a relatively short period of time.

A square foot of solar collector operating at an efficiency of 40% in a solar flux of 180 watts/m^2 can deliver 200,000 Btu per year to year-round uses such as domestic water heating, about half this amount to space heating. Put another way, 10,000 ft^2 of such collectors can deliver the equivalent of 0.945 barrels per day to water heating, or about 0.47 bbl/day to space heating, with limited storage. If we assume that collector systems that can operate reliably at an efficiency of 40% can be delivered *and* installed at $20/ft^2 in volume production (or $15/ft^2 at 30% efficiency, etc.), then this translates to a capital cost of about $423,000/bbl.-day delivered to space heating, $211,700/bbl-day delivered to water heating. (Note however, that if a solar system is replacing a 70% efficient oil furnace, this capital cost for space heating is equivalent to $296,000/bbl-day *replaced.*)

While most people working in the field envision 40% to 80% solar heating systems, Lovins claims that a 100% system based on seasonal storage would be cheaper. For a 20 bbl/year (4 kilowatt average) heating requirement in a 180 watts/m^2 average solar flux, the collector area is 600 ft.^{2} (55.7 m^2) to supply 100% of requirements at 40% collector efficiency with seasonal storage. In winter, however, requirements would rise to 8Kw to 10Kw while collector output could drop to around 30Kw, creating a need for 5Kw to 7Kw from seasonal storage. This would necessitate a storage capacity of 20 to 25 kilowatt-months. As Lovins observes, 100 m^3 of water can provide 8Kw-months of storage at a ΔT of 50°C (at 100% storage efficiency). Thus, the storage required would be 250 m^3 at 100% efficiency, or 360 m^3 at 70% efficiency (about 7 m^3 per m^2 of collector). At a cost of $35 per m^3 (optimistically), this amounts to $12,600 to $15,050. Thus the total system cost would be in the neighborhood of $26,000, or about $470,000/bbl.-day. (This contrasts sharply with Lovins' estimate of $50,000 to $70,000/bbl.-day.) Of course, collector area can be traded for storage volume, or the heat load curve flattened somewhat by careful insulation (which should be done anyway for cost effectiveness), but the cost per barrel per day delivered remains in the same general range.

This is hardly to suggest that solar heating and other renewable sources should not be vigorously pursued. In view of expected future fuel price increases and the need to conserve conventional fluid fuels, a substantial Federal program of R&D, supports and incentives is warranted to speed commercialization of solar heating and other technologies.

Several of the points raised by Lovins should be reinforced:

- conservation measures will generally be the most cost-effective investment;
- solar-designed housing with passive systems have a significantly lower cost than that given above for "add-on" active systems;
- "soft" solar applications appear more promising than "hard" ones;
- insofar as is practicable, it is important to try to match supplies to end-use demand rather than vice-versa.

On Electrical Generation

Comments on Lovins' analysis of electric generating costs notwithstanding, it is the case that electricity is generally more expensive than direct fuel use in most inter-substitutible applications; recent estimates suggest this to be true also for synthetic gas. Inappropriate applications of electricity are costly and wasteful, and policy should attempt to avoid such through supply/demand matching and new end-use technology.

On the other hand, electricity since its introduction has been more expensive on the average than fluid fuels, just as natural gas has been more expensive than coal. Much of the dramatic growth in the use of these fuels must be attributed to their convenience—a factor that is hard to quantify.

Future growth in electricity use can be expected for several reasons, including:

- increased demand of electrical end uses;
- convenience, particularly in inter-fuel switching;
- institutional problems for direct coal use;
- limits to non-electrical energy source production.

This raises important questions about efficiency. Certainly, the development of new fluid fuels, natural or synthetic, particularly for transportation, must be a central part of a cogent energy policy, as must be the early development of renewable sources for applications that would otherwise be supplied electrically (or the development of direct use of nuclear heat for process applications).

There are critical policy decisions required to avoid electric substitution in inappropriate applications. Industrial switching to electricity is continuing because of convenience in procurement and application. (Netschert reports of factories using resistance heating for steam generation!) With natural gas in short supply and oil suspect since the embargo, many industries may prefer to switch to electricity and pay the price premium rather than become embroiled in fuel procurement problems or in Federal, State and local regulations over coal use—leaving utilities to act as their surrogate in procurement and regulatory matters.

Most current estimates project year 2000 electricity use to be 2.5 times or greater than today's. This represents a considerable onus to further improve the efficiency of electric generation, distribution and end-use. Major efforts on the part of industry and government will be required to implement cogeneration and total energy systems, improve generating efficiency (topping and bottoming cycles, HTGR, MHD, etc.), improve T&D efficiency (transmission, storage) and improve end-use efficiency (insulation, heat pumps, high-temperature heat pumps, solar-assisted heat pumps, storage; lighting, refrigeration, air-conditioning, water heaters, electric motors, etc.)

To accomplish this will require innovative and flexible management—a sense of purpose tempered by a realistic view of what is achievable. As Pickering has said: "practical conservatism goes hand in hand with abstract radicalism, because meaningful novelty is effectively postponed by both parties". The road we take can neither be impractical, nor the same stuff in larger doses.

Comments on Cogeneration

A major increase in use of cogeneration—whether electricity as a byproduct of industrial steam generation or steam and heating as a byproduct of utility electrical generation—is undeniably desirable for the increased efficiency and cost savings that result. Von Hippel and Williams state that

60% of all industrial steam production (greater than 100,000 lbs/hr) is suitable for cogeneration. Thus, if all of this 60% were produced in the year 2000 by fluidized-bed, gas-turbine (Stal-Laval type) cogeneration systems, they estimate an electric output of 3.6 quads/year (at 3413 Btu/Kwh, or about 10^{12} Kwh/year) could be achieved—about half the present U.S. electrical consumption. The potential for savings from utility industry waste heat are even greater.

The barriers impeding major deployment of cogeneration are (like much of the energy problem) institutional:

- geographic mismatches of primary production and "byproduct" demand (note in this regard that Lovins' suggestion to use only hydro and industrial cogeneration for electrical production involves serious geographical mismatches);
- the current policy of ex-urban siting of large power plants and resistance to urban siting of even relatively small plants (e.g. community resistance to Harvard Medical School's total energy facility);
- regulatory problems (rates, environmental requirements, etc.);
- inadequate financial incentives for cogeneration (beyond the producer's own needs) or for construction of district heating systems;
- motivating private industry to take on the increased burden of cogeneration production beyond its own needs.

FIGURE 4.1
Comparison of Cogeneration and Separate Generation

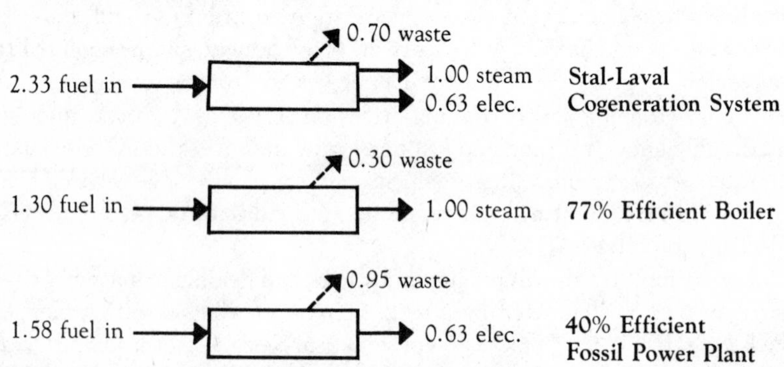

In the last regard it should be noted that industrial cogeneration, while more efficient, is hardly a "something-for-nothing" proposition. As Figure 4.1 shows, the Stal-Laval cogeneration system requires 20% less fuel for steam and electricity production than the equivalent separate boiler and power plant (assuming that waste heat from the power plant is not utilized). However, in switching to this system, the industrial user is faced with an 80% increase in fuel procurement and a 130% increase in emissions as well as the increased capital cost. It is difficult to convince an industrial concern to accept this burden unless it plans to use a significant fraction of the power, or is given sufficient financial inducement (such as a high tax credit or high rate of return), or has the utility cover part of the construction costs. In fact, given current institutional problems of coal use, it may be difficult enough to get the industrial producer to switch from fluids to coal, let alone to install cogeneration.

Nevertheless, cogeneration is desirable and every effort needs to be made to remove the barriers to utility and industrial cogeneration.

As a final point, it should be noted that cogeneration and other total energy system concepts transcend the boundaries of "hard" and "soft," as does their desirability. A "hard" example is the proposed district heating system for Uppsala, Greater Stockholm and Sodertälje using the Forsmark nuclear plant and heat pump feedback at the tail end.

Transportation

The transport sector is given only light attention by Lovins. This is a crucial area because of the potential for efficiency, conservation (as Lovins points out) and increased mass transit, and because of the difficulties in meeting mid- to long- term demand.

Lovins' suggestion for a fuel alcohol industry 10 to 14 times the size of the present U.S. beer and wine industry does little to help. Szego and Kemp estimate that land managed and harvested in a manner similar to a Southern tree farm would be sufficient to support 1Mw of electrical production per square mile. To produce one-third of current transportation fuel demand (19 quads) in this manner would require about 100,000 square miles—about 20% of current U.S. harvested farm acreage or about 35% of U.S. national forest acreage. Variations in crop, climate, insolation, water supply and long-term productivity would likely increase required area significantly. Biomass conversion (particularly of wastes) needs better attention, but the scale and manner of Lovins' approach seems inappropriate in an era when food supplies are critical.

Nevertheless, faced with declining oil supplies, a minor penetration of electric vehicles before the turn of the century and uncertain prospects for synthetics, supply/demand matching in the transportation sector looms as a major difficulty.

Capital Investment Requirements for Soft and Hard Paths

While condemning the cost of the "hard" path as being prohibitively consumptive of capital, Lovins does little to provide an overall estimate of the cost of the "soft" path—apparently because he feels it is clearly cheaper. What follows is a "rough cut" at such an estimate.

Lovins' illustrative soft path shows a year 2000 demand of 95 quads, composed of 25 quads coal, 35 quads of oil and natural gas and 35 quads from "soft" technologies. Since Lovins provides no breakdown by end-use, we have developed an illustrative one (see Table 4.2) by adaptation from recent reports of I.E.A. and Von Hippel and Williams (cogeneration accounts for the 6 quad "discrepancy" between supply and demand).

Table 4.3 shows the distribution of oil, gas, coal and soft technology supply to meet these year 2000 end-use demands, following Lovins' suggestions as closely as possible. The results show that monumental changes are required even by 2000 (with just 37% "soft" energy as compared to 100% by 2025):

- 60% of all industrial process steam boilers are converted to coal-fired, fluidized-bed, gas turbine cogeneration systems;
- 12½% of industrial steam boilers are coal-fired and 27½% are solar-powered;
- all industrial heating is converted to gas;
- all cooking and drying is converted to gas;
- virtually all residential and commercial space is equipped with high-efficiency solar heating (70% of space heat, 83% of water heat) with natural gas backup;
- 55% of cooling is solar with electric backup;
- electricity demand is 23% *higher* than 1975 (57% cogenerated, 15½% wind, 15½% hydro, 12% coal);
- 40% of transportation demand is provided by biomass alcohol.

TABLE 4.3

ILLUSTRATIVE ENERGY INPUTS FOR YEAR 2000 SOFT PATH SCENARIO, QUADS

Oil		**Gas**		**Coal**	
Transport	12.5	Space Heat	3.5	Indus. Cogen.*	17.7
Feed	7.0	Water Heat	0.5	Process Steam	2.1
Misc.	0.5	Cooking, Drying	1.7	Electricity	3.0 (1.0)
	20.0	Indust. Heat	9.0	Feed	1.7
		Misc.	0.3	Misc.	0.5
			15.0		25.0

'Soft' Technology		(*% of end-use*)	**Electricity**		
Transp. Biomass	9.0	(41%)	Cogen.	4.8	
Solar Space Heat	8.0	(70%)	Coal	1.0	
Solar Water Heat	2.5	(83%)	Wind	1.3	
Solar Air Cond.	2.5	(56%)	Hydro	1.3	
Solar Process Steam	4.5	(27%)		8.4	(25.2)
Wind Electric	1.3 (4.0)				
Hydroelectric	1.3 (4.0)				
	29.1 (34.5)				

*60% *of process steam is cogenerated using Stal Laval-type systems.*

TABLE 4.2

ILLUSTRATIVE YEAR 2000 END-USE REQUIREMENTS FOR THE 'SOFT PATH,' QUADS

Industrial			**Commercial**			**Residential**			**Transport**	
	Direct	Elec.		Direct	Elec.		Direct	Elec.	Direct	Elec.
Process Steam	16.5	—	Space Heat	3.5	—	Space Heat	8.0	—	21.5	0.2
Electric Drive	—	4.7	Water Heat	0.5	—	Water Heat	2.5	—	21.5	0.2
Electrolytic	—	0.3	Cooking	0.2	—	Cooking	1.0	—		(0.5)
Heat	9.0	—	Refrig.	—	0.3	Drying	0.5	—		
Feed	7.5	—	Air Cond.	1.1	0.3	Refrig.	—	0.3		
Misc.	0.5	0.3	Lighting	—	0.6	Air Cond.	1.4	0.4		
	33.5	5.3	Feed.	1.2	—	Lighting	—	0.3		
		(16.0)	Misc.	0.3	0.2	Misc.	0.5	0.5		
				6.8	1.4		13.9	1.5		
					(4.2)			(4.6)		

Of course the matching of supply and demand and the mix of soft technology can be varied, but this does not substantially affect the pattern or scale of deployment. In fact, the changes required seem hardly less staggering than for Lovins' hard path, particularly if the associated support facilities and infrastructure changes are included.

TABLE 4.4
ILLUSTRATIVE ESTIMATE OF 'SOFT' PATH CUMULATIVE CAPITAL COST TO THE YEAR 2000 (1976 DOLLARS)

Item	Unit Cost $/bbl-day	Year 2000 Supply, Quads	Delivery Factor[1]	Cumulative Cost to 2000, billions
Industrial Cogeneration	49,000[2]	9.9 steam	0.75	276
		4.8 elec.	0.95	
Transportation Biomass	30,000[3]	9.0	(1.0)	127
Solar Space Heating	423,000	8.0	0.7	1118
Solar Water Heating	211,700	2.5	0.7	175
Solar Air Conditioning	250,000[4]	2.5	0.7	207
Solar Process Heat	250,000	4.5	0.75	398
Wind Electric	400,000	1.3	0.9	221
Hydroelectric	130,000	1.3[5]	0.9	18
			Subtotal	2540
Conventional Coal, Oil and Gas Supply				400
Retirement of Central Electric Capacity[6]				180
			Total	$ 3120

Notes:

(1) *Since unit costs are given in $/bbl-day delivered and supply is given in primary quads, this factor corrects for the delivery efficiency of conventional supply. Cumulative cost is calculated as unit cost times supply times delivery factor times 472,000 bbl-yr/quad-day.*

(2) *Assuming a capital cost of $100 million (Lovins' figure) for a 65 Mwe, 105 Mwt system; daily delivery of 1268 bbl-equivalent steam, 785 bbl-equivalent electricity (90% capacity factor, 5% loss factor).*

(3) *FEA, Lovins high estimate.*

(4) *Assumed incremental cost for a building also equipped with solar heating.*

(5) *Only 0.3 quads of this is new capacity.*

(6) *Assuming an average of $300 per kilowatt of unamortized cost on 600 Gwe of capacity and associated transmission operating, under construction or on order.*

Table 4.4 provides a rough estimate of the total capital cost of the soft path to the year 2000—$2540 billion (1976 dollars) for the 30 to 35 quads of soft supply plus about $400 billion for conventional fossil supply and (very roughly) $180 billion for early retirement of central electric plant and transmission—or a total of about $3120 billion. There is considerable uncertainty in this figure: for example, biomass cost may be low for transportation-grade fuels, conversion and delivery; solar heating costs would be lower for passive systems (but could continue to be higher for active systems); it is assumed that no "soft" energy equipment requires replacement in this period; future cost decreases in soft technology may be underestimated; some early retirement and conversion costs are neglected. In short, the uncertainty in the $3.1 trillion 'soft path' cost to 2000 is probably on the order of ± $1 trillion. For example, if solar system costs were $10/ft.² (or equivalent) installed, and wind turbine costs were $275,000/bbl-day, the soft path cost would be reduced to $2102 billion (1976 dollars).

This estimated capital cost amounts to about 5½% of cumulative GNP for the period 1975-2000 (at 3.1% real GNP growth)—a figure that undoubtedly precludes its feasibility in terms of capital formation. In fact, it is more expensive than Lovins' hard path and much more costly than a reasonable moderate path.

FIG. 4.2

ESTIMATES OF CUMULATIVE 1975–2000 ENERGY FACILITIES INVESTMENT

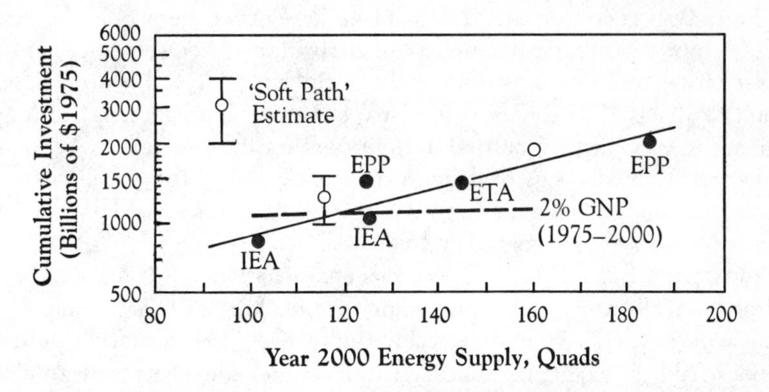

Figure 4.2 repeats the estimates of cumulative energy system investment costs discussed in Section 3.2. Even allowing for uncertainties and neglected cost items, it would appear that the cumulative 1975–2000 cost of a moderate (100 to 125 quad) path would be in the neighborhood of 1.0 to 1.25 trillion dollars, and certainly should not exceed 1.5 trillion.

This can be illustrated in another fashion. Consider a moderate scenario of 115 quads demand in 2000 (60 quads direct, 55 quads electric). The incremental electricity demand at end-use is:

$(55\text{-}20.5) \times 1/3 \times 0.9 = 10.35$ quads.

At $140,000/bbl-day, new generating capacity and associated T&D would cost

$10.35 \times 140{,}000 \times 472{,}000 = \$684$ billion.

Assuming a cost of $20,000/bbl-day for the 60 quads of direct supply (new and replacement, mostly fossil), this would cost

$60 \times 20{,}000 \times 472{,}000 = \$566$ billion.

This amounts to a cumulative cost of $1.25 trillion for this 115 quad path (in reasonable agreement with the IEA, ETA and EPP estimates)—about 2.2% of cumulative 1975-2000 GNP. A similar analysis yields a 1975-2000 cost on the order of 2.0 trillion dollars for Lovins' "hard path."

In conclusion, it should be noted that the foregoing analysis is not intended to recommend against a realistic deployment of soft technologies. Obviously, some soft technologies can provide less expensive energy than competing hard technology supply, even at higher capital cost. This applies, for example, to the use of solar hot water heating or to the substitution of solar space heating for electric resistance heating in many regions. Finally, we note, as Lovins, that the most cost-effective investment is very frequently increased efficiency and conservation.

Lovins: Dr. Forbes's calculations of the total capital cost of what he takes to be a soft energy path begins with his construction of what he says is an illustrative version of one. The data he presents concern only the year 2000, not a 50-year path, and their derivation is not shown beyond a blanket reference to the studies by the Weinberg group and by von Hippel and Williams. The former is a relatively weak-conservation future (e.g. there is very little industrial retrofit) arbitrarily assumed to be half-electrified, and is thus of doubtful relevance. The latter, while a good starting point on conservation (89 q in 2000 vs. my estimated 95), makes no effort to match supply to end-use structure.

Insofar as I can deduce Dr. Forbes's end-use structure, it has some odd features. (His accounting conventions are unclear, so it is possible that some of these oddities may be accounting artifacts.) He assumes excessive industrial heat requirements at high temperatures and meets them all with 9 q of gas (no coal), has too little solar process steam (only a quarter of process requirements below about 300°C) and no solar direct heat (e.g. crop drying), has probably too much biomass (9 q, whereas I claim only 6 q from residues), has far too much electricity for needs requiring it, has therefore too much cogeneration (arbitrarily limited to a single system), and has retired all but about 50 GW(e), all coal-fired, of present thermal power stations rather than retiring non-nuclear stations only through normal attrition as I assumed. He therefore counts a $180 billion penalty against the soft path for premature retirement of unamortized generating and transmission facilities—the former to be replaced by one type of cogeneration system, the latter presumably by storage assessed against his wind systems. (Premature retirement to save oil and gas would also presumably be required in a nuclear-oriented path, but does not show on his hard-path or "moderate-path" accounts.) No district heating is assumed.

Details of the calculations also appear to be inconsistent: for example, the 17.7 q going to cogeneration should yield 9.9 q of process steam according to Table 4.4 in EPDC, 7.6 q according to Figure 4.1 in EPDC, and 10.1 q according to Table 4.3's note that 4.5 q of solar process heat represents 27% of end-use for that purpose. (It is not clear which value is assumed.) The general effect of these peculiarities, and especially of the poor matching of supply to end-use thermodynamic structure, appears to be to raise computed total cost. This is particularly true, as noted below, for low-temperature solar heating systems, where Dr. Forbes's failure to consider conservation measures inflates total costs by more than $1 trillion. For the moment, however, let us accept for purposes of argument his idiosyncratic end-use structure and examine his assumptions about whole-system costs (1976 $ per bbl/day delivered enthalpy).

First a few general comments bearing on pp 7–8 of Dr. Forbes's testimony to your Committee [of which EPDC may be considered a backup paper]. Any reader of my published cost computations can readily ascertain that they are carefully stated in constant 1976 $, contrary to Dr. Forbes's unspecified claim that they are not. My computations of delivered energy cost take full account of end-use efficiency, again contrary to Dr. Forbes's suggestion that I ignore this important factor. The same is true of infrastructure in direct-fuel systems. As for lifetime of physical capital (and extracted resource deposits), this is reflected in amortization rates and therefore in delivered energy costs, particularly for oil and gas, and more than reflected in the accelerated depreciation allowed for power stations. I have assumed in my analysis that soft technologies are built to last at least as long as competing hard technologies and can thus use the same fixed charge rate. (In fact there is substantial doubt that, say, nuclear stations will last anywhere near 40 years, but I assume they will as a conservatism.) For the case of short-lived investments such as some oil and gas field equipment, one would use a higher fixed charge rate.

On p 8 [of his testimony] Dr. Forbes gives his estimate of about $140,000/bbl per day for a nuclear-electric system. His "derivation," on p 40 of EPDC, is a wholly undocumented statement of "a reasonable estimate"; it follows my methodology, except in neglecting to consider the capacity factor of distribution equipment, but it substitutes generally lower costs and a higher capacity factor than those I document. This technique of asserting "my number is better than your number" while not addressing the sources or alleged faults of the numbers I document is unedifying and unworthy. Accordingly, rather than stating as a fact, "Lovins overestimates the costs of some hard technologies," Dr. Forbes should state, "Lovins presents detailed and documented calculations of nuclear capital cost; I think many of the data are too high, but think you should take my word for it, and do not propose to cite any sources of my own or to present any arguments that Lovins's sources (chiefly the Bechtel data base) are wrong."

I turn now to Dr. Forbes's own calculations, presented in EPDC pp 42–4 and 55, of the capital cost of soft and transitional technologies, given in the order of Table 4.4 of EPDC.

Industrial cogeneration. Dr. Forbes considers a Stal-Laval system using a coal-fired fluidized-bed gas turbine producing 65 MW(e) delivered plus 105 MW(t) delivered process steam. In fact I do not consider such a system, let alone estimate its cost at $100 million as he claims. On p 119 of *SEP* I estimate $40 million (Stal-Laval data) for a system delivering 60 MW(e) plus 105(t) of *district-heating hot water*, and hence nominally producing about 65 MW(e) + 120 MW(t). Dr. Forbes has included in his $100 million about $20–25 million for a district heating grid, whereas the process steam would be used on the factory premises, and has counted the entire electricity distribution system ($35 million) as marginal even though total electricity use in his year-2000 estimates has increased only 23% from its 1975 value, and hence at most 23% of distribution investment could be counted as marginal. Assuming, as Dr. Forbes has apparently done, that plant costs are the same for raising steam as for heating water, plant cost ($40 million) plus 0.23 × $35 million (a maximum) plus $10 million for the fuel cycle (at $3000/bbl per day) equals $58 million as a maximum capital cost—42% less than the value Dr. Forbes wrongly ascribes to me.

(Dr. Forbes's calculation contains further numerical discrepancies I cannot resolve; he seems to be assuming that a barrel of crude oil contains 6.12 GJ rather than the more conventional 5.80, hence that a bbl/day is 70.8 kW rather than 67.1. I must also assume that the assumed cogeneration system does not operate as in Figure 4.1 of EPDC so that 4.8 q of electricity corresponds to 7.6 q of steam, for Table 4.4 of EPDC gives 9.9 q of steam.)

Transportation biomass. Dr. Forbes cites $30,000/bbl per day as "FEA, Lovins high estimate" without noting it is their high estimate for pyrolysis of municipal waste, where one must deal with very heterogeneous material containing metals, chlorine-rich plastics, etc. The high estimate I cite (from FEA data) for bioconversion of farm and forestry residues is about $20,000/bbl per day including collection investment. As a weighted average of the whole range of bioconversion projects, it seems conservative even including the whole local distribution system for the fuel output (*SEP* p 125). Dr. Forbes's value should thus be reduced by at least a third.

Solar space and water heating. Dr. Forbes assumes collector efficiency 0.40, delivered heat 68 W/m² for water heating and 34 for space heating, and 1977 installed system price of $20/ft² collector (1976 $ throughout), yielding $423,000/(bbl per day) for space heating and $212,000 for water heating. Professor Bethe assumes respectively 0.42–0.50, 58 W/m² for combined space and water heating, and $20/ft² (though he describes a Los Alamos system at $12–15/ft² today, yielding $171,000/(bbl per day)). I assume 0.42 efficiency including tank losses, 76 W/m² for combined space and water heating with the collector working into a large tank load, documented installed system prices (less seasonal storage) of about $14/ft² in the late 1970s and $9/ft² in the mid-1980s, and corresponding delivered heat investments (again omitting seasonal storage tanks at an assumed price of $26/m³) of about $110,000/(bbl per day) in the late 1970s and $66,000 in the mid-1980s.

Disagreements so far arise from (a) Dr. Forbes's considering only current, not future, solar equipment prices, whereas I consider the maturation of a now fragmented solar market—hence fewer middlemen as manufacturers franchise and market directly—and the more widespread use of today's best designs; (b) Dr. Forbes's considering separate, thus inefficient, space and water heaters, made even more inefficient by the absence of a large thermal load for the space-heating collectors to work into during much of the year; (c) Dr. Forbes's emphasis on unnecessarily costly types of solar equipment rather than on the types of high-quality, contractor-installed devices that can already be obtained by careful shopping in areas that have relatively mature solar markets (see discussion and citations in *Annual Review of Energy*, "Soft Energy Technologies," 3: 477, 1978). Discrepancies in collector efficiencies and conversion factors account for a further 10% in system price.

Barbara Greene's compendium of Northern California active solar systems (LBL-5229, August 1976) lists five buildings with contractor-installed collectors priced under $20/ft², the lowest being $2.57/ft² materials cost for the Hewlett-Packard plant retrofit at Sunnyvale; if this had been built not by H-P maintenance staff but by small job contractors at the *highest* price shown in the table ($8.44/ft² in the $16.88/ft² Sonoma house), it would still have totalled only $11/ft², compared with $16.88, $15.75, and $17.55/ft² shown for some houses. (Contrary to Dr. Forbes's assertion that uninstalled flat plates sell for about $20–25/ft², Dr. Greene's entries for owner-installed collectors give uninstalled prices including $3.12, $5.33, $3.94, and (1969 $) $3.75/ft².) Dr. Greene's survey shows hot-water-heater-only solar systems contractor-installed at prices/ft² including $15.83, $15.19, $15, $17.56, and $14.29, with materials costs running as low as $1.09/ft² (tank-type) or $3.00 (flat plate)—yet costs and prices are normally higher per ft² for water-heating than for space—or space-and-water-heating systems because the former have a higher ratio of fixed to variable costs. It may also be of interest that Bruce Anderson

(TEA, Harrisville, NH 03450) routinely contractor-installs high-quality field-erected active solar systems for \$10–12/ft²; that the Los Alamos integrated collectors were estimated to install commercially at \$8.30/ft² in new buildings (LA-UR-75-1335, 1975) with further improvement possible by learning; and that analysis of the 1977 Office of Technology Assessment report *Application of Solar Technology to Today's Energy Needs* shows complete installed system prices of \$10.4–11.5/ft² for the Reynolds "Torex 14" system. In short, Dr. Forbes has chosen values favoring his case from a very wide range of empirical values. I consider the existence of lower values—assuming good quality and normal commercial margins throughout—to be a proof that though it is possible to pay as much as Dr. Forbes assumes, it is not necessary and, in a competitive market place, is therefore not a reasonable basis for calculating the costs that can be realistically achieved.

For seasonal heat storage, Dr. Forbes assumes a storage volume of about 7 m^3/m^2 collector, roughly a 16-fold overestimate, and an "optimistic" tank price of \$35/m^3, roughly a 35% overestimate for a typical case (more for large tanks). The volume overestimate arises from failure to understand how seasonal storage works—a point on which Professor Bethe likewise challenged but later accepted my analysis. This misunderstanding arises largely from Dr. Forbes's assumption that a typical house requires an average heating load of 4 kW and a winter-peak heating load of 8–10 kW, whereas heat-conserving improvements that are now economically attractive can reduce these figures to about 0.5 kW and 1–1.2 kW respectively (for space *plus* hot water), thus making the required tank volume 0.4–0.5 m^3/m^2 for typical U.S. installations (less in sunny areas) and the cost relatively low both for the tank and for the reduced collector area. In particular, storage price becomes about \$12/m^2 collector, not Dr. Forbes's \$245; collector area per typical house becomes less than 7 m^2, not 56; and storage investment per house becomes about \$100, not \$13–15,000. Total solar investment per house thus becomes (1976 \$ around 1985) about \$566 rather than about \$26,000, and investment per bbl/day supplied heat becomes about \$76,000 (assuming rather high storage volumes and costs) rather than Dr. Forbes's weighted water-and-space-heat average of \$373,000, an overestimate of roughly fivefold accounting for more than \$1 trillion in investment (assuming his heat requirements to be correct).

In fact the error is even larger, for Dr. Forbes, as he states in EPDC, has neglected passive (and neighborhood) systems, which would lower the average cost of the solar mix. If, for example, 12 million new dwelling units (of order two-thirds of new construction to 2000) were passive-solar at a marginal capital cost of \$400/kW, and if three-fourths of 40 million retrofit units were on a neighborhood scale at a liberal \$40,000/bbl per day (see the *Annual Review of Energy* paper cited above and the Hollands &

Orgill paper whose summary conclusions are reprinted in Volume 2 of the Senate record), then with active retrofits costing $70,000/bbl per day, the unit-weighted average investment would be only $43,000/bbl per day, or 8.7 times below Dr. Forbes's estimate, saving some $1144 billion.

Nor does Dr. Forbes provide satisfactory citations for his calculations. He gives no source for his $35/m³ storage-price estimate, ignoring my ACES, Crouch & Adams, and OTA citations for lower values. For collector systems he states merely: "Flat plate collector systems for solar space and water heating are currently selling at prices averaging around $20 to $25 per square foot uninstalled, varying from under $10/ft² uninstalled to over $40/ft² installed[(41)]." His note 41 states only: "For example, see advertisements and case studies in *Solar Age.*" This is hardly a suitable rebuttal to my citation of extensive reviews by IIASA, OTA, LBL, and other groups. (See *Annual Review of Energy* article for further discussion.)

Solar cooling. Dr. Forbes gives a marginal cost of $250,000 per bbl/day but cites no source. The CONAES Solar Panel has estimated that for non-residential buildings (presumably the great bulk of the 2.5 q assumed) and for 1990 (a representative base year for pricing most of the installations to 2000) the price would be about $2500/ton or about $74,000 per bbl/day displaced. Even in 1975 the CONAES estimate is only $124,000, half Dr. Forbes's value. The CONAES residential estimate for 1990 is $201,000 per bbl/day; a weighted average of this figure (20%) and the 1990 non-residential estimate (80%) would be $99,400 per bbl/day, or 40% of Dr. Forbes's undocumented estimate. [Virtually all cooling should in any case be done passively at very low cost.]

Solar process heat. Dr. Forbes gives an investment of $250,000 per bbl/day but again cites no source. The CONAES 1975 estimate for hot water is equivalent (at 180 W/m² insolation and 0.4 First Law efficiency) to $60,000 per bbl/day. I cite a 315°C Winston-collector system with a retail price today of $118/m² and an efficiency of 0.44 (*SEP* p 126, n 31). If installation price is $22/m², total investment is $120,000 per bbl/day, or 48% of Dr. Forbes's estimate. I suggest in *Annual Review of Energy* that even this $140/m² is probably too high.

Wind electricity. Dr. Forbes cites $400,000. His testimony gives as sources only an estimate "on the order of $300,000 to $400,000 per barrel per day for the National Research Council-type vertical-axis wind turbine, and about $500,000 . . . for the ERDA/NASA-type horizontal-axis wind turbine (with reasonable allowance for future cost reduction)." EPDC gives some details:

- DAF 200-kW(e) Darrieus: run-on installed price $180,000 (I cited $150,000 on the basis of discussions with the Canadian government's staff and consultants); capacity factor 0.3 (same as mine): T&D investment \$150/kW (I assumed about \$126 because local siting avoids high-voltage transmission, and at most a third of the 600-VAC–and–under distribution is marginal); 10% losses (I assume 4% because of local use); result, \$3889/kW or \$275,000 per bbl/day (my result, \$3037/kW or \$204,000 per bbl/day).

Not content with this value, Dr. Forbes adds a storage cost of \$13,400 per bbl/day, not an excessively unreasonable value except that in practice, wind would be used as a fuel- and water-saver in a hydro-peaked grid and thus would not require storage. (In fact he cites, but does not document, storage costs of \$50,000-100,000 per bbl/day.) Further, Dr. Forbes then combines this unnecessary \$200/kW storage with further-reduced capacity factors—0.25 and 0.20—to obtain investments of \$390,000 and \$490,000 per bbl/day (values I can only reproduce by assuming the \$200 for storage is per kW peak installed, not stored, in which case it *is* unreasonably high). Not only would these values become \$258,000 and \$319,000 on my documented assumptions, but such low capacity factors correspond to using the wind machines in sites where one would not build them at all, and are thus grossly unreasonable distortions of a realistic wind regime. By stating that "the number of locations at which capacity factors of 30% could be achieved in the U.S. would be limited" but not saying how limited, Dr. Forbes gives the incorrect impression that 30% is a best-case value and that most machines would do far worse. In fact many would do better (especially on the Great Plains) and those that did significantly worse would not be built at all. The Grandpa's Knob site was indeed carefully selected, but not through extended measurements; today, wind experts realize that a longer series of wind data is necessary and would have suggested better sites. (In any case Vermont is not among our most favorable wind areas.)

- Dr. Forbes mentions an estimate of \$5000/kW *average* (1971 \$) for the Smith-Putnam machine, which he converts to 1976 \$ with an apparent deflator of 8%/y, yielding (with 10% losses) \$548,000 per bbl/day. This high figure reflects not only a relatively low capacity factor but also an excessive construction cost (see below).
- Likewise Dr. Forbes relies upon the \$5000/kW *peak* cost of the ERDA/NASA Plum Brook prototype, which is generally recognized as having been an exorbitantly priced and very badly designed disaster—and a pathological case with high first costs bearing no relation to production machines. As I shall show in a moment, Dr. Forbes's apparently generous assumption that production machines can be made, with storage, for 40% of the Plum Brook cost (yielding, on his assumptions of 10% losses, 35% capacity factor, and 70.8

kW/bbl per day, $445,000 per bbl/day—though Dr. Forbes calculates $483,000 per bbl/day) is not generous enough to make up for the grossly excessive assumed base price.

My article in Volume 3 of the *Annual Review of Energy* reviews, with extensive citations, the following installed prices, both carefully estimated and empirical (marked +):

MW-range designs by Lockheed, GE, other ERDA contractors	$400-600/kW peak
+ 175-kW updated Gedser machine at Cuttyhunk Island	about 750/kW peak
2-MW Tvind machine (much labor is not costed)	350/kW peak
10-kW Smith Chalk-wheel machine:	
+ homemade version (material costs)	88/kW peak
mass produced with tower	500/kW peak
200-kW Gedser machine if built today	1050-1350/kW av.
4-kW typical horizontal-axis machine	about 600/kW peak
+22.5-kW Riisager El-Windmolle machines, installed in 1978 (1978 $) in Denmark	760/kW peak
+20-kW Borre machines, installed in 1978 (1978 $)	600/kW peak
+ few-kW machine homemade from old car parts (material costs)	about 50/kW peak
200-kW DAF Darrieus (run-on installed price)	750/kW peak
+ 200-kW DAF Darrieus, offering price of no.2 unit	1000/kW peak
+200-kW DAF Darrieus, installed price complete, no. 1 unit	1275/kW peak

[In spring 1978, Charles Schachle of Moses Lake, Washington sold a 3-MW machine to Southern California for a turnkey *$356/kWp (rated at 17.9 m/s) and offered subsequent units for *$205/kWp installed. He plans to market 25-kW machines in 1979–80 at a factory cost (including tower) of about $240/kWp.]

From these values it appears that $0.4 \times 5000 - 200$ (storage) = $1800/kW peak is a far from reasonable estimate for production machines. From the very detailed Danish (Gedser-update) and ERDA-contractor studies, $400-600/kW peak or about $1050-1700/kW average (sent out) appear realistic. Unfortunately, Dr. Forbes does not consider any of these machines (save his own estimates for the DAF machine) and ignores both my U.S. and my Danish references.

In summary, whole-system wind-electric investments from about $80,000 to at most about $200,000 per bbl/day appear realistic. Even including Dr. Forbes's superfluous storage at his $13,400 per bbl/day, $120,000 (corresponding to the upper bound for the Gedser update) seems a generous value—30% of Dr. Forbes's inflated estimate.

Hydroelectric. Dr. Forbes assumes $130,000 per bbl/day but cites no source. Since the additions would be microhydro sets, and since the whole 15 GW (at 0.65 average capacity factor) could be accommodated within the scope of existing dams, large and small, estimates by MITRE, the New England Regional Commission, and several New England state governments suggest that the marginal investment should be toward the lower end of the $300-1200/kW peak price range. At $500/kW peak, 0.65 capacity factor, 5% losses, and a generous $75/kW for marginal transmission and distribution (in practice none should be needed), typical whole-system investments is $62,000 per bbl/day, or 48% of Dr. Forbes's undocumented estimate.

Summary. If Dr. Forbes's Table 4.4 in EPDC is recalculated with the capital costs just derived, but using his estimates of quads required and of delivery factors, the total investment for soft technologies and cogeneration to 2000 becomes not $2540 billion but $741 billion for soft and transitional technologies, a 71% reduction corresponding to his average 3.4-times overestimate. Some 64% of the difference between Dr. Forbes's Table 4.4 total and the revised total is due to the dominant term, solar space and water heating, where the weighted-average unit investment changes from $373,000 per bbl/day to $43,000 per bbl/day (for a mix of active single-building, active neighborhood, and passive single-building systems in the ratio 1:3:1.2). Further recalculation taking account of the inappropriateness of much of Dr. Forbes's assumed end-use structure and technological mix would further reduce the investment required. Specifically, if decreased electrification—hence decreased cogeneration—were cancelled by increased district heating, biomass decreased by a third, microhydro increased threefold, solar process heat of 4.5 q increased by a like amount of $60,000 per bbl/day water preheat, and solar space and water heating prorated to end-use requirements of 0.5 kW per dwelling unit—lest retrofits exceed the housing stock—total investment would decrease to about $711 billion plus whatever is required for fossil fuels.

These calculations do not try to recompute Dr. Forbes's unexplained $400 billion estimate for conventional fossil-fuel supply; this would be a very complex task owing to the phasing of old and new investments. Conservation investments are omitted from supply investments, consistent with Dr. Forbes's wish that conservation not be considered an exclusive feature of only the soft path. Cash-flow advantages of soft over hard technologies (owing to the short lead time and low operating cost of the former) are ignored. No debit is included for premature retirement of electrical capacity because none is assumed save for nuclear power, where 80-90% of the investment can be salvaged if desired by non-nuclear retrofit and where in most cases the capital cost of existing plants will already have been recovered through accelerated depreciation allowance and investment tax credit.

The above discussion appears to justify my contention, which Dr. Forbes denied, that his calculated capital costs for soft technologies either cannot be justified or, where documented at all, are based on pathological cases not representative of realistic costs. (His testimony illustrates this practice with the Plum Brook and Putnam wind machines and with the technically incompetent New England solar "tests.")

In short, Dr. Forbes's cost estimates do not begin to approach generally accepted professional standards of accuracy and objectivity, and lack the realistic and detailed technical basis that he claimed in testimony.

Index of Authors, Institutions, and Titles

Index of Subjects